SENSORY ANALYSIS OF FOODS

Second Edition

SENSORY ANALYSIS OF FOODS

Second Edition

Edited by

J. R. PIGGOTT

*Department of Bioscience and Biotechnology, Food Science Division,
University of Strathclyde, Glasgow, UK*

ELSEVIER APPLIED SCIENCE
LONDON and NEW YORK

ELSEVIER SCIENCE PUBLISHERS LTD
Crown House, Linton Road, Barking, Essex IG11 8JU, England

Sole Distributor in the USA and Canada
ELSEVIER SCIENCE PUBLISHING CO., INC.
52 Vanderbilt Avenue, New York, NY 10017, USA

First edition 1984
Second edition 1988

WITH 71 TABLES AND 62 ILLUSTRATIONS

© 1988 ELSEVIER SCIENCE PUBLISHERS LTD

British Library Cataloguing in Publication Data

Sensory analysis of foods.—2nd ed.
  1. Food. Sensory analysis
  I. Piggott, J. R. (John Raymond).
  664'.07

ISBN 1-85166-231-6

Library of Congress Cataloging-in-Publication Data

Sensory analysis of foods/edited by J. R. Piggott.—2nd ed.
  p.   cm.
Includes bibliographies and index.
ISBN 1-85166-231-6
  1. Food—Sensory evaluation. I. Piggott, J. R. (John Raymond),
1950–
TX546.S45   1988
664'.07—dc19

No responsibility is assumed by the Publisher for any injury and/or damage to persons or property as a matter of products liability, negligence or otherwise, or from any use or operation of any methods, products, instructions or ideas contained in the material herein.

Special regulations for readers in the USA

This publication has been registered with the Copyright Clearance Center Inc. (CCC), Salem, Massachusetts. Information can be obtained from the CCC about conditions under which photocopies of parts of this publication may be made in the USA. All other copyright questions, including photocopying outside of the USA, should be referred to the publisher.

All rights reserved. No part of this publication may be reproduced, stored in a retrieval system, or transmitted in any form or by any means, electronic, mechanical, photocopying, recording, or otherwise, without the prior written permission of the publisher.

Printed in Great Britain by Galliard (Printers) Ltd, Great Yarmouth

PREFACE

Reviews of recent improvements in understanding of the senses and in the theory and practice of sensory analysis and sensory evaluation have not been readily accessible to food scientists and technologists, so the purpose of this book is to provide an overview of present knowledge and practice in sensory analysis. It is not intended to be a laboratory manual, but to provide the practising sensory scientist with a review of progress around the world, and the beginner with a firm foundation on which to build his own expertise. I hope it will be of value to all those involved in the management and operation of sensory testing in industrial, government and university laboratories.

Individual chapters have been contributed by acknowledged leaders in their fields, who have surveyed the literature and brought their own unrivalled experience to its interpretation. The first two chapters examine the chemical senses of taste and smell, by which sensory analysts and consumers perceive food flavours. Texture perception and assessment is discussed in the next chapter, and this is followed by a chapter describing vision and appearance assessment. Texture assessment might appear to be straightforward, but this chapter shows how the textural characteristics which people perceive in foods are often not simply related to easily measured mechanical properties. Colour vision has been well researched and systems for instrumentally assessing colour have been used for more than fifty years, but the next chapter points out that there are still pitfalls involved in trying to substitute instrumental measurement for the human senses. Difference tests and methods are described in the next chapter, which shows how such apparently simple tasks mask complexities which are still not completely understood. This is followed by a review of ranking and scaling methods, an area where the work done in the behavioural sciences is beginning to have an impact in food science. Descriptive methods are the most ambitious sensory tests in current use, and have perhaps shown the greatest increase in use in recent years, moving from being difficult experimental methods to being routine analytical tools.

The next chapter moves out of the laboratory, and is devoted to consumer studies; it recalls that food which is not acceptable, however well analysed it is, is not food but waste. The final two chapters describe some of the important statistical procedures available for analysis of sensory data—first what can be described as the conventional descriptive and inferential methods, and second the less conventional variants of multidimensional scaling. Each chapter includes a full list of references to enable the interested reader to follow up subjects of interest.

This book also demonstrates where information and understanding is lacking. We must not forget that sensory analysis is simply an analytical tool—it is only useful insofar as the results it produces are useful. Sensory analysis at present stands on a plateau, separated by an abyss from the chemical sciences on the one side and from market research on the other. While there are signs in the following chapters of bridge-building, it must be the task of the years to come to complete the bridges and show how the composition of food affects its sensory properties and hence its acceptability to the consumer.

In compiling this second edition of *Sensory Analysis of Foods*, I have taken the opportunity of revising and updating it, in order to take account of the changes and advances which have occurred in the past years. I have once again relied on the experience and skill of the contributors of the individual chapters to explain their own fields; the quality of their contributions demonstrates that this faith was justified. I am pleased to acknowledge also the assistance of colleagues, too numerous to mention individually, for invaluable discussions and suggestions, and of the publishers for their patience and help.

J. R. PIGGOTT

CONTENTS

Preface	v
List of Contributors	ix
1. The Sense of Taste K.-H. PLATTIG	1
2. The Sense of Smell J. A. MARUNIAK	25
3. Texture Perception and Measurement J. G. BRENNAN	69
4. Colour Vision and Appearance Measurement D. B. MACDOUGALL	103
5. Sensory Difference Testing and the Measurement of Sensory Discriminability J. E. R. FRIJTERS	131
6. Scaling and Ranking Methods D. G. LAND and R. SHEPHERD	155
7. Current Practices and Application of Descriptive Methods J. J. POWERS	187

8. Consumer Studies of Food Habits 267
   H. L. MEISELMAN

9. Statistical Analysis of Sensory Data 335
   G. L. SMITH

10. Preference Mapping and Multidimensional Scaling . . 381
    H. J. H. MACFIE and D. M. H. THOMSON

Index 411

LIST OF CONTRIBUTORS

J. G. Brennan

    Department of Food Science and Technology, University of Reading, Whiteknights, PO Box 226, Reading RG6 2AP, UK

J. E. R. Frijters

    Department of Marketing and Marketing Research and Department of Food Science, Wageningen Agricultural University, De Leeuwenborch, Hollandseweg 1, 6706 KN Wageningen, The Netherlands

D. G. Land

    Taint Analysis and Sensory Quality Services, 8 High Bungay Road, Loddon, Norwich NR14 6JT, UK

D. B. MacDougall

    Agricultural and Food Research Council, AFRC Institute of Food Research—Bristol Laboratory, Langford, Bristol BS18 7DY, UK

H. J. H. MacFie

    Agricultural and Food Research Council, AFRC Institute of Food Research—Bristol Laboratory, Langford, Bristol BS18 7DY, UK

J. A. Maruniak

    College of Arts and Sciences, Division of Biological Sciences, University of Missouri—Columbia, 213 Lefevre Hall, Columbia, Missouri 65211, USA

H. L. MEISELMAN

Behavioral Sciences Division, Department of the Army, US Army Troop Support Command, Natick Research, Development and Engineering Center, Natick, Massachusetts 01760-5020, USA

K.-H. PLATTIG

Institut für Physiologie und Biokybernetik der Universität Erlangen-Nürnberg, Universitätstrasse 17, D-8520 Erlangen, Federal Republic of Germany

J. J. POWERS

Department of Food Science and Technology, Food Science Building, University of Georgia College of Agriculture, Athens, Georgia 30602, USA

R. SHEPHERD

Agricultural and Food Research Council, AFRC Institute of Food Research—Norwich Laboratory, Colney Lane, Norwich NR4 7UA, UK

G. L. SMITH

Ministry of Agriculture, Fisheries and Food, Torry Research Station, PO Box 31, 135 Abbey Road, Aberdeen AB9 8DG, UK

D. M. H. THOMSON

Department of Food Science and Technology, University of Reading, Whiteknights, PO Box 226, Reading RG6 2AL, UK

Chapter 1

THE SENSE OF TASTE

K.-H. PLATTIG

Institut für Physiologie und Biokybernetik der Universität Erlangen-Nürnberg, Federal Republic of Germany

1. INTRODUCTION

The classic sensory modalities of olfaction and taste represent, together with the cutaneous modalities of mechano-, thermo- and nociception, the group of 'lower senses' (von Skramlik, 1926). In addition, olfaction and taste are named 'chemical senses'. The term 'lower senses', however, might be misleading; there is no suggestion that the phylogenetically 'old' chemical senses are of minor importance for the organism. Quite the contrary, olfaction and taste (being sensors for food intake and sexual reproduction) as well as the cutaneous senses are very important in the development of vertebrates to the human stage (Boudreau, 1980).

The only justification for the term 'lower senses' is that the 'higher senses', vision and audition, reveal a higher and more rational clarity of sensory perception. This rational clarity is lacking for the chemical and cutaneous modalities to such a degree that some people (for instance anosmics), having lost their ability to *smell*, only complain 'I cannot *taste*!'. In fact, lack of rationality might also enable more direct access to deeper layers of personality, as food scientists, perfumers and others know.

The term 'chemical senses' means that the primary sensory process, initiating a taste or smell sensation, should be a chemical binding of substances with taste or smell properties to the surfaces of the membranes of the respective receptor cells or their processes.

Audition and vision act over long distances to receive information for the organism. Likewise, smell also makes use of molecules of odorants, which may approach the organism borne by the wind over quite a long distance. In contrast, taste is a close-range or contact sense, operating not on a long-distance basis but only when in contact with the source of taste.

Its receptors in man and vertebrates are mainly situated on the tongue surface, but in part (although to a very minor degree) they are spread over all the oral cavity and down the oesophagus and even on the trachea and the larynx.

2. ANATOMICAL STRUCTURES OF THE TASTE SYSTEM: RECEPTOR CELLS, TASTE BUDS, NERVOUS PATHWAYS AND CENTRAL PROJECTIONS

The receptive bases for the perception of the four taste qualities—sweet, sour, bitter and salty—are taste receptor cells. In most vertebrates and also in man about 30 to 70 single cells of three or four different types (Murray and Murray, 1960; compare Fig. 3 and below) are situated in the taste buds. The taste buds, numbering between 0 and 200, are on certain papillae (nipple-like elevations) on the surface of the tongue or spread simply over the oral, oesophageal or tracheal mucosa. On the human tongue there are three types of taste papillae: fungiform, foliate and vallate papillae.

Filiform papillae, a fourth kind of human tongue papilla, lack taste buds completely; they only consist of thin, horny processes of the otherwise uncorned tongue surface and have a purely mechanical function of holding certain food constituents. These corneous filiform papillae are much more pronounced in animal tongues such as those of cattle, in which they help to hold the grass for tearing it out.

Taste buds are found in only about 40–50% of the fungiform papillae (Arvidson, 1976; Arvidson and Friberg, 1980). Some of the human taste papillae, especially the vallate ones, have more than 100 taste buds (Fig. 1 and Table 1).

2.1. Localisation of the Taste Papillae on the Tongue

Fungiform papillae are spread over almost all the tongue, except for a strip of about 2–3 cm right and left of its mid-line on the dorsum or back of the tongue (which has only filiform papillae). Fungiform papillae are up to 3 mm high and about 0·3–2 mm diameter. The vallate papillae are larger, up to 5–7 mm in diameter. Contrary to the diffuse distribution of fungiform papillae all over the edges of the tongue and even on its base, the vallate papillae are situated in a V-shaped borderline (linea terminalis), which separates the anterior two-thirds of the tongue surface from its base. The apex of this V, looking towards the rear (pharynx), is formed by a blind opening (foramen caecum). During foetal development the thyroid gland,

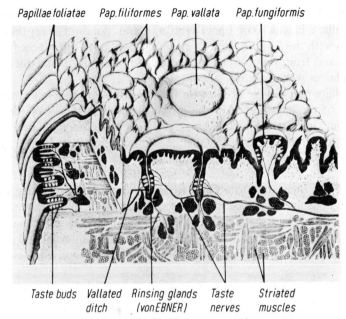

FIG. 1. Taste papillae on the human tongue from surface and sectioned view. (From Rohen, 1978.)

at this early stage with an excretory duct, develops from this opening towards the neck region. The duct is lost on the change to purely endocrine functions.

In man, in this terminal line there are usually only about 7–14 rather large vallate papillae resembling fungiform papillae in appearance. The

TABLE 1
HUMAN TONGUE PAPILLAE AND THEIR TASTE BUDS IN ADULTS

	Number of papillae	Taste buds per papilla	Taste buds in all papillae
Circumvallate (*P. vallatae*)	8–12 (7–14)	100–200	1 000–1 500
Foliate (*P. foliatae*)	15–20	≃ 10	150–200
Fungiform (*P. fungiformes*)	≃ 100	0–4	300–400
Filiform (*P. filiformes*)	≃ 1 000	0	0

main differences are that vallate papillae are larger and surrounded by a deep ditch, in which von Ebner's rinsing glands pour out a very thin serous saliva with the assumed task of cleaning away taste substances from this ditch and from the taste buds (located in both walls of the ditch, in the papillar as well as on the lateral side (Fig. 1)).

Foliate papillae are situated on the side edges of the tongue, mostly in front of the lateral ends of the terminal line, and they, like the vallate papillae, have von Ebner's rinsing glands. Foliate papillae are very well developed in rodents and also in human infants, but their number and size decrease in adults (Fig. 1 and Table 1).

2.2. Taste Buds and Taste Cells: Receptor, Basal and Supporting Cells

Taste buds are immersed in the flat epithelia of the papillae, but also of other parts of the mouth, pharynx and oesophagus. They are reminiscent of little barrels, onions or oranges, up to about 70 μm in height and 40 μm in diameter (Fig. 2) having an opening, the taste pore, to the oral or

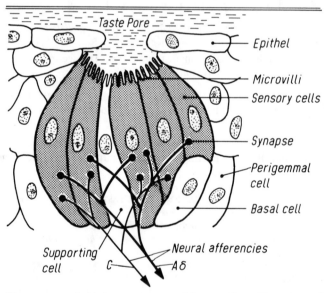

FIG. 2. Human taste bud, its structure and innervation. The microvilli of the sensory cells protrude into a fluid-filled space in the taste pore. Only two afferent fibres are drawn, while actually about 50 fibres branch within just one taste bud, which has its cells (about 40–70) assembled like the slices of an orange. Efferent fibres as described in the text are not shown. (Modified according to Altner, 1978.)

pharyngeal cavity and basal synaptic nerve connections. In an adult there are about 2000 taste buds, of which about 1000 are situated on the vallate papillae, each of which may have up to 200 taste buds. Out of about 100 fungiform papillae only about 45–50% have 1–4 taste buds, while about half of them lack taste buds completely. In small infants anatomists have estimated up to about 10 000 taste buds.

Within the taste buds, there are sensory receptor cells to be differentiated from supporting (indifferent) and basal cells (Fig. 2). Beidler and Smallman proved as early as 1965, by marking with ^{3}H-thymidine, that there is a permanent regeneration of taste cells with a half-time of 250 ± 50 h, which depends on cell divisions of the basal cells. This division of the basal cells makes the young taste receptor cells move from the periphery of the taste bud to its centre, and the older taste cells in the central portion of the taste buds die away and are then pressed out from the taste bud into the mouth through the taste pore.

The taste pore is a small channel of about 2 μm diameter, in which small processes of the active taste cells protrude to the oral cavity. These microvilli of about 0·1–0·2 μm diameter and 1–2 μm length are assumed to bear chemical receptor sites, to which taste molecules may bind for initiating the primary processes of taste after the stimulating substances have passed through the watery oral (salivary) mucus by diffusion (Beidler, 1954, 1961, 1971).

The main task of the receptor cells appears to be binding of taste molecules and initiation of the primary sensory processes; the main task of the basal cells seems to be the renewal of the set of taste cells. It is not clear whether the only task of the supporting (sustentacular = indifferent) cells is to be a mechanical support to the basal and gustatory cells, or anything else (Fig. 3).

Having no nerve axons of their own, but a synaptic connection to afferent gustatory fibres right at their cell bases, the taste cells are secondary sensory receptor cells. Olfactory cells in contrast end without a synapse in their own axon (filum olfactorium); therefore they are primary receptor cells. During foetal development and after taste nerve degeneration, the ingrowing taste nerves seem to provoke the formation of taste cells and buds (Guth, 1958, 1971). All taste cells situated together in one taste bud are connected by basal synapses to about 50 afferent taste fibres running towards the central nervous system. These synapses have to be renewed in each renewal of the sensory cells, but the precise mechanism of this permanent regeneration of synapses is quite unclear.

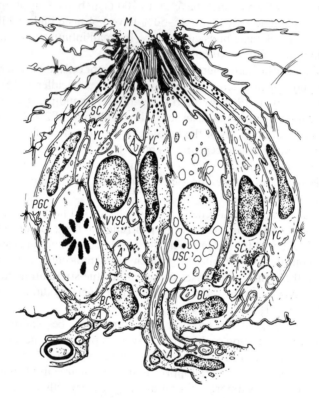

Fig. 3. Cellular organisation of a taste bud in a foliate papilla in rabbit according to a scheme of Andres drawn after electron microscopic observations. There are very young sensory cells (VYSC), young (ripening) sensory cells (YC) and (ripe or active) sensory cells (SC), dying (or dead) sensory cells (DSC) and basal cells (BC). Sustentacular or supporting cells are not shown in this picture (as 'indifferent' cells they might just be precursors of VYSCs). A, Axons of neural afferencies; PGC, perigemmal cell. Murray and Murray (1967) differentiated within the receptor cells 'dark' and 'light' cells or type 1 and type 2 cells from type 3 cells, which coincide with the age development stages of Andres. (Drawing according to electron microscopic observations by Andres (1975), Murray and Murray (1967) and Murray (1971).)

2.3. Neural Pathways of Taste and its Cortical Projection

The nerve supplies for the taste buds are: (I) the telodendria (terminal branchings) of myelinated nerve fibres of about 1–6 μm diameter (Aδ-fibres and B-fibres in the scheme of Erlanger and Gasser‡; conduction velocity 3–30 m s^{-1}), and (II) unmyelinated fibres of 0·3–1 μm diameter (C-fibres according to Erlanger and Gasser (1924, 1937), conducting at about 0·5–2 m s^{-1}). Nerve fibre (II) might be an efferent system, probably carrying information of the vegetative nerve system to the taste cells, while nerve fibre (I) appears to carry the gustatory afferent information to the central nervous system. These taste afferences reach the brain mainly via the seventh and the ninth cranial nerves (N. VII = n. facialis for the anterior two-thirds of the tongue, N. IX = n. glossopharyngeus for the posterior one-third or base of the tongue; the areas are separated by the linea terminalis and the vallate papillae), but the pharyngeal and oesophageal taste buds are mainly connected to the tenth cranial nerve (N.X = n. vagus). To reach the facial nerve the taste fibres from the anterior two-thirds of the tongue initially join the lingual nerve, which is part of the trigeminal nerve (N.V = fifth cranial nerve), and then the chorda tympani, connecting the lingual nerve to the facial nerve. The lingual nerve runs between the chewing muscles, and releases behind one of them (musculus pterygoideus medialis) the chorda tympani, which enters the middle ear (cavum tympani) through a tiny fissure in the petrous bone.

The chorda tympani, bearing afferent taste and efferent salivation and lacrimal information, runs on the inner side of the tympanic membrane between the middle ear ossicles and joins the facial nerve through another somewhat larger bone channel. The cell bodies of all taste afferents from the anterior one-third of the tongue and also those of the taste buds in the palate, approaching N. VII via the great superficial petrosal nerve (n. petrosus superficialis major), are situated in the ganglion geniculi of the facial nerve (Fig. 4). The central processes of these gustatory neurons join the facial nerve to the nucleus tractus solitarii in the brain stem (rhombencephalon: medulla oblongata and mesencephalon), which is a common relay station not only for the trigemino-facial, but also for the vagal nerves from pharynx, larynx and oesophagus (Yamamoto, 1983).

The first synaptic relay, after the initial gustatory synapses at the taste

‡ Erlanger & Gasser brought a neurophysiological classification of nerve cells and fibres according to the fibres' conduction velocity, which corresponds to the fibre diameter; Erlanger-Gasser A α are the thickest, followed by A β, -γ, -δ, B, and C as the thinnest fibres with velocities ranging between 120 and about 0·3 m s^{-1} and diameters between 22 and about 0·3 μm.

cell bases, is in this nucleus tractus solitarii. Here the second neuron of the gustatory pathway starts crossing in the lemniscus medialis (medial or sensory lemniscus = band of Reil) up to the nucleus ventralis thalami (arcuate nucleus of the thalamus), in which the third neuron starts up to the cerebral cortex. The cortical taste projection is situated in Brodmann area 43, behind the parietal gyrus postcentralis, right adjacent to the somatosensory projection of the mouth and tongue region.

3. TASTE QUALITIES AND THEIR NEURAL CODING (SUBJECTIVE AND OBJECTIVE PHYSIOLOGY OF TASTE)

To naive observation it is usually not clear that original taste sensations are in reality very complex and superimposed by information coming from other sensory modalities. In terms of physiology we would say that taste, playing an important role in an oral metering for control of appetite and food intake (Katsuki et al., 1977; Booth, 1981), does not act alone, but is part of a complex oro-facial sense. This includes olfactory and somatosensory information on the contents of the oral cavity. The participating somatosensory information consists not only of thermo- and mechano-receptive data, but also of pain or pain precursors, which for instance are elicited by pepper or capsaicine (in chillies).

Anosmic people, who might have lost the ability to smell—for instance by damaging the fine olfactory nerve fibres, e.g. in a traffic accident, by viral or bacterial toxins or by chemical substances, sometimes even if applied for therapeutic purposes—very often feel only the subjective complaint that 'food does not taste good anymore, there's no taste at all', while a very brief check-up of taste abilities proves that sweet, sour, bitter and salty can be tasted very well. It is the flavour of the food constituents, which comes up from the rear of the mouth through the choanae (posterior or pharyngeal nares) to the nose and its olfactory receptors, that is not perceived in these cases.

3.1. Qualities, Subqualities and Thresholds of Taste

For the basic taste qualities—sweet, sour, bitter and salty—some substances, with their threshold concentrations, are collected in Table 2.

Subqualities of taste may be electric, alkaline and metallic tastes. Metallic taste is elicited by some metal salts, and it is said to be elicited also by some formed metals, which, however, according to their surface electric potentials might usually in reality produce electric taste. Alkaline taste

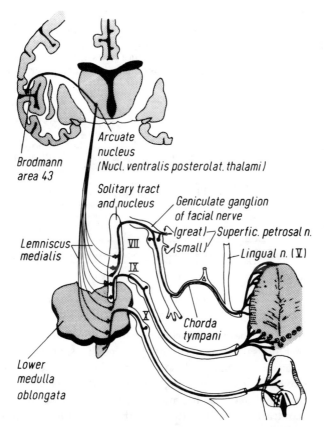

FIG. 4. The neural pathway of taste with three neurons between tongue, respective oral cavity, and cortical representation in the parietal lobe (gyrus postcentralis = Brodmann area 43). Further details in the text. (Pattern of the picture taken from Netter, 1972.)

(soapy) is also linked to electric taste, since it is caused by cathodal DC stimulation, while the anode elicits a sour taste. Potassium carbonate (potash), being in common use for the production of soap, is the original substance described as producing soapy or alkaline tastes.

3.2. Threshold and Quality Conditions of Taste

Suprathreshold taste sensations are elicited by different substances in very different concentrations (Table 2). Also the threshold concentrations can vary between individuals by up to about two orders of magnitude, but even

TABLE 2

SOME TASTE-SUBSTANCES AND THEIR THRESHOLD CONCENTRATIONS IN (MOL LITRE^{-1}) OR EQUIV. LITRE^{-1} (FOR ACIDS), AND NUMBERS OF MOLECULES (ML^{-1} OF SOLUTION) ELICITING UNSPECIFIC THRESHOLD RESPONSES IN MAN
(Calculated according to von Skramlik (1926) and other authors)

Sweet	mol litre^{-1}	molecules ml^{-1}	Bitter	mol litre^{-1}	molecules ml^{-1}
Glucose	0·08	4·8 × 10^{19}	Strychnine hydrochloride	0·000 001 6	9·6 × 10^{14}
Sucrose	0·017	1·0 × 10^{19}	Quinine	0·000 008	4·8 × 10^{15}
Lactose	0·065	3·9 × 10^{19}	Nicotine	0·000 019	1·1 × 10^{16}
Saccharine	0·000 023	1·4 × 10^{16}	Caffeine	0·000 7	4·2 × 10^{17}
BeCl$_2$	0·000 3	1·8 × 10^{17}	MgSO$_4$	0·004 6	2·8 × 10^{18}

Sour	equiv. litre^{-1}	molecules ml^{-1}	Salty	mol litre^{-1}	molecules ml^{-1}
HCl	0·000 9	5·4 × 10^{17}	NH$_4$Cl	0·004	2·4 × 10^{18}
Acetic acid	0·001 8	1·1 × 10^{18}	NaF	0·005	3·0 × 10^{18}
Tartaric acid	0·001 2	3·6 × 10^{17}	MgCl$_2$	0·015	9·0 × 10^{18}
Citric acid	0·002 3	4·6 × 10^{17}	KCl	0·017	1·0 × 10^{19}
Malic acid	0·000 5	1·6 × 10^{18}	NaCl	0·03	1·8 × 10^{19}

intra-individually there are considerable fluctuations dependent on time of day, hunger and satiety or hormonal influences; taste perception develops changing sensitivities and preferences throughout the life span (Schiffman et al., 1976, 1979a, b; Plattig et al., 1980; Cowart, 1981). So far, however, a drastic threshold decrease has not been found in relation to the female cycle, in contrast to the three to four orders of magnitude reported by Le Magnen (1950) for the olfactory threshold for the musky odour of exaltolide, for about 24–36 h around the time of ovulation.

Therefore Table 2 can give only very rough values, which might hold for the so-called 'specific' or 'recognition' or 'sensation threshold'. Specific threshold means that a substance is clearly recognised in regard to its quality. It is preceded by the 'unspecific' or 'detection threshold', when lower concentrations are applied, which means that something tasting different from neutral is detected, but definite determination of the quality in this state is impossible.

Moreover, substances may elicit different qualities if the stimulating concentrations are changed (Table 3). Potassium chloride normally tastes salty, but starts out sweetish. With increasing concentration it becomes strongly sweet with some bitter, then rather bitter and only finally salty with still some addition of bitter and even sour. Sodium chloride also starts

TABLE 3
TASTE QUALITIES ELICITED IN MAN BY DIFFERENT CONCENTRATIONS OF NaCl AND KCl
(From Renquist, 1919, 1920)

Concentration ($mol\ litre^{-1}$)	NaCl	KCl
0·009	No taste	Sweet
0·01	Faintly sweet	Strongly sweet
0·02	Sweet	Sweet, faintly bitter
0·03	Sweet	Bitter
0·04	Salty, faintly sweet	Bitter
0·05	Salty	Bitter, salty
0·1	Salty	Bitter, salty
0·2	Purely salty	Salty, bitter, sour
1·0	Purely salty	Salty, bitter, sour

sweet, and only in higher concentrations appears to be purely salty. This phenomenon has been very well known for a long time and depends not only on the stimulating concentration, but also on the size of the stimulated area on the tongue surface. Some neurophysiological addition (or inhibition) of single receptor excitation via single papilla interactions must occur (Miller, 1971), but it is quite unclear whether the regeneration of the taste cells might also play a role in this regard.

The old experience that the four taste qualities can be elicited from different parts of the human tongue with different ease or difficulty (e.g. Henkin and Christiansen, 1967) is demonstrated in Fig. 5. The tip of the tongue is particularly sensitive for sweet, the following lateral edges for sour and then salty and the base of the tongue especially for bitter. This distribution, however, means a particular sensitivity only in statistical terms for each quality: stimulation of single human papillae (Plattig and Innitzer, 1976) reveals that bitter can also be elicited from the tip of the tongue and sweet from the posterior edges, so that Fig. 5 only gives a higher probability of qualities.

However, it might be of practical importance—not only for medicine— that any disturbance of the chorda tympani or of the facial nerve, for instance in otitis media or in basal skull fractures, may predominantly decrease the sensory perception for sweet, sour and salty, while simultaneously the bitter sensation from the base of the tongue might even be increased. Bitter perception, in contrast, may be decreased in all cases of damage to the glossopharyngeal nerve.

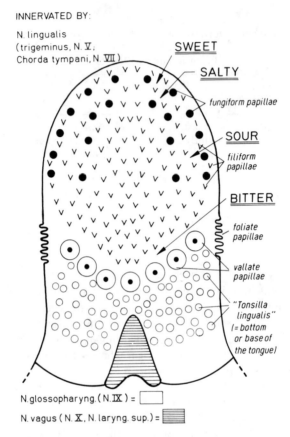

FIG. 5. Scheme of the tongue surface showing the distribution of the taste papillae, the innervation and the areas of maximal sensitivity for the single taste qualities.

3.3. Neural Coding of Taste

Taste intensities are coded for gustatory perception by the magnitude of receptor potentials and by the action potential frequencies of the gustatory afferences, using the common principle of all sensory (and also motor and visceral) nerves (Maes and Erickson, 1982). The coding of taste qualities, however, in the responses of taste cells and nerves is, together with that of smell, less clear than for most other sensory modalities (Sato, 1982). Von Békésy (1964a, 1966) postulated quality-specific tongue papillae (one papilla should respond only to sour and another only to sweet and so on), but this could not be proved (Plattig and Innitzer, 1976). The cells even in only one taste bud apparently respond to substances of different quality, but to different degrees as shown in Fig. 6 from experiments in hamster and rat. The receptor potentials of one single taste cell responding to (1) NaCl, (2) quinine-HCl, (3) sucrose and (4) HCl are shown in track I, individual responses of eight taste cells to the same stimuli 1–4 in tracks II and III, and neural discharge rates in tracks IV and V. Especially in the discharge rates, but also in the cell excitation levels certain taste profiles can be recognised, which appear to be very typical for the four different qualities. This is in agreement with the across-fibre pattern hypothesis of Pfaffmann (1955, 1964) and Pfaffmann et al. (1976).

The across-fibre pattern hypothesis postulates, for the central nervous system, the necessity of evaluating several taste fibres to obtain the information on the one quality of the present stimulation (Schiffman and Erikson, 1974; Schiffman et al., 1981; van der Molen, 1982), and it makes use of the graded or only relative specificity of the taste cells and the taste fibres (Tomita and Pascher, 1964; Ichioka and Hayashi, 1974).

Whether the convergence of the taste cells to the taste neurons of about 4–8:1, axonal transport in the taste nerves (Oakley et al., 1980) and/or the permanent regeneration of taste cells and the continuous reformation of synapses between taste cells and taste nerves might participate in this relative specificity remains still unclear. However, it appears to be quite clear that the receptive fields of the nerve cells in the nucleus tractus solitarii have considerable overlap on the surface of the tongue, which even crosses over more than one taste papilla, and there is considerable peripheral interaction among single papillae (Miller, 1971; Yamamoto, 1983).

3.4. Primary Processes in the Taste Cells

As mentioned earlier the precise nature of the receptive and transductive processes in human taste is unclear. There are only a few findings in man, most experiments having been done in other vertebrates, and these primary processes of taste cell activation may not only depend on species and kind of stimulus, but also vary from cell to cell within one species. It is widely accepted that all of gustation is initiated by adsorption and chemical binding of the tastant molecules to quality-specific cellular acceptor molecules, which seem to be located on the surface of the apical taste cell membrane, i.e. that of the microvilli protruding into the taste pore (Beidler, 1954, 1961; Sato, 1980; Faurion, 1982; Faurion and MacLeod, 1982; Avenct et al., 1988). It can be assumed that the acceptor molecules change certain ultrastructural features, after binding stimulating molecules, so that changes of membrane permeability occur which might initiate the excitation of the taste cells.

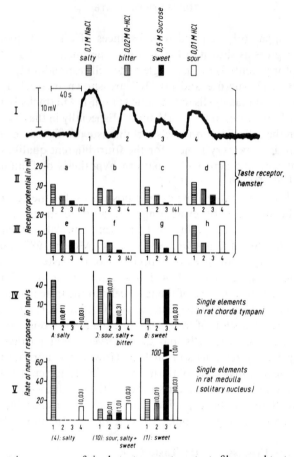

FIG. 6. Electric responses of single taste receptors, taste fibres and taste relay cells (or fibres) of the solitary nucleus to different gustatory stimuli. All responses are to (1) NaCl (0·1 M), (2) quinine-HCl (Q-HCl 0·02 M), (3) sucrose (0·5 M) and (4) HCl (0·01 M), applied to the tongue surface with water rinses between stimuli and recorded by micropipette electrodes, unless stated otherwise. Details of the tracks I-V are as follows. I, Time course of receptor potential in millivolts of a hamster taste receptor. (From Kimura and Beidler, 1961.) II + III, Response amplitude in millivolts of eight (a-h) hamster taste receptors. (Values from Kimura and Beidler 1961.) IV, Frequency response in pulses s^{-1} of three single fibres of rat chorda tympani. Deviations from the concentrations given above: (2) Q-HCl reduced to 0·01 M, (3) sucrose only in J is reduced to 0·3 M, (4) HCl is 0·03 M in A + B, but in J it is 0·01 M as usual. (Data from Pfaffmann, 1955.) V, Frequency response in pulses s^{-1} of three single units in the rat solitary nucleus. Deviations from the concentration given above: (2) Q-HCl reduced to 0·01 M, (3) sucrose elevated to 1·0 M, (4) HCl elevated to 0·03 M. (Data from Pfaffmann et al., 1961.) In tracks IV and V the single fibres and nerve cells are denominated as 'salty', 'sweet', 'sour-salty-sweet' elements according to the predominant responses to the respective qualities.

Teeter et al. (1987) listed at least three possible hypotheses for the events within taste cells following taste stimulation: (1) changes in membrane conductance; (2) release of intracellular second messengers; or (3) changes in the phase-boundary potential associated with the fixed charges at the external surface of their apical membranes. Also specific proteins have been suggested as the acceptor molecules, especially for sweet and bitter (c.f. Sato, 1987). Sugars may, in mammals, initiate transepithelial movements of Na^+ ions through amiloride sensitive channels (Schiffman et al., 1983; DeSimone et al., 1984) and, in the case of insects, also a specific glucosidase has been proposed as an acceptor (Hansen et al., 1975; Hansen, 1978). Furthermore, acetylcholin-esterase together with esterases and phosphatases are found in taste cells of vertebrates (Baradi and Bourne, 1959; Baradi and Brandes, 1963; Kurihara and Koyama, 1972; Price, 1973); they may interfere with taste cell excitation.

3.4.1. Adaptation of Taste Cells and Fibres

Rinsing the tongue with taste solutions of constant concentration elicits first an overshooting response, after which the pulse rate of taste nerve potentials decreases to a steady state. This steady state might be zero after low stimulus concentrations, or some stationary discharge. Corresponding to that adapted state of nerve discharges, the subjective sensation may reduce to zero or stay at some level above zero, but apparently there is some additional central adaptation or habituation (Cohen et al., 1955, 1957; Gent, 1979; Sato, 1980).

3.4.2. Taste Modification and Enhancement

Taste modifiers or taste enhancers might help in elaborating the hidden receptor mechanisms (Faurion et al., 1980a, b). The taste enhancer known for the longest time is glutamate, which has been in use in Eastern Asia for a long time, but has also been used in Western countries for about 10 or 15 years to an increasing degree. Glutamate enhances especially the taste of meat and meaty preparations, owing to its content of nucleotides.

Gymnemic acid or its potassium salt, from the Indian plant *Gymnema sylvestre*, blocks only the sweet sensation without influencing the other taste qualities. Sugar tastes like sand when gymnemic preparations are brought into the mouth previously or simultaneously with it.

The taste-modifying protein miraculin from Miracle fruit (*Synsepalum dulcificum*), originally grown only in Western Africa, changes the sour taste to sweet. The native Africans used to chew these red, olive-sized miracle berries or extracts of them, and then the sour taste of citrons changed to a sweet taste like that of oranges. This is caused by impregnation of the inner oral surface by miraculin; the mechanism assumed by Kurihara and Beidler (1968, 1969) is that postulated sweet receptors are raised out of the epithelial surface to enable a permanent binding between them and xylose or arabinose constituents of the miraculin molecule. This hypothesis has to be proved further, and the action of these taste modifiers might be compared to the competitive blocking action of curare at the motor endplate on a more sophisticated base. Miraculin binds to the sweet receptors only if acids produce a kind of shrinking of the receptor membrane. This shrinking causes an elevation of the sweet receptors, and the content of the xylose or arabinose sugars of the miraculin molecule is then responsible for eliciting the sweet sensations.

4. ELECTRIC TASTE

All sensory modalities can be stimulated not only in their natural ways but also in an 'inadequate way'. Strong mechanical influences like boxing or pushing the eye may elicit visual sensations such as seeing stars. This very rough method of inadequate mechanical stimulation is improved by the application of electric current, which has the very great advantage of easy and convenient control of the stimulating intensity and its time course.

For defining the role of inadequate stimulation, and for differentiating inadequate from adequate stimulation, it has to be noted that inadequate stimulation needs much higher stimulus intensities in terms of energy than adequate stimulation. Therefore the adequate stimulus, which best fits the particular receptor cells or a particular sensory pathway or sensory channel, is that which elicits sensory excitation with the lowest amount of stimulating energy.

Electric taste is very well known to anyone who has ever tried to lick the two poles of an electric battery, in which case mostly sour is perceived. It is less well known (but was mentioned in Section 3.1) that only the positive pole (anode) of an electric DC source elicits sour taste, while the negative pole (cathode) of an electric battery usually elicits soapy or alkaline taste (Bujas and Chweitzer, 1938; von Békésy, 1964*b*; Bujas, 1971; Bujas *et al.*, 1979). This DC stimulation of taste receptors on the human or animal tongue can even be used for threshold determinations as in electrogustometry (Krarup, 1958; von Békésy, 1965; Plattig, 1969).

In the case where pulses of interrupted DC current are used, of which the

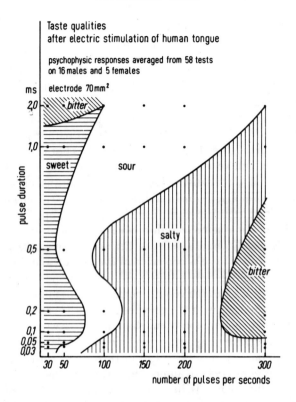

FIG. 7. Different taste qualities elicited by electric pulses of different frequency (abscissa) and duration (ordinate) applied to the anterolateral edge of human tongue via a silver–silver chloride electrode of 10 × 7 mm. The psychophysical responses were averaged from 58 tests on 16 males and 5 females. (From Plattig, 1971.)

frequency and single pulse duration are changed, different taste qualities can be elicited from the tip of the tongue (Fig. 7). One has to take into account, however, that in this case the electric pulses go through the taste papilla right down to the taste nerves, so that the originally stimulated substrates are mainly the taste nerves rather than the taste receptors (Thumfart et al., 1978). For DC electrogustometry, however, the direct current gives rise to electrolytic processes by which protons and anions are liberated from saliva and its constituents, so that in this case an adequate stimulation is finally initiated by the DC electric stimulation.

5. NORMAL AND PATHOLOGICAL STATES OF TASTE PERCEPTION

Functional testing of taste perception by adequate stimulation or by electrogustometry might usefully assist in finding out whether people have normal or decreased functions of taste. Decreased perception of taste is called hypogeusia, while the term ageusia means complete loss of taste. Both might occur following lesions of the neural pathways as described above, but also in the case of certain diseases such as hormonal disorders or after a prolonged intake of certain drugs (Henkin, 1967, 1971, 1972, 1974, 1976; Henkin and Christiansen, 1967; Henkin and Shallenberger, 1970; Henkin et al., 1967, 1972, 1975). In addition, it should be taken into account that hypogeusic or ageusic patients usually lose their appetite and the ease of intestinal functions, since secretion of saliva or of gastric or intestinal juices depends greatly on taste reflexes. Certain aversive taste stimuli may elicit vomiting, if their taste is perceived or even if not, and this latter case usually indicates that peripheral taste mechanisms and subcortical taste reflexes perform well, but the central perception does not.

A final problem deals with the genetics of taste. There are certain taste differences in man, inherited in Mendelian fashion, best known as a taste blindness to the bitterness of phenylthiocarbamide (PTC), or other compounds containing an N—C=S group (Fox, 1932; Harris and Kalmus, 1949a, b; Harris et al., 1949; Kalmus, 1958; Kalmus and Lewkonia, 1973; Fischer and Griffin, 1959; Fischer et al., 1962, 1965, 1966). Thresholds for N—C=S substances are in a characteristic way bimodally distributed with some differences between Western and Afro–Asian populations.

Finally, may I quote the hopeful statement of Ziporyn (1982): 'The period of neglect of the chemosenses may finally be over'.

REFERENCES

Altner, H. (1978). Physiology of taste. In: *Fundamentals of Sensory Physiology*, R. F. Schmidt (Ed.), Springer-Verlag, New York. pp. 218–27.

Andres, K. H. (1975). *Neue morphologische Grundlagen zur Physiologie des Riechens und Schmeckens* (New morphological results for the physiology of smell and taste). *Arch. Oto-Rhino-Lar.*, **210**, 1–41.

Arvidson, K. (1976). Scanning electron microscopy of fungiform papillae in the tongue of man and monkey. *Acta Otolaryngol.*, **81**, 496–502.

Arvidson, K. and Friberg, U. (1980). Human taste: responses and taste bud number in fungiform papillae. *Science*, **209**, 807–8.

Avenct, P., Hofmann, F. and Lindemann, B. (1988). Transduction in taste receptor cells requires cAMP-dependent protein kinase. *Nature*, **331**, 351–4.

Baradi, A. F. and Bourne, G. H. (1959). Histochemical localization of cholinesterase in gustatory and olfactory epithelia. *J. Histochem. Cytochem.*, **7**, 2–7.
Baradi, A. F. and Brandes, D. (1963). Electron microscopy of alkaline phosphatase in papilla foliata. *J. Histochem. Cytochem.*, **11**, 815–17.
Beidler, L. M. (1954). A theory of taste stimulation. *J. Gen. Physiol.*, **38**, 133–9.
Beidler, L. M. (1961). Taste receptor stimulation. In: *Progress in Biophysics and Biophysical Chemistry, Vol. 12*, J. A. V. Butler, H. E. Huxley and R. E. Zirkle (Eds), Pergamon Press, Oxford, pp. 107–51.
Beidler, L. M. (Ed.) (1971). *Chemical Senses 2—Taste (Handbook of Sensory Physiology, Vol. IV/2)*, Springer-Verlag, Berlin. 410 pp.
Beidler, L. M. and Smallman, R. L. (1965). Renewal of cells within taste buds. *J. Cell. Biol.*, **27**, 263–72.
Békésy, G. von (1964a). Duplexity theory of taste. *Science*, **145**, 834–5.
Békésy, G. von (1964b). Sweetness produced electrically on the tongue and its relations to taste theories. *J. Appl. Physiol.*, **19**, 1105–13.
Békésy, G. von (1965). Temperature coefficients of the electrical thresholds of taste sensations. *J. Gen. Physiol.*, **49**, 27–35.
Békésy, G. von (1966). Taste theories and the chemical stimulation of single papillae. *J. Appl. Physiol.*, **21**, 1–9.
Booth, D. A. (1981). The physiology of appetite. *Brit. Med. Bull.*, **37**, 135–40.
Boudreau, J. C. (1980). Taste and the taste of foods—A review and a report on a symposium. *Naturwiss.*, **67**, 14–20.
Bujas, Z. (1971). Electrical taste. In: *Chemical Senses 2–Taste (Handbook of Sensory Physiology, Vol. IV/2)*, L. M. Beidler (Ed.), Springer-Verlag, Berlin, pp. 180–99.
Bujas, Z. and Chweitzer, A. (1938). Les modifications électroniques d'excitabilité pour le goût électrique. *Compt. Rend. Soc. Biol.*, **127**, 1071–2.
Bujas, Z., Frank, M. and Pfaffman, C. (1979). Neural effects of electrical taste stimuli. *Sens. Proc.*, **3**, 353–65.
Cohen, M. J., Hagiwara, S. and Zottermann, Y. (1955). The response spectrum of taste fibers in the cat: a single fiber analysis. *Acta Physiol. Scand.*, **33**, 316–32.
Cohen, M. J., Landgren, S., Ström, L. and Zottermann, Y. (1957). Cortical reception of touch and taste in the cat. *Acta Physiol. Scand.*, **40** (Suppl. 135), 1–50.
Cowart, B. J. (1981). Development of taste perception in humans: sensitivity and preference throughout the life span. *Psychol. Bull.*, **90**, 43–73.
DeSimone, J. A., Heck, G. L., Mierson, S. and DeSimone, S. K. (1984). The active ion transport properties of canine lingual epithelia *in vitro*. *J. Gen. Physiol.*, **83**, 633–56.
Erlanger, J. and Gasser, H. S. (1924). The compound nature of the action current of nerve as disclosed by the cathode ray oscillograph. *Am. J. Physiol.*, **70**, 624–66.
Erlanger, J. and Gasser, H. S. (1937). *Electrical Signs of Nervous Activity*. University of Pennsylvania Press, Philadelphia, Oxford University Press, Oxford and London.
Faurion, A. (1982). Etude des mécanismes de la chimioreception du goût sucré. Thèse de Doctorat d'Etat, Univ. Pierre et Marie Curie, Paris 389 pp.
Faurion, A. and MacLeod, P. (1982). Sweet taste receptor mechanisms. In: *Nutritive Sweeteners*, Birch, G. G. and Parker, K. J. (Eds), Applied Science Publishers, London, pp. 247–73.

Faurion, A., Bonaventure, L., Bertrand, B. and MacLeod, P. (1980a). Multiple approach of the sweet taste sensory continuum: psychophysical and electrophysical data. In: *Proceedings of the 7th International Symposium on Olfaction and Taste and of the 4th Congress of the European Chemoreception Research Organization*, H. van der Starre (Ed.), Information Retrieval Ltd, London, p. 86.

Faurion, A., Saito, S. and MacLeod, P. (1980b). Sweet taste involves several distinct receptor mechanisms. *Chem. Senses*, **5**, 107–21.

Fischer, R. and Griffin, F. (1959). On factors involved in the mechanism of 'taste-blindness'. *Experientia*, **15**, 447–8.

Fischer, R., Griffin, F. and Mead, E. L. (1962). Two characteristic ranges of the taste sensitivity. *Med. Exp. (Basle)*, **6**, 177–82.

Fischer, R., Griffin, F., Archer, R. C., Zinsmeister, S. C. and Jastram, P. S. (1965). Weber ratio in gustatory chemoreception; and indicator of systemic (drug) reactivity. *Nature*, **207**, 1049–53.

Fischer, R., Griffin, F. and Rockey, M. A. (1966). Gustatory chemoreception in man: multidisciplinary aspects and perspectives. *Perspec. Biol. Med.*, **9**, 549–77.

Fox, A. L. (1932). The relationship between chemical constitution and taste. *Proc. Nat. Acad. Sci. (Washington)*, **18**, 115–20.

Gent, J. F. (1979). An exponential model for adaptation in taste. *Sens. proc.*, **3**, 303–16.

Guth, L. (1958). Taste buds on the cat's circumvallate papillae after reinnervation by glossopharyngeal, vagus and hypoglossal nerves. *Anat. Rec.*, **130**, 25–37.

Guth, L. (1971). Degeneration and regeneration of taste buds. In: *Chemical Senses 2–Taste (Handbook of Sensory Physiology, Vol. IV/2)*, L. M. Beidler (Ed.), Springer-Verlag, Berlin, pp. 63–74.

Hansen, K. (1978). Insect chemoreception. In: *Taxis and Behavior (Receptors and Recognition, Ser. B, Vol. 5)*, G. L. Hazelbauer (Ed.), Chapman and Hall, London, pp. 233–92.

Hansen, K., Bührer, H. and Wiecorek, H. (1975). α-Glucosidases as sugar receptor proteins in flies. *Proceedings European Chemoreception Research Organisation Symposium Wädenswil*, Information Retrieval Ltd, London, pp. 79–88.

Harris, H. and Kalmus, H. (1949a). Genetical differences in taste sensitivity to phenylthiourea and to anti-thyroid substances. *Nature*, **163**, 878–9.

Harris, H. and Kalmus, H. (1949a). The measurement of taste sensitivity to phenylthiourea (PTC). *Ann. Eugen.*, **15**, 24–31.

Harris, H., Kalmus, H. and Trotter, W. R. (1949). Taste sensitivity to phenylthiourea in goitre and diabetes. *Lancet*, **3**, 1038.

Henkin, R. I. (1967). On the mechanism of the taste defect in familial dysautonomia. In: *Olfaction and Taste II*, T. Hayashi (Ed.), Pergamon Press, Oxford, pp. 321–35.

Henkin, R. I. (1971). Disorders of taste and smell. *J. Am. Med. Ass.*, **218**, 1946.

Henkin, R. I. (1972). (I) Loss of smell and taste after respiratory infection, (II) Precaution and treatment of hypogeusia due to head and neck irradiation. *J. Am. Med. Ass.*, **220**, 870–1.

Henkin, R. I. (1974). Salt taste in patients with essential hypertension and with hypertension due to primary hyperaldosteronism. *J. Chron. Dis.*, **27**, 235–44.

Henkin, R. I. (1976). Taste dysfunction and penicillamine (Reply to a letter of Sidney Lerner, M.D.). *J. Am. Med. Ass.*, **236**, 250–1.

Henkin, R. I. and Christiansen, R. L. (1967). Taste localization on the tongue, palate, and pharynx of normal man. *J. Appl. Physiol.*, **22**, 320–6.
Henkin, R. I. and Shallenberger, R. S. (1970). Aglycogeusia: The inability to recognize sweetness and its possible molecular basis. *Nature*, **227**, 965–6.
Henkin, R. I., Keiser, H. R., Jaffe, I. A., Sternlieb, I. and Scheinberg, I. H. (1967). Decreased taste sensitivity after D-penicillamine reversed by copper administration. *Lancet*, **ii**, 1268–71.
Henkin, R. I., Talal, N., Larson, A. L. and Mattern, C. F. T. (1972). Abnormalities of taste and smell in Sjogren's syndrome. *Ann. Int. Med.*, **76**, 375–83.
Henkin, R. I., Larson, A. L. and Powell, R. D. (1975). Hypogeusia, dysgeusia, hyposmia following influenza-like infection. *Ann. Otol. Rhinol. Laryngol.*, **84**, 672–82.
Ichioka, M. and Hayashi, H. (1974). Spatio-temporal nerve impulse patterns in rat chorda tympani fibres in correlation with four primary taste qualities. *Proc. Jap. Acad.*, **50**, 392–5.
Kalmus, H. (1958). Improvements in the classification of the taster genotypes. *Ann. Human Genet.*, **22**, 222–30.
Kalmus, H. and Lewkonia, I. (1973). Relation between some forms of glaucoma and phenylthiocarbamide tasting. *Brit. J. Ophthalmol.*, **57**, 503–6.
Katsuki, Y., Sato, M., Takagi, S. F. and Oomura, Y. (1977). *Food Intake and Chemical Senses*, University of Tokyo Press, Tokyo, 614 pp.
Kimura, K. and Beidler, L. M. (1961). Microelectrode study of taste receptors of rat and hamster. *J. Cell. Comp. Physiol.*, **58**, 131–9.
Krarup, B. (1958). Electro-Gustometry: a method for clinical taste examinations. *Acta Otolaryngol. (Stockholm)*, **49**, 294–305.
Kurihara, K. and Beidler, L. M. (1968). Taste-modifying protein from miracle-fruit. *Science*, **161**, 1241–3.
Kurihara, K. and Beidler, L. M. (1969). Mechanism of the action of taste-modifying protein. *Nature*, **222**, 1176–9.
Kurihara, K. and Koyama, N. (1972). High activity of adenyl cyclase in olfactory and gustatory organs. *Biochem. Biophys. Res. Commun.*, **48**, 30–4.
Le Magnen, J. (1950). Nouvelles données sur le phénomène de l'exaltolide. *Compt. Rend. Acad. Sci. (Paris)*, **230**, 1103–5.
Maes, F. W. and Erickson, R. P. (1982). Intensity coding in rat taste organ. *Proceedings Vth Congress European Chemoreception Research Organisation, Regensburg*.
Miller, I. J. (1971). Peripheral interaction among single papillae inputs to gustatory nerve fibers. *J. Gen. Physiol.*, **57**, 1–25.
Molen, J. N. van der (1982). A study on coding and variability in taste responses of the blowfly. Diss. (rer. nat.) Gröningen, 149 pp.
Murray, R. G. (1971). Ultrastructure of taste receptors. In: *Chemical Senses 2— Taste. (Handbook of Sensory Physiology, Vol. IV/2)*, L. M. Beidler (Ed.), Springer-Verlag, Berlin, pp. 31–50.
Murray, R. G. and Murray, A. (1960). The fine structure of the taste buds of rhesus and cynomalgus monkeys. *Anat. Rec.*, **138**, 211–33.
Murray, R. G. and Murray, A. (1967). The fine structure of taste buds of rabbit foliate papillae. *J. Ultrastruct. Res.*, **19**, 327–53.
Netter, F. H. (1972). *The CIBA Collection of Medical Illustrations, Vol. 1: Nervous System*, CIBA, Basel, 168 pp.

Oakley, B., Chu, J. S., Jones, L. B. and Sloan, H. E. (1980). Taste bud structure and function depend upon axonal transport. *Proceedings of the 7th International Symposium on Olfaction and Taste and of the 4th Congress of the European Chemoreception Research Organization*, H. van der Starre (Ed.), Information Retrieval Ltd, London, p. 286.

Pfaffmann, C. (1955). Gustatory nerve impulses in rat, cat and rabbit. *J. Neurophysiol.*, **18**, 429–40.

Pfaffmann, C. (1964). Taste, its sensory and motivating properties. *Am. Scientist*, **52**, 187–206.

Pfaffmann, C., Erickson, R. P., Frommer, G. P. and Halpern, B. P. (1961). Gustatory discharges in the rat medulla and thalamus. In: *Sensory Communication*, W. A. Rosenblith (Ed.), MIT Press, Cambridge, Mass., pp. 455–73.

Pfaffmann, C., Frank, M., Bartoshuk, L. M. and Snell, T. C. (1976). Coding gustatory information in the squirrel monkey chorda tympani. In: *Progress in Psychobiology and Physiological Psychology*, J. M. Sprague and A. N. Epstein (Eds), Academic Press, New York, pp. 1–27.

Plattig, K.-H. (1969). Über den elektrischen Geschmack. Reizstärkeabhängige evozierte Hirnpotentiale nach elektrischer Reizung der Zunge des Menschen. (Habilschr. Erlangen 1968). *Z. Biol.*, **116**, 161–211.

Plattig, K.-H. (1971). Elektrischer Geschmack, *Umschau*, **71**, 64.

Plattig, K.-H. and Innitzer, J. (1976). Taste qualities elicited by electric stimulation of single human tongue papillae. *Pfluegers Arch.*, **361**, 115–20.

Plattig, K.-H., Kobal, G. and Thumfart, W. (1980). Die chemischen Sinne Geruch und Geschmack im Laufe des Lebens—Veränderungen der Geruchs- und Geschmackwahrnehmung. *Z. Gerontol.*, **2**, 149–57.

Price, S. (1973). Bitter stimuli and phosphodiesterase on tongue epithelium. *Nature*, **241**, 54–5.

Renquist, Y. (1919). Über den Geschmack. *Scand. Arch. Physiol.*, **38**, 97–201.

Renquist, Y. (1920). Der Schwellenwert des Geschmacksreizes bei einigen homologen und isomeren Verbindungen. *Scand. Arch. Physiol.*, **40**, 117–24.

Rohen, J. W. (1978). *Funktionelle Anatomie des Nervensystems*, F. K. Schattauer, Stuttgart, 380 pp.

Sato, M. (1987). Taste receptor proteins. *Chem. Senses*, **12**, 277–83.

Sato, T. (1980). Recent advances in the physiology of taste cells. *Prog. Neurobiol.*, **14**, 25–67.

Sato, T. (1982). Characteristics of rat taste cell responses to four basic taste stimuli. *Proceedings of the 16th Symposium on Taste and Smell, Japanese Association for Study of Taste and Smell*, Chuo University, Tokyo.

Schiffman, S. S. and Erickson, R. P. (1974). The range of gustatory quality: psychophysical and neuro approaches. *Proceedings 1st Congress European Chemoreception Research Organisation, Paris*, p. 10.

Schiffman, S. S., Moss, J. and Erickson, R. P. (1976). Thresholds of food odors in the elderly. *Exp. Aging Res.*, **2**, 389–98.

Schiffman, S. S., Hornack, K. and Reily, D. (1979a). Increased taste thresholds of amino acids with age. *Am. J. Clin. Nutr.*, **32**, 1622–7.

Schiffman, S. S., Orlandi, M. and Erickson, R. P. (1979b). Changes in taste and smell with age: biological aspects. In: *Sensory Systems and Communications in the Elderly*, J. M. Ordy and K. Brizzee (Eds), Raven Press, New York, pp. 247–68.

Schiffman, S. S., Cahan, H. and Lindley, M. G. (1981). Multiple receptor sites mediate sweetness: evidence from cross-adaptation. *Pharmacol. Biochem. Behav.*, **15**, 377–92.

Schiffman, S. S., Lockhead, E. and Maes, F. W. (1983). Amiloride reduces the taste intensity of Na and Li salts and sweeteners. *Proc. Nat. Acad. Sci.* (*Washington*), **80**, 6136–40.

Skramlik, E. von (1926). *Handbuch der Physiologie de niederen Sinne. Bd. 1: Die Physiologie des Geruchs- und Geschmacks-sinnes*, G. Thieme, Leipzig, 532 pp.

Teeter, J., Funakoshi, M., Kurihara, K., Roper, S., Sato, T. and Tonosaki, K. (1987). Generation of the taste cell potential. *Chem. Senses*, **12**, 217–34.

Thumfart, W., Plattig, K.-H. and Laumer, R. (1978). Klinische Erfahrungen mit der Impulsgustometrie, einer neuen elektrischen Geschmacksprüfung im vergleichenden klinischen Einsatz. *Laryng. Rhinol.*, **57**, 134–42.

Tomita, H. and Pascher, W. (1964). Über die Geschmacksfunktion nach Ausfall der sensorischen Zungennerven. *HNO* (*Berlin*), **12**, 163–9.

Yamamoto, T. (1983). Neural mechanisms of taste function. In: *Frontiers of Oral Physiology, Vol. 4*, Y. Kawamura (Ed.), Karger, Basel, pp. 102–30.

Ziporyn, T. (1982). Taste and smell: the neglected senses. *J. Am. Med. Ass.*, **247**, 277–9; 282–5.

Chapter 2

THE SENSE OF SMELL

J. A. MARUNIAK

*Division of Biological Sciences, University of Missouri,
Columbia, Missouri, USA*

1. INTRODUCTION

This chapter will review the role of the sense of smell in the sensory processing of foods. While most people equate smell with olfaction, it is now known that there are contributions from the gustatory and trigeminal systems that make up or influence those sensations commonly referred to as smell. Nonetheless, olfaction alone probably does contribute the most crucial sensory information for assessing the quality and palatability of foods and is therefore of overriding importance in determining what a lay person would call the 'taste', and the professional would call the 'flavour', of foods. It should be remembered that people refer to the taste of foods when talking about a complex sensation evoked by stimulation of all the oral and nasal chemosensory systems because the brain blends the information from the individual systems into a single perceptual gestalt.

In our attempts to understand the function of smell in the sensory analysis of foods we will begin by examining the process of eating and identify the role played by the sense of smell at each stage. Generally, the first stage in the analysis of food by chemosensory systems occurs before the item is placed in the mouth. At this stage, and probably only at this stage, the food is analysed almost entirely by olfaction. Volatile odours diffusing from the food are drawn through the nose and detected by the olfactory receptor cells therein. The resulting neural activity of the receptor cells is processed by the brain and an odour sensation or smell is engendered for that food. Assuming a favourable response to the smell of this food, the item is placed into the mouth and the second stage of chemosensory analysis of foods begins.

While the main physiological function of the mouth is to prepare food for chemical digestion, a number of things happen during this process that are important for chemosensory analysis. First, food is usually chewed into smaller pieces and moved around the mouth in a process called mastication. At the same time it is mixed with saliva and its temperature tends to equilibrate with oral temperature. Mastication, therefore, increases the surface area and exposure of new surfaces of the food for chemical digestion and in the process causes an increase in the release of odours, tastants and pungent compounds from the food. Compounds in the food bolus partition into the saliva and air-spaces in the mouth according to their solubilities and vapour pressures. Clearly the volatility, especially of polar odours, is enhanced in the warm, humid environment of the mouth. Chewing movements cause tastants and pungent compounds in saliva to stimulate the gustatory and trigeminal receptors throughout the mouth and cause food volatiles to pass to the back of the mouth and into the respiratory airstreams of the nasopharynx. During expirations these volatiles are carried up into the nose and stimulate the olfactory receptor cells and nasal trigeminal receptors via the so-called 'retronasal' route.

The sensory analysis of the food item by oral and nasal chemosensory systems ends, for the most part, after the food bolus is swallowed. It is therefore, as described, a two-stage process. The rest of this chapter will review what is known about the details and mechanisms of this two-stage process with a predominant emphasis on the role of nasal chemoreception, particularly olfaction.

The location of the nasal chemoreceptors of man that are known to be involved in the analysis of foods are shown in Fig. 1. The trigeminal chemosensory system responds to many odorants (Tucker, 1963, 1971; Silver and Moulton, 1982; Silver et al., 1986) and mediates an array of physiological reflexes which can affect the olfactory system (for a review see Silver and Maruniak, 1981). The trigeminal system is clearly involved in the perception of noxious stimuli within the nasal cavity, but there is evidence from normal and anosmic humans, pigeons, and dogs that the trigeminal system contributes to the sense of smell, particularly intensity (Allen, 1937; Cain, 1974, 1976; Doty et al., 1978; Walker et al., 1979; Cain and Murphy, 1980; Silver and Maruniak, 1981). A third nerve in the nasal cavity of man, the terminal nerve, may also be chemosensitive (Brookover, 1914, 1917; Pearson, 1941, 1942; Demski and Northcutt, 1983; Jennes, 1986; for a review see Wysocki, 1979). Its receptive field within the human nasal cavity is not clear, although in other mammals it projects to areas near the external nares (Bojsen-Møller, 1975). Its cell bodies and fibres appear to be the substrate for an extensive gonadotropin releasing hormone (GnRH) system

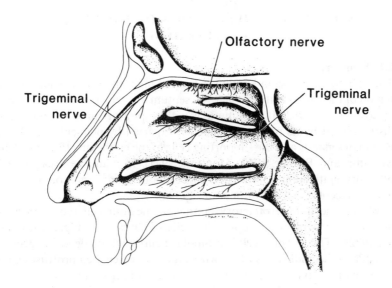

FIG. 1. A representation of the lateral wall of the human nasal cavity showing the nasal turbinates and distributions of olfactory and trigeminal nerves.

that innervates the olfactory system and may be involved in coordinating olfaction with reproductive behaviour and physiology (Jennes and Stumpf, 1980; Phillips et al., 1980; Schwanzel-Fukuda and Silverman, 1980; Demski and Northcutt, 1983; Macrides and Davis, 1983; Meredith, 1983; Witkin and Silverman, 1983; Jennes, 1986).

While man does not possess a functional accessory olfactory system (Wysocki, 1979; Johnson et al., 1985), it should, at least, be mentioned during a discussion of the human olfactory system since much of what is known about the anatomy and physiology of the olfactory system is based upon animals in which the accessory olfactory system plays a major role in chemosensory behaviour (for reviews see Wysocki, 1979; Meredith, 1983; Wysocki et al., 1985). In addition, many other mammals possess a chemoreceptive septal organ (see Marshall and Maruniak, 1986). The sense of smell in man therefore results from stimulation of chemoreceptors of the olfactory, trigeminal and, possibly, terminal nerves. Olfaction contributes by far the predominant component and will therefore receive most of our attention. Readers are referred to several reviews for more extensive information on terminal and trigeminal nerve chemoreception (Silver and Maruniak, 1981, Demski and Northcutt, 1983; Keverne et al., 1986).

2. STRUCTURE AND FUNCTION OF THE OLFACTORY EPITHELIUM

2.1. Structure

The olfactory mucosa in humans is located high-up in the dorsoposterior region of the nasal cavity on the roof, septum and superior turbinates (Fig. 1; Lovell *et al.*, 1982). It is contiguous with respiratory mucosa and can be distinguished by its pale yellow colour imparted by pigment granules in Bowman's glands and the supporting cells of the epithelium (Bloom, 1954; Moulton and Biedler, 1967; Moulton, 1971). In man, the area of each olfactory epithelium is estimated to be about $1\,cm^2$, and contain three million receptor cells (Moran *et al.*, 1982a).

Within the human olfactory epithelium lie four cell types: receptor neurons, microvillar cells, supporting cells and basal cells (Fig. 2; Moran *et al.*, 1982b). The receptor cell is a bipolar neuron whose dendritic process extends to the epithelial surface where it ends in a ciliated protrusion, the olfactory knob (Graziadei, 1971a). In man it has been estimated that 10–30 cilia extend from each olfactory knob into the mucous layer (Ohno *et al.*, 1981). The microvillar cell has only recently been described in humans (Moran *et al.*, 1982b). It is thought to be a second type of sensory neuron whose function is presently unknown. Microvillar cells are flask-shaped and similar to other bipolar sensory neurons. They possess a cytoplasmic process that appears to be an axon growing from the basal pole of the cell toward the lamina propria and short microvilli extend from their apical surface into the mucous layer. A different kind of microvillar cell is the only type of receptor cell in the vomeronasal organs of the accessory olfactory system of many species.

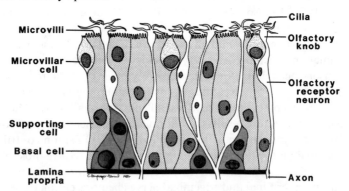

FIG. 2. A representation of the structure of the human olfactory epithelium.

Supporting cells are columnar cells whose cytoplasm extends from the basement membrane to the epithelial surface where their surfaces are formed by microvilli which extend up into the mucous layer (Fig. 2). They are joined to other supporting cells and to olfactory receptor cells by tight junctions which extend around the edge of all cells close to the epithelial surface, presumably providing a barrier to penetration of the epithelium by foreign matter. The functions of the supporting cells are not clear. Like glial cells in the central nervous system, they may provide secretory, supportive, nutritive or electrically insulative or buffering properties essential for the functioning of receptor cells. It is clear that they help produce mucus and probably clear odours from the olfactory epithelium (Moran et al., 1982a; Jafek, 1983; Getchell et al., 1987). Graziadei (1971b) has suggested that these cells may be the structural substrate within which receptor cells maintain orderly orientations with respect to each other. In frogs and turtles he has observed that receptor cells are sometimes arranged in interlocking rings centred around groups of supporting cells. It is not known if these orderly arrangements of receptor cells are present in humans or if they have any functional significance.

Lying close to the basement membrane are the basal cells (Fig. 2). These are cells which undergo division and give rise to new receptor neurons (Moulton et al., 1970; Thornhill, 1970; Graziadei and Metcalf, 1971; Graziadei et al., 1980). The average lifespan of receptor cells for a number of species has been estimated to be about a month (Moulton, 1975; Graziadei and Monti-Graziadei, 1979; Samanen and Forbes, 1984). Ultrastructural analysis has indicated that olfactory receptor neurons can undergo neurogenesis in all vertebrates studied, including primates (Graziadei et al., 1980). Such plasticity in the receptive substrate for olfaction has interesting implications for any theory of quality coding, since smell sensations seem to remain constant in spite of the continuous renewal of receptor neurons and their axonal connections with the olfactory bulbs.

The olfactory epithelium is perforated in places by ducts of the Bowman's glands which originate beneath it in the lamina propria. These glands, found only in olfactory mucosa, are compound tubulo-alveolar structures and appear to be combined mucous-serous glands (Bloom, 1954; Bang, 1964; Bojsen-Møller, 1964; Graziadei, 1971a). It is thought that the mucous sheet covering the epithelium is composed primarily of secretions from the Bowman's glands (Moulton and Beidler, 1967; Getchell et al., 1987). In man it moves at a variable speed of from 0 to 14 mm min^{-1} over the mucosal surface (Andersen et al., 1971; Simon et al., 1977; Puchelle et al., 1981). Since odorants must dissolve in the mucus before reacting with olfactory receptor cells (Getchell et al., 1980), mucus movement may be one of the main routes by which odorants are cleared from the nasal cavity. Recently Getchell et al. (1987) have shown increased mucus secretion by Bowman's glands and supporting cells in response to odour stimulation. (For more detailed descriptions of the olfactory epithelium see Moulton and Beidler, 1967; Graziadei, 1971a; Vinnikov, 1974; Moulton, 1976.)

2.2. Function

The site of the odour-receptor cell interaction is thought to be the cilia of the receptor cells (Atema, 1973, 1975; Ottoson, 1973; Menco et al., 1976, Bronshtein and Minor, 1977; Getchell et al., 1980; Rhein and Cagan, 1980; Adamek et al., 1984, Lancet, 1986; Nakamura and Gold, 1987). Cilia have a very large surface area, are the first part of the receptor cell an odour molecule would encounter, and therefore are the most logical structures upon which the molecular receptive sites could be located. The molecular receptive sites are thought to be protein, lipid, lipoprotein, glycoprotein, or phospholipid (Davies and Taylor, 1954; Davies, 1965, 1971; Ash and Skogen, 1970; Dodd, 1971; Koyama and Kurihara, 1972; Gennings et al., 1977; Sperber, 1977; Cagan and Zeigler, 1978; Fesenko et al., 1978, 1979; Price, 1978, 1981; Chen and Lancet, 1984; Baldaccini et al., 1986).

It has only been in the past few years that much progress has been made in elucidating the molecular mechanisms by which odours interact with, and are transduced by, the olfactory receptor neurons. Our present state of knowledge can be summarised as follows. The nature of the initial event in transduction, the interaction of an odour molecule with an olfactory receptor cell, remains vague. Essentially, there is a dichotomy between those that argue for, and those that argue against, the existence of odour-specific proteinaceous receptor sites on the cilia.

The most recent arguments against specialised receptor sites emphasise that there is no reason to invoke the existence of such sites to explain most of the extant data. Rather, factors such as the differential absorption and metabolism of odours, and odour-induced changes in receptor cell metabolism and membrane fluidity can account for the known responses of olfactory receptor neurons, given a certain amount of variability in membrane lipids and non-odour receptor proteins (Price, 1984; Kashiwayangi and Kurihara, 1985). Studies of the responses of a non-olfactory neuroblastoma cell line to odours have been particularly compelling. These reveal that the neuroblastoma cells respond to stimulation by a variety of odours in a manner that is very similar to the responses of olfactory receptor neurons (Kashiwayangi and Kurihara, 1985). This suggests that no specialised receptors are necessary to account for the responses of olfactory receptor neurons to odours.

Arguments for the existence of specialised receptors for odours tend to emphasise special cases where data cannot be adequately explained by other models. For example, no one has been able to rationalise satisfactorily the marked qualitative differences in the odours of enantiomers such as + and − carvones without invoking the existence of specialised receptor sites (Friedman and Miller, 1971; Leitereg et al., 1971; Russell and Hills, 1971; Ohloff, 1980; Ohloff et al., 1986). Further, the results of a number of studies on the effects of reagents that chemically modify olfactory receptor neurons, on subsequent responses to odours, appear to be unexplainable without the existence of specialised odour receptor sites (Getchell and Gesteland, 1972; Menevse et al., 1978; Delaleu and Holley, 1983; Shirley et al., 1983; Schafer et al., 1984; Mason et al., 1985). Finally, Ohloff (1986) argues that the large changes in odour quality and intensity of a molecule, that sometimes result from only minor structural changes, indicate that specialised odour receptive sites must be involved.

Regardless of whether specialised odour receptor sites mediate the initial event in transduction, the ensuing events are becoming clear. Olfactory receptor neurons, like many other sensory and non-sensory cells, appear to use common second messenger pathways to communicate signals transduced on their surface membranes to the appropriate parts of the cell. Thus when an odour interacts with a responsive olfactory receptor neuron it initiates changes in the G-proteins that span the neuronal membrane (Heldman and Lancet, 1986; Lancet, 1986; Pace and Lancet, 1986; Bruch and Kalinoski, 1987, Sklar et al., 1987). These G-proteins initiate either the formation of cyclic nucleotides (cAMP or cGMP) or the increase of calcium ions inside the cell, both of which activate protein kinases (Menevse et al., 1977; Pace et al., 1985; Heldman and Lancet, 1986; Lancet, 1986; Nakamura and Gold, 1987; Vodyanoy and Vodyanoy, 1987). Activation of a protein kinase allows phosphorylation of probably one type of the membrane's proteinaceous ion channels resulting in a localised conductance change in the receptor neuron (Nakamura and Gold, 1987; Vodyanoy and Vodyanoy, 1987).

In general, stimuli that affect sensory receptors cause an increase in membrane conductance which results in depolarisation of the receptor cell (Schmidt, 1978). Presumably the same is true for olfactory stimuli and receptor cells. Experimental evidence suggests that the ion(s) responsible for depolarisation of olfactory receptors is potassium (Vodyanoy and Murphy, 1983), or sodium and potassium (Takagi et al., 1968; Minor, 1971; Getchell, 1986).

A rather extensive literature has been published characterising the responses of olfactory receptor cells (Gesteland et al., 1963, 1965; Lettvin and Gesteland, 1965; Døving, 1966; Mathews and Tucker, 1966; Altner and Boeckh, 1967; Shibuya and Tucker, 1967; O'Connell and Mozell, 1969; Shibuya, 1969; Kafka, 1971; Mathews, 1972a; Blank, 1974; Duchamp et al., 1974; Holley et al., 1974; Getchell and Getchell, 1975; Baylin, 1979; Revial et al., 1982; Sicard and Holley, 1984; Sicard, 1986). In purified air conditions, olfactory receptor cells generate impulses at a 'resting rate' that is relatively low for sensory receptors, of about 0.05–$2.0 \, s^{-1}$. When receptor cells are exposed to an odour they may increase, decrease or show no change in their rate of impulse generation compared to their resting rates. Almost all of these data were collected in non-mammalian vertebrates and they show that individual olfactory receptor cells seem to respond to a large, though apparently unique, set of odorants. A recent report on mouse olfactory receptor cells suggests that the responses of mammalian receptor cells may be more specific than those of other vertebrates (Sicard, 1986). In any case this broad responsiveness has hindered researchers in finding correlations between receptor cell responses and the physical or chemical properties of odours. Indeed, studies have shown that even non-odorous compounds such as urea, glucose, sucrose and testosterone can elicit olfactory responses when presented under the right conditions (Getchell, 1969; MacLeod et al., 1979).

Insects also have broadly responsive receptor cells but, in addition, possess receptor cells which are specialised for the detection of biologically important odours such as pheromones and food odours (Schneider, 1971). There appear to be only two reports of such specialised receptor cells in vertebrates. Minor and Vasileva (1980) and Minor et al. (1980) have found highly specialised olfactory receptor cells in the rat for mediating responses to sex pheromones.

2.3. Structure-Activity Relationships and the Olfactory Code

According to Beets (1978): 'The structure of any molecular species represents, under specific physical conditions, the only source of information carried by the molecules... Consequently, the question whether a relationship between molecular structure and odour exists is meaningless. The only legitimate question is whether it is simple enough to be detected'. The relationship between molecular structure and odour quality is the olfactory code. There are many theories relating odour quality to various molecular properties. Whereas some workers attempt to provide a molecular basis for all odour qualities, others try to explain only a subset of qualities. In most cases these theories have not been very successful in predicting the odorous qualities of molecules other than those from which the theory derives. Nonetheless, all deserve some consideration since they represent diverse and serious attempts to systematise what seems to be a particularly complex relationship between structure and function. In addition, not all odour–receptor interactions may occur through the same mechanism and therefore a unified theory of olfactory coding may not be possible.

In the puncturing theory of Davies (1962, 1969, 1971), olfactory receptor neurons are depolarised by ions flowing through holes punched in the receptor membrane by odour molecules. This theory, therefore, falls into the group that holds that specialised odour receptive sites do not exist. Odour quality would depend on variations in current flow which result from, and are regulated by, the time for diffusion of an odour through the membrane and the time for the consequent hole to heal. Cherry et al. (1970) attempted to test the puncturing theory by studying the interactions between small molecules and artificial lipid membranes. They found no correlation between the threshold value for an odour and the resistance changes induced by the odour in their lipid membranes and therefore could not support the theory.

The infrared theory of olfaction was proposed by Wright and maintains that molecular vibration is the most important parameter in determining odour quality (Wright, 1954, 1974; Wright and Burgess, 1969, 1971, 1975). In Wright's words: 'There is now persuasive evidence that specific patterns of low-frequency molecular vibration determine the olfactory responses that molecules exhibit' (Wright, 1977). Critics have said that such vibrations merely provide correlations with other more meaningful structural parameters (Beets, 1978). Nonetheless support for the theory still exists (Belmont and Ambarek, 1985). A more important insufficiency in this theory is its inability to account for differences in odours of enantiomers (which would have identical vibrations) and particularly those enantiomers where one has an odour while the other does not (Friedman and Miller, 1971; Leitereg et al., 1971; Russell and Hills, 1971; Ohloff, 1980).

The hydrogen bonding theory of olfaction as proposed by Randebrock (1971) holds that the ability of many odours to form one or more hydrogen bonds provides the mechanism for transmitting odour quality. Randebrock suggests that odour–receptor interactions are mediated by membrane-helical proteins. While such proteins are present in the surface membranes of all cells, in the olfactory receptor cells the hydrogen bonding sites of each protein could uniquely interact with an odour and transmit its quality.

The oxidative theory of olfaction (Zinkevich and Minor, 1969; Zinkevich and Treboganov, 1973; Vinnikov et al., 1979) has its origins in the observations that odours are partially oxidised in the olfactory mucosa (Zinkevich and Minor, 1969) by molecular oxygen (Zinkevich and Treboganov, 1973). It holds that odour sensations originate in the enzyme-catalysed oxidations of odours in which molecular oxygen acts as a cofactor that permits odour–receptor interactions.

The stereochemical theory of odour quality is the most popular and best supported theory for how odour quality relates to molecular structure. It holds that the three-dimensional shape of a molecule, or part of a molecule, is the most important molecular parameter that the olfactory system uses to assign odour quality (Pauling, 1946; Amoore, 1952). For example, Amoore has observed a strong correlation between the psychophysical assessment of odour quality for 107 odours and the size and shape of the odour molecule (Amoore, 1965; Amoore et al., 1967). Ohloff (1980, 1986) has extensively studied the variation of odour quality with changes in odour molecular parameters and concludes that stereochemistry is of overriding importance. Beets (1964) has suggested that only part of a molecule may be responsible for the quality of an odour. It is now known that molecules with very different overall structures but similar partial structures can have similar odour qualities (Beets and Thiemer, 1970; Boelens, 1974; Ohloff, 1986). For example, Polak et al. (1978) were able to attribute an 'earthy' note to the part of a decalin molecule containing a five- or six-membered carbon ring with α-methyls or methylenes on both sides of a hydroxy group. Interestingly, they found that some subjects were able to focus on this 'earthy' note while sampling the pure compound, and believed they were smelling a mixture. Wright (1982) has reconciled his vibration theory with the growing support for the importance of stereochemistry by proposing that the odour–receptor interaction involves only partial odour structures, and that vibration and stereochemistry combine to determine the interaction.

While the proponents of the various versions of the stereochemical theory believe that odour quality depends mostly on the shape of parts or all of the molecule, many also believe that for small molecules odour quality depends mainly on the nature of the functional groups (Polak, 1973; Beets, 1975). Klopping (1971) was the first to propose that small molecules such as methyl cyanide, methyl isocyanide, dimethyl ether and dimethyl sulphide must derive their odour quality primarily from the electronic properties of their functional groups. He argued that their odour qualities are so different, yet shapes so similar, that the functional groups must predominate over shape in importance for quality. Beets (1975) believes that molecular shape can play only a minor role in the quality coding of small molecules and suggests that hydrogen bonding must be the critical parameter in interactions of receptors with such small molecules. Wright (1982) also believes that small molecules are unique in that they can more easily reach and transmit their vibrational characteristics to receptor sites.

Finally, there is evidence that even for large molecules both stereochemistry and functional groups can play a role in determining odour quality. Ohloff and Giersch (1980) have described a group of mostly flowery smelling odours up to 250 MW which possess hydroxy and carbonyl functional groups in the hydrophobic region of the molecule. If the two functional groups were separated by less than 3 Å the molecules possessed an odour, but they were odourless if the separation was more than 3 Å. They assumed that the functional groups interacted with a complementary system in the receptor site through intramolecular hydrogen bonding (see Randebrock's hydrogen theory above). Another class of odours that seem to act according to these criteria are those possessing the caramel odour (Ohloff and Giersch, 1980). Further evidence for the importance of functional groups for odour quality comes from an interesting report by Schafer and Brower (1975). They found that a panel of 73 organic chemists were reasonably successful in identifying the functional groups of 36 unknown and unfamiliar odorants. Their correct identification of functional groups ranged from 86% to 50% for odours containing amines, sulphur, esters, phenols and carboxylic acids, respectively. Aldehyde and ketone groups were recognised 42% of the time; hydrocarbons, 36%; alcohols, 25%; ethers, 20%; and halides, 16%. In a follow-up study, when the best subjects were tested for ability to identify sterically hindered functional groups, success was significantly lower than with unhindered groups.

Most of these theories have used a comparison of odour molecular properties with psychophysical evaluation of quality to speculate about the underlying biochemical and physiological mechanisms. Other theories of olfactory quality coding have arisen from electrophysiological studies of receptor cells. In such studies no one has been able to find simple relationships between receptor cell responsiveness and particular classes of chemical compounds or their odours. As a result, theories deriving from these studies have emphasised that the code for odour quality lies in the responses of populations rather than of individual receptor cells. Adrian

(1950) envisaged that such population responses could result in spatial and temporal patterns of receptor activity as odours passed through the nose. Spatial patterns of receptor cell responses arise from two sources. First, odours may be distributed differently across the epithelium by their interactions with the mucus-lined nasal passages as they travel from the external to the internal nares: the 'chromatographic model' of olfaction (Mozell, 1964a, b, 1966, 1970, 1971; Mozell and Hornung, 1981). Second, receptor cells having different sensitivities may be distributed differentially across the epithelium (Adrian, 1950; Moulton, 1967; Mozell, 1971). This mechanism has been termed 'inherent patterning' (Moulton, 1976). In fact, differential receptor sensitivities across the olfactory receptor sheet have been demonstrated in many, but not all, species tested (frog—Mustaparta, 1971; rat—Thommesen and Døving, 1977; salamander—Kubie and Moulton, 1979; Kubie et al., 1980; Mackay-Sim and Kubie, 1981; Mackay-Sim et al., 1982; char—Thommesen, 1982) and Erickson and Caprio (1984) report no differential distribution of sensitivity in the catfish. Electrophysiological and autoradiographic studies in the salamander reveal that receptors with similar response affinities seem to be grouped together on the olfactory epithelium so that each odour elicits a characteristic topographic pattern of receptor responses that is constant over a range of concentrations (Kubie and Moulton, 1979; Kubie et al., 1980; Mackay-Sim and Kubie, 1981; Nathan and Moulton, 1981; Mackay-Sim et al., 1982; Mackay-Sim and Shaman, 1983).

In addition to spatial variables, temporal patterns of receptor responses may also contribute to olfactory coding. Mozell (1966) found that when different odours passed through the frog's nose they stimulated the same region of the epithelium with different latencies. Adrian (1950) had observed similar temporal differences in responses of cells in the olfactory bulb. Thus, the quality of an odour may be based upon both where and when it elicits receptor responses. As Mozell (1971) has stated: 'These two possible mechanisms for discrimination, selective receptor sensitivity and analysis across regions, need not be mutually exclusive, and by operating together they can generate a combined code with many more permutations than could either alone'.

Anatomical studies have not revealed any systematic regional differences in receptor cell morphology which might underlie the inherent patterning phenomenon (Graziadei and Monti-Graziadei, 1979; Breipohl et al., 1982). However, studies in two amphibians reveal an anterior-to-posterior gradient in epithelial thickness, with the anterior being thicker and containing more receptor cells than the posterior (Dubois-Dauphin et al.,

1980; Mackay-Sim and Patel, 1984). In addition, a gradient in receptor cell neurogenesis has been reported, with the posterior, thinner epithelium undergoing turnover much more rapidly than the anterior (Mackay-Sim and Patel, 1984).

Further support for the possible importance of receptor activity patterns in the coding of odour quality can be found in the existence of odour evoked patterns of activity in the olfactory bulbs. These spatial patterns of activity probably arise, in part from the topographic projection of similar patterns from the epithelium (Adrian, 1950; Le Gros Clark, 1951; Land, 1973; Land and Shepherd, 1974; Sharp *et al.*, 1975; Skeen, 1977; Costanzo and O'Connell, 1978, 1980; Dubois-Dauphin *et al.*, 1980; Kauer, 1981). However, even at the level of the olfactory bulbs, neurons are rather broadly responsive to odours (see following section). Projections beyond the olfactory bulbs take the form of a non-classical topographic pattern (Haberly, 1985). Haberly (1985) concludes that the olfactory system does not have specialised feature detector cells like other sensory systems. Instead of cells being tuned to primary odours, he believes that odour quality is encoded at all levels in ensemble form.

In summary, physiological and anatomical studies suggest that within the olfactory epithelium, odour quality is encoded by rather broadly tuned receptor cells whose responses create a topographic and temporal ensemble pattern which comprises the unprocessed representation of odour quality to the brain. Such spatio-temporal patterns could arise from the distribution of odour molecules 'imposed' by their physical interactions with nasal passages and mucus, from breathing and sniffing patterns, and from genetically determined differences in distributions of receptor cells. However, with the present state of knowledge none of the theories of olfactory coding fully explain our odour sensations and in none of the proposed theories have the roles of mixtures of odours, odour concentration and flow rate been clearly delineated.

3. STRUCTURE AND FUNCTION OF THE OLFACTORY BULBS

In man, the axons of the olfactory receptor neurons from each nasal cavity pass to the ipsilateral bulb in about 10–20 bundles. In the bulbs they synapse superficially with the dendrites of mitral and tufted cells in structures called glomeruli. The axons from approximately 26 000 receptor cells converge on each glomerulus which is innervated by about 24 mitral and 70 tufted cells (Allison and Warwick, 1949). Mitral and tufted cells are the output cells of the olfactory bulbs and their bodies lie within deeper layers of the bulbs. Their axons pass from the bulbs in the lateral olfactory tract (LOT) to the olfactory cortex.

Electrophysiological studies of mitral cells reveal that they have higher spontaneous and odour-evoked firing rates than receptor cells and also show more rapid and longer lasting odour-induced adaptation (Mozell, 1962; Døving, 1964; Mathews, 1972b; Potter and Chorover, 1976; Baylin and Moulton, 1979; Chaput and Holley, 1979; Mair, 1982a, b). Individual mitral cell responses have proven to be more difficult to classify than those of receptor cells. Mitral cells respond to many odours and may be excited, or suppressed, or respond with a temporal combination of these (Walsh, 1956; Mancia et al., 1962; Døving, 1964; Higashino et al., 1969; Pfaff and Gregory, 1971; Mathews, 1972a, b; Kauer, 1974, 1977; Boulet et al., 1978; Mair, 1982a). Recently, Chaput and Holley (1985) found that in awake freely-breathing rabbits, the overall olfactory bulb response to odours was a reorganisation of spontaneous levels of neural activity without an increase in total activity. Part of the complexity of mitral cell responses may be a function of odour concentration: some mitral cells appear to be 'concentration tuned', firing at a relatively constant rate to an optimum odour concentration, but showing suppression above and below this concentration (Kauer, 1974, 1977; Kauer and Shepherd, 1977). Meredith (1986) attributes the complexity of mitral cell responses to the actions of lateral inhibitory circuits in the bulb. In any case, attempts to correlate similarities of mitral cell responses with psychophysical or molecular properties of odours have been compounded by the complexity of the responses.

The extreme convergence of receptor cells onto the mitral and tufted cells (about 250 receptor cells: one output cell) further complicates attempts to understand olfactory coding. It is speculated that such convergence allows the olfactory system to detect extremely low odour concentrations by increasing the probability that an output cell will respond even when only a small number of receptor cells are stimulated (Holley and Døving, 1977; Van Drongelen et al., 1978).

In addition to the mitral and tufted cells there are eight types of interneurons which must contribute heavily to the processing of information within the olfactory bulb (Schneider and Macrides, 1978). Periglomerular and granule cells make direct synaptic contact with the mitral and tufted cells and are thought to modify their activity through potent inhibition (Yamamoto et al., 1963; Nicoll, 1972; Getchell and

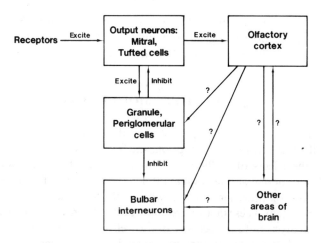

FIG. 3. A representation of the functional interconnections of the neurons of the olfactory bulbs.

Shepherd, 1975a,b; Macrides, 1977; Nakashima et al., 1978; Jahr and Nicoll, 1982; Meredith, 1986). The other interneurons probably interact only with the granule cells (Price and Powell, 1970; Schneider and Macrides, 1978). Some of these may indirectly affect mitral and tufted output by inhibiting granule cell inhibition (Yamamoto et al., 1963; Youngs et al., 1976). Figure 3 is a schematic representation of the cells of the olfactory bulb and their possible involvement and interactions in processing olfactory information.

Anatomical studies in a number of species have shown that the output of the olfactory bulbs passes via the lateral olfactory tract to the olfactory cortex of the telencephalon on the ventral surface of the forebrain (Broadwell, 1975; Scalia and Winans, 1975; Devor, 1976; Haberly and Price, 1977; Skeen and Hall, 1977; Davis et al., 1978; DeOlmos et al., 1978). The olfactory bulbs in turn receive centrifugal fibres from these cortical areas (Davis et al., 1978; DeOlmos et al., 1978; Haberly and Price, 1978). Additionally while the olfactory bulb projections are only to the telencephalon, the projections from the rest of the brain back to the bulbs come from all areas of the neuraxis, including the telencephalon, diencephalon, and rhombencephalon (Shipley and Adamek, 1984). No other primary sensory relay structure receives such diffuse projections from the brain. In addition, the olfactory cortical areas are interconnected by a substantial system of association fibres (Haberly and Price, 1978). These extensive and sometimes reciprocal interconnections between the bulb and the rest of the brain are thought to coordinate olfactory processing with the behavioural, endocrinological and nutritional status of the individual (Davis et al., 1978; Shipley and Adamek, 1984).

4. OLFACTORY MODULATORS

For the remainder of this chapter some of the factors which may routinely affect olfactory sensitivity, perception, and function will be discussed. These include the effects of sniffing technique, mixing of stimuli, adaptation, hunger, autonomic and hormonal influences, and aging. Figure 4 shows a compilation of these factors and where and how they might affect the olfactory system. These factors are important to the researcher studying the sensory analysis of foods because of their effects on the sense of smell and the consequent variability they may introduce into experiments.

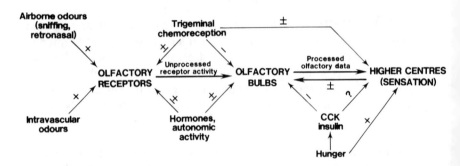

FIG. 4. Flow diagram of the olfactory system showing where and how the various modulators exert their effects. Plus signs indicate facilitatory actions; minus signs indicate suppressive actions.

4.1. Sniffing

The olfactory receptors in most mammals, including man, are located in the upper recesses of the nasal cavity away from the major flow of respired air. Under normal restful breathing conditions only about 5% of inspired air reaches the olfactory receptors, but during sniffing this increases to about 20–30% (Becker and King, 1957; DeVries and Stuvier, 1961; Shevrygin, 1973; Hornung et al., 1987). Sniffing, therefore, appears to be an adaptation for enhancing odour transport to the recessed olfactory epithelia.

For human subjects sniffing can be defined as the pattern of inspiration and expiration through the nose during odour sampling. In a comprehensive investigation of human sniffing, Laing (1982) observed wide variations among the sniffing patterns of untrained subjects but found that individual patterns were remarkably consistent over time. The average sniff among subjects was 200 ml in 0·4 s giving an inhalation rate of 30 litre min^{-1}. For individuals, sniffing rate, volume and duration remained constant between experiments and were not affected by the type of odour or its pleasantness. However, the duration of a sniffing episode, but not the initial sniff rate or volume, decreased as the odour concentration increased (Laing, 1982). Furthermore, natural sniffing patterns were as effective as experimenter-instructed sniff patterns in detection and intensity estimation tasks (Laing, 1983). For investigators testing human panels these data indicate that sniffing techniques used by individuals apparently do not usually affect performance.

There is some debate about whether human threshold or intensity values vary with inhalation rate. LeMagnen (1944/1945) obtained thresholds that were 20 times lower when the inhalation rate was increased from 12 to 96 litre min^{-1}. Rehn (1978) observed a twofold increase in perceived intensity when inhalation rate changed from 6 to 60 litres min^{-1}. However, Teghtsoonian *et al.* (1978) found no differences in perceived intensities of odours whether sniffing was 'weak' or 'strong'. They proposed a perceptual constancy model to explain the invariance of odour intensity with inhalation rate. The data of Youngentob *et al.* (1986) provide support for this model. They found that by increasing the resistance while subjects were sniffing, the perceived intensity of an odour decreased even though the concentration was unchanged. Thus the perceived effort of sniffing affected the perceived intensity.

Certainly there is a lower limit of inhalation rate necessary to cause deflection of inspired air up into the region of the receptor cells, and there is probably a higher inhalation rate above which performance no longer improves. As Tucker (1963) has concluded from his animal studies: 'It appears that the effect of increasing the nasal flow rate is to increase the odour concentration at the receptors being recorded from, until ultimately the concentration can go no higher'. Laing's data (1982, 1983) showed that natural inhalation rates were sufficient to obtain optimum performance by humans. Since the maximum inhalation rate varied among individuals from 12 to 71 litre min^{-1}, olfactometers capable of delivering odours at these flow rates should be used in human tests. Along this line, Doty *et al.* (1986) reported that the capacity of sniff bottles used in psychophysical tests (65 ml to 285 ml) significantly affected detection thresholds.

4.2. Mixture Effects

For those studying the role of the sense of smell in the sensory analysis of foods there are two general categories of mixtures that are important: homosensory and heterosensory mixtures. A homosensory mixture would be one that is composed of two or more odours. A heterosensory mixture might contain an odour and a tastant, or odour and pungent compound. We will first examine how the components of heterosensory mixtures interact and affect perception of flavour.

When food is in the mouth during the second stage of the chemosensory analysis of foods, responses of both oral and nasal chemoreceptors contribute to its flavour. The olfactory receptor neurons are stimulated by odours travelling the retronasal route. Whenever an odour stimulus in the mouth reaches the olfactory receptor neurons via the retronasal route, it is perceived as an oral and not nasal sensation. This is referred to as the taste–smell illusion (Hyman *et al.*, 1979; Murphy and Cain, 1980; Burdach *et al.*, 1984). There are very few data comparing the qualitative differences in perception of odours when they reach the nose through sniffing vs via the retronasal route (Burdach *et al.*, 1984). It does, however, appear that both engender odour sensations through the same basic mechanism (Burdach *et al.*, 1984; Voirol and Daget, 1986). Nonetheless, there seem to be some differences, since as Rozin (1982) points out, there are foods such as Limburger cheese that have a good flavour but do not smell good.

A number of studies show that both gustatory and trigeminal stimulation can affect the perceived intensity of a simultaneously presented odour. There is general agreement that the overall intensity of a heterosensory gustatory–olfactory presentation is about the same or only slightly less than the sum of the individual intensities (Murphy *et al.*, 1977; Murphy and Cain, 1980; Gillan, 1983; Hornung and Enns, 1984; Enns and Hornung, 1985). For heterosensory trigeminal–olfactory presentations, however, the perceived intensity depends on the concentration of the trigeminal component (Cain, 1974, 1976; Cain and Murphy, 1980; Lawless *et al.*, 1985).

Trigeminal chemical stimulation has been traditionally considered to evoke noxious or irritating sensations such as pungency, tickle, sting, burn, cool, warmth, and pain (Moncrieff, 1955; Doty, 1975; Cain, 1976). However, there is now considerable electrophysiological evidence that odours and concentrations generally considered to be non-irritating can stimulate trigeminal as well as olfactory chemoreceptors (Beidler and Tucker, 1956; Tucker, 1961, 1963, 1971; Silver and Moulton, 1982). In addition, Doty has clearly shown that anosmic humans can detect many odours commonly used in olfactory research and formerly thought not to stimulate the trigeminal system (Doty, et al., 1978). Thus there are many single compounds that excite both the olfactory and trigeminal chemoreceptors and thereby act as if they were a heterosensory mixture (see Cain, 1976). In fact, single compounds can evoke any and all combinations of olfactory, trigeminal and taste sensations.

When an odour and relatively non-irritating concentrations of a non-odorous trigeminal stimulus such as carbon dioxide are presented together there is a linear increase in the perceived intensity of the odour over that obtained for the odour alone (Cain, 1974, 1976; Cain and Murphy, 1980). However, as the trigeminal stimulant becomes more and more irritating, the perceived intensity of the odour is reduced. Thus, when capsaicin (the potent hot trigeminal irritant in chili peppers) is presented to the mouth at the same time as an odour is presented to the nose the perceived intensity of the odour is substantially reduced (Lawless et al., 1985). The mechanism for this may lie partly in trigeminal modulation of olfactory bulb excitability since the bulbs are more excitable when trigeminal neural activity is blocked (Stone et al., 1968; Stone and Rebert, 1970). Olfactory sensitivity may be further altered by the changes in respiration, nasal blood flow, patency and mucous secretion known to result from trigeminal stimulation (Allen, 1937; Alarie, 1966; James and Daly, 1969; Tucker, 1971; Ulrich et al., 1972). These effects on the nasal cavity probably result from autonomic nervous system activity elicited by irritating trigeminal stimulants.

We will next address the effects of mixing two or more odours, i.e., a homosensory mixture, on the perception of a mixture and its individual components. There is general agreement that the perceived intensity of a mixture of odours is almost always less than the sum of the intensities of the components (Jones and Woskow, 1964; Laing et al., 1984). Laing et al. (1984) report that a rare exception may occur when both components of a binary mixture are weak and their intensities synergise.

There has been much interest in how the overall quality of a mixture of odours is related to its components. In general, the quality of binary mixtures has been found to be most like that of the component possessing the strongest unmixed perceived intensity (Cain, 1975; Laing et al., 1984). In many cases the odour with the strongest intensity can completely suppress perception of a less intense odour (Cain, 1975; Laing et al., 1984). Lawless (1987) reported an interesting technique of unmasking the suppressed odour in such a mixture. He found that by pre-adapting subjects to the more intense odour of a binary mixture, the odour of the less intense component could be unmasked.

When a mixture consists of odours of equal unmixed intensities, both are perceived and contribute to the overall quality (Laing and Willox, 1983; Laing et al., 1984). The idea that two odours having different qualities will blend to form a mixture with a new unique quality seems to no longer hold for most mixing conditions (Berglund and Lindvall, 1982; Laing et al., 1984). It seems then, that the flavour chemist and the olfactory system can be most creative in blending odours of weak and equal intensities.

In summary, it appears that as the brain processes the responses of the oral and nasal chemosensory systems to food in the mouth, it blends them into a perceptual gestalt whose intensity is most often equal to or less than the sum of the intensities that would be elicited if each component stimulus were presented individually to each chemosensory system. The effects of such CNS blending of components on the quality or flavour of foods is not, at present, well understood.

4.3. Adaptation

Everyone is familiar with the powerful effects of olfactory adaptation on the perception of odours in the home, workplace, restaurants and other environments. The sense of smell is like most of the other senses in that it shows adaptation to prolonged stimulation. Thus as a human or other animal is alerted to or becomes aware of the continued presence of an odour, the response is attenuated, sometimes to subliminal levels. Selective attention can often return the perception. The process has been widely studied for the sense of smell and experiments reveal that olfactory adaptation is an exponential function, i.e., that the perceived intensity of a constant stimulus decreases exponentially (Ekman, 1958; Ekman et al., 1967; Cain, 1970; Pryor et al., 1970; Steinmetz et al., 1970; Engen, 1971; McNulty, 1974; Overbosch, 1986; Overbosch et al., 1986). Everyone is also familiar with the seemingly quick recovery of the olfactory system from adaptation. However, recovery has not been as well-studied. Ekman et al. (1967) state only that recovery of olfactory sensitivity is very fast upon termination of the adapting odour. Doty et al. (1986) found that the duration of the intertrial interval (ITI) (and thus length of time for recovery) in their study did not affect odour detection thresholds. The shortest ITI they used was 8 s. Other studies suggest that recovery after long odour exposures is prolonged (Koster, 1972; Chaput and Panhuber, 1982).

A few factors have been identified that affect adaptation. Overbosch (1986) found that individuals adapt at different rates, although the shapes of their adapting curves remain the same. Further, he reported that adaptation proceeds more quickly, but complete adaptation takes longer, for higher concentrations of an odour. In the case of irritating trigeminal stimuli there is evidence that adaptation either does not occur or is very prolonged (Ekman et al., 1967; Engen, 1986).

The most important neural substrates for adaptation appear to be in the brain rather than at the level of the olfactory receptor cells (Getchell, 1986). There is a general consensus that receptor neurons adapt too slowly to account for the exponential decreases in perceived intensities seen in psychophysical studies (Ottoson, 1956, Getchell and Shepherd, 1978; Van As et al., 1985; Getchell, 1986). This conclusion is based on studies of olfactory receptor neurons which show that their response to a prolonged stimulus is an initial peak of activity (the phasic response) followed by a protracted plateau (the tonic response) for the duration of stimulation. Even though the activity of receptor neurons may not mirror the course of perceptual adaptation, the phasic/tonic nature of their responding may be crucial. Given the complex inhibitory circuity that exists in the olfactory bulbs (see Section 3) the initial phasic response of the olfactory receptor neurons may set up the bulbar pattern of excitation/inhibition that is involved in identification and quantification of the odour while the tonic response drives bulbar adaptation.

Studies performed on awake freely-breathing rabbits suggest that the overall activity of the olfactory bulb output neurons does not change much from spontaneous levels during either short or long exposures to odours (Chaput and Panhuber, 1982; Chaput and Holley, 1985). In response to odours, the spontaneous level of activity becomes limited to inspirations, and expirations are characterised by almost no neural activity. Even when the rabbits were continuously exposed for an hour to a fairly high concentration of odour, the overall level of activity in the bulbs did not change much (Chaput and Panhuber, 1982). Thus it appears that changes in overall levels of olfactory bulb activity can not account for adaptation. However, Chaput and Panhuber (1982) did observe changes in the odour and inspiration synchronised bulbar activity that might underlie perceptual adaptation. During their hour-long continuous exposure of rabbits to odour they observed that some of the output neurons lost their inspiration synchronised activity and reverted back to an activity pattern more characteristic of odour-free spontaneous activity patterns.

In summary, the data indicate that the activity of the olfactory receptor neurons is probably not directly responsible for the shifts in perception that are termed adaptation. The more likely neural substrates for olfactory adaptation appear to be at the level of the olfactory bulbs or higher.

4.4. Hunger

The effects of hunger on the human sense of smell are not clearly understood. Some experiments show that hungry subjects are more sensitive to odours whereas other experiments show no differences (Goetzl and Stone, 1947; Janowitz and Grossman, 1949; Goetzl et al., 1950; Hammer, 1951; Schneider and Wolf, 1955; Berg et al., 1963). There is also some debate about the influence of satiety on the hedonics of food odours. Cabanac (1971) found that food-related odours are perceived as pleasant when a subject is hungry and unpleasant when the subject is satiated, but Mower et al. (1977) found that the hedonics of food odours changed with satiety only in some subjects.

One probable contributing factor to post-prandial shifts in olfaction is the effect of bloodborne odours on perception. Although odours are normally carried to the olfactory receptors in the air currents associated with breathing, sniffing and eating, they can also reach the receptors after being injected into the circulatory system (Bednar and Langfelder, 1930; Marco et al., 1956; Dishoeck and Versteeg, 1957; Teatini and Pincini, 1961; Antonelli, 1962; Guttich, 1964; Maruniak et al., 1983a,b). The responses resulting from such bloodborne odours have been termed intravascular or haematogenic olfaction. Human and animal studies reveal that intravascular odours are carried to the olfactory receptors by two mechanisms. First, they can diffuse out of the bloodstream in the lungs and stimulate the olfactory receptors as would airborne odours. Second, they can diffuse from capillaries underlying and surrounding the olfactory epithelium and move to the olfactory receptor sites by mechanisms similar to those responsible for nutrient and metabolite exchange (Maruniak et al., 1983a). A similar mechanism operates within the gustatory system where bloodborne tastants apparently reach receptors by direct diffusion from lingual capillaries (Bradley, 1973). Thus, post-ingestional mixture and adaptation effects of bloodborne odours may help account for differences in olfactory sensitivity to food odours and their pleasantness before and after a meal. Consequently, bloodborne odours may modulate the olfactory system, thereby subliminally influencing hunger, feeding and perhaps other behaviours involving the sense of smell.

In rats, electrical activity in the olfactory bulbs evoked by food odours is greater when they are hungry than when satiated, but there are no differences in responses to non-food odours (Pager, 1977). Cain (1975) has duplicated these findings with injections of insulin, a hormone whose blood levels increase after a meal. Since the olfactory bulbs contain the highest concentrations of insulin receptors in the brain (Havrankova et al., 1978), insulin levels may exert a continuing modulation on the olfactory system.

Another hormone and putative CNS neurotransmitter, associated both with digestion and satiety, is cholecystokinin (CCK), released by the small intestine during a meal. Injected CCK can eliminate or reduce food intake in hungry rats, pigs, sheep, monkeys and humans (Gibbs *et al.*, 1973, 1976; Della-Fera and Baile, 1979; Falasco *et al.*, 1979; Anika *et al.*, 1981, Kissileff *et al.*, 1981). It appears to be one of the factors that regulate appetite following ingestion of food and may act directly on the brain (Morley, 1982). Reports from several laboratories suggest that CCK may also be involved in modulating olfaction. Very high concentrations of CCK and its receptors have been found in the olfactory bulbs of rats (Rehfeld, 1978; Saito *et al.*, 1980; Vanderhaeghen *et al.*, 1980; Seroogy *et al.*, 1985) and the concentration of CCK receptors increased when rats were fasted (Saito *et al.*, 1981). Dupont *et al.* (1982) compared brain levels of CCK in rats eating bland or highly palatable diets and reported higher levels in the anterior olfactory nuclei (which receives olfactory bulb output) of those animals which ate the highly palatable (and presumably more odorous) diet. These investigators concluded that CCK may be involved in olfaction as a neurotransmitter in the anterior olfactory nucleus.

4.5. Autonomic and Hormonal Influences

Some hormones may affect the sense of smell by altering nasal patency. Others may directly modulate electrical activity in the olfactory bulbs or other olfactory areas in the brain. It is well known that histamine dramatically decreases nasal patency while epinephrine increases it. These effects are accomplished by radical alterations in the degree of engorgement of nasal vascular sinuses (Tucker and Beidler, 1956; Jackson, 1965). Additionally, the aperture of the nasal cavity decreases in response to cold, parasympathetic activity, strong trigeminal stimulation, and local irritation of the nasal mucosa by allergens (Tucker and Beidler, 1956; Tucker, 1963; Jackson, 1970; Cole *et al.*, 1983; Maruniak and Silver, unpublished data from the pigeon). Sympathetic simulation or decongestants increase the nasal aperture by vasoconstriction within the nasal mucosa (Beickert, 1951; Stoksted and Thomsen, 1953; Tucker and Beidler, 1956; Beidler, 1957; Jackson, 1965). Sympathetic activity is increased during exercise, some affective states, electrical stimulation of the superior cervical ganglion, noise, pain and pressure (Tucker and Beidler, 1956; Beidler, 1957; Cole *et al.*, 1983). In rabbits, sympathetic activity decreases these responses (Tucker and Beidler, 1956). Presumably these effects are attributable solely to variations in the access of odours to receptors caused by changes in nasal patency. When possible, then, attempts should be made during experiments to eliminate any factors which might induce changes in the activity of the autonomic nervous system.

Besides these acute changes in nasal patency there are regular cyclical and reciprocal changes between the nasal cavities called the nasal cycle. Nasal cycles are present in about 80% of humans and each cycle lasts from 30 min to 5 h (Heetderks, 1927; Stoksted, 1952; Eccles, 1977; Cole and Haight, 1986). The nasal cycle seems to be controlled by sympathetic activity (Beickert, 1951; Stoksted and Thomsen, 1953) and is characterised by vasoconstriction causing one side of the nasal cavity to dilate and secrete more mucus while the other side is constricted and has diminished mucous secretion (Hasegawa and Kern, 1977). In human psychophysical studies the nasal cycle is generally not considered to affect performance if odour presentations are made to both nostrils simultaneously since combined flow rate remains constant throughout the cycle.

There has been much interest in the effects of reproductive hormones on olfactory sensitivity because of the intimate relationship between odours and reproductive activity in man and many mammals. It is generally believed that women have lower thresholds for most odours than men (Moncrieff, 1946; Schneider and Wolf, 1955; Koelega and Koster, 1974; Doty et al., 1986). Because of this, many studies have focused on the variations of female olfactory sensitivity with the oestrous or menstrual cycles and concomitant oestrogen levels. Most, but not all investigators, report a peak in olfactory sensitivity at about the time of ovulation when oestrogen levels are highest, and one or more peaks in sensitivity at other times of the cycle (LeMagnen, 1952; Vierling and Rock, 1967; Henkin, 1974; Amoore et al., 1975; Doty et al., 1981). In animals, behavioural studies and electrophysiological recordings during the oestrous cycle have shown a peak olfactory sensitivity during pro-oestrus when oestrogen levels are highest (Pietras and Moulton, 1974; Schmidt and Schmidt, 1980; Rittner and Schmidt, 1982). Additionally, oestrogen administration is reported to increase olfactory sensitivity in women (Schneider et al., 1958). Oestrogen administration also enhances odour detection in rats, increases electrical activity in the olfactory bulbs of fish and cats, and increases electrical activity of other olfactory regions of the brain (Terasawa and Timiras, 1967; Oshima and Gorbman, 1969; Innes and Michal, 1970; Phillips and Vallowe, 1975; Cartas-Heredia et al., 1978). These studies suggest that oestrogens augment olfactory activity within the brain and thereby enhance olfactory sensitivity.

A second means by which oestrogens might increase sensitivity is by changes in the nasal mucosa. The nasal mucosa in rats and rabbits is thicker and more vascularised after oestrogen treatment and there are significant histological changes (particularly in the supporting cells) in the nasal mucosa of pre-ovulatory monkeys (Taylor, 1961; Saini and Breipohl, 1977). In humans, Mair *et al.* (1978) found that olfactory sensitivity changed during the menstrual cycle only for large water-insoluble odours and concluded that changes in mucous secretion could be responsible for changes in sensitivity.

Cyclic variations in olfactory sensitivity attributable to androgens have not been found in men or male animals (Pietras and Moulton, 1974; Doty *et al.*, 1981). For this reason many researchers prefer to use male subjects in order to eliminate possible experimental variability caused by female reproductive cyclicity. While such a strategy is acceptable in animal and some human studies it might lead to sex-biased errors when inferring generalities about human olfaction from data gathered in all male subject groups.

4.6. Aging

One of the curses of growing old is an inevitable dulling of the senses. The sense of smell is no exception and age frequently brings diminished acuity (Schiffman, 1977; Schiffman and Pasternak, 1979; Schemper *et al.*, 1981; Schiffman and Leffingwell, 1981; Stevens and Lawless, 1981; Eskenazi *et al.*, 1986; Stevens and Cain, 1987). While many of the elderly complain about the flat taste of foods (Cohen and Gitman, 1959), many also report that eating is one of their last remaining pleasures (Van Toller and Dodd, 1987).

Eskenazi *et al.* (1986) found that olfactory sensitivity peaks for most people in their mid 30s. By age 80 olfactory acuity has declined by about 20% (Van Toller and Dodd, 1987). In keeping with the evidence presented in Section 4.2 that nasal and retronasal odours evoke sensations via a common mechanism, aging has been reported to affect retronasal and nasal olfaction equally (Stevens and Cain, 1986*b*).

The mechanisms by which aging takes its toll on the senses is not well understood. For olfaction, it has been reported that there is an average loss of about 1% of the olfactory receptor neurons each year throughout life (Smith, 1942). While the loss of 80% of the receptor neurons would explain the olfactory deficits of an 80-year-old, it does not explain why proportionate deficits are not seen in 30-, 40-, 50- and 60-year-olds.

Oral and nasal chemosensory systems can be differentially affected by aging. Aging appears generally to affect olfaction more than gustation

(Stevens et al., 1984; Stevens and Cain, 1985). Apparently, advancing age can diminish components of both oral and nasal trigeminal responses (Stevens et al., 1982; Stevens and Cain, 1986a). For a given individual, ageing may negatively impact either olfactory or trigeminal sensitivity without affecting the other (Stevens et al., 1982).

For scientists studying the sensory analysis of foods, the effects of ageing offer a challenging fertile ground for making positive contributions to the quality of life of our senior citizens. Given that eating is one of the few pleasures of the elderly (Van Toller and Dodd, 1987), yet food often is perceived to have lost much of its flavour (Cohen and Gitman, 1959), we should be able to help by fortifying their foods with flavour additives. Schiffman (1977) reports that the elderly prefer food that has been enriched by the addition of exogenous flavours. Because of the disproportionate impact of ageing on olfaction compared to gustation, it has been suggested that attempts to improve the palatability of foods for the elderly focus on the addition of odorous rather than gustatory flavourings (Stevens et al., 1984).

REFERENCES

Adamek, G. D., Gesteland, R. C., Mair, R. G. and Oakley, B. (1984). Transduction physiology of olfactory receptor cilia. *Brain Res.*, **310**, 87–97.

Adrian, E. D. (1950). Sensory discrimination with some recent evidence from the olfactory organ. *Brit. Med. Bull.*, **6**, 330–1.

Alarie, Y. (1966). Irritating properties of airborne materials to the upper respiratory tract. *Arch. Environ. Health*, **13**, 433–49.

Allen, W. F. (1937). Olfactory and trigeminal conditioned reflexes in dogs. *Am. J. Physiol.*, **118**, 532.

Allison, A. C. and Warwick, R. T. (1949). Quantitative observations on the olfactory system of the rabbit. *Brain*, **72**, 186–97.

Altner, H. and Boeckh, (1967). Uber das Reaktionsspektrum von Rezeptoren aus der Riechschleimhaut von Wasserfroschen (*Rana esculenta*). *Z. Vergleich Physiol.*, **55**, 299–306.

Amoore, J. E. (1952). The stereochemical specificities of human olfactory receptors. *Perfumery Essent. Oil Record*, **43**: 321–3.

Amoore, J. E. (1965). Psychophysics of odor. *Cold Spring Harbor Symp. Quant. Biol.*, **30**, 623–37.

Amoore, J. E., Palmieri, G. and Wanke, E. (1967). Molecular shape and odor: pattern analysis by PAPA. *Nature*, **216**, 1084–7.

Amoore, J. E., Popplewell, J. R. and Whissell-Buechy, D. (1975). Sensitivity of women to musk odor: No menstrual variation. *J. Chem. Ecol.*, **1**, 291–7.

Andersen, I., Lundquist, G. and Proctor, D. (1971). Human nasal mucosal function in a controlled climate. *Arch. Environ. Health*, **23**, 408–15.

Anika, S. M., Houpt, T. R. and Houpt, K. A. (1981). Cholecystokinin and satiety in pigs. *Am. J. Physiol.*, **9**, R310–18.

Antonelli, A. A. (1962). Electrophysiologische untersuchungen am Olfaktionssystem de Katze wahrend intrahamatischer injektion von Essenzlosungen. *Z. Laryngol. Rhinol. Otol. Ihre Grenzgeb.*, **41**, 822–8.

Ash, K. O. and Skogen, J. D. (1970). Chemosensing: selectivity, sensitivity, and additive effects on a stimulant-induced activity of olfactory preparation. *J. Neurochem.*, **17**, 1143–53.

Atema, J. (1973). Microtubule theory of sensory transduction. *J. Theor. Biol.*, **38**, 181–90.

Atema, J. (1975). Stimulus transmission along microtubules in sensory cells: an hypothesis. In: *Microtubules and Microtubule Inhibitors*, M. Borgers and M. de Brabander (Eds), Elsevier, New York, pp. 247–57.

Baldaccini, N., Gagliardo, A., Pelosi, P. and Topazzini, A. (1986). Occurrence of a pyrazine binding protein in the nasal mucosa of some vertebrates. *Comp. Biochem. Physiol.*, **84B**, 249–53.

Bang, B. G. (1964). The mucous glands of the developing human nose. *Acta Anat.*, **59**, 297–314.

Baylin, F. (1979). Temporal patterns and selectivity in the unitary responses of olfactory receptors in the tiger salamander to odor simulation. *J. Gen. Physiol.*, **74**, 17–36.

Baylin, F. and Moulton, D. G. (1979). Adaptation and cross-adaptation to odor stimulation of olfactory receptors in the tiger salamander. *J. Gen. Physiol.*, **74**, 37–55.

Becker, R. F. and King, J. E. (1957). Delineation of the nasal air streams in the living dog. *AMA Arch. Otolaryngol.*, **65**, 428–36.

Bednar, M. and Langfelder, O. (1930). Uber das intravenose (hamatogene) Riechen, *Mschr. Ohrenheilk, Laryngo-Rhinol.*, **64**, 1133–9.

Beets, M. G. J. (1964). A molecular approach to olfaction. In: *Molecular Pharmacology*, J. Ariens (Ed.), Academic Press, New York, pp. 3–51.

Beets, M. G. J. (1975). Pharmacological aspects of olfaction. In: *Methods in Olfactory Research*, D. G. Moulton, A. Turk and J. W. Johnson (Eds), Academic Press, New York, pp. 445–70.

Beets, M. G. J. (1978). Odor and stimulant structure. In: *Handbook of Perception VIA. Tasting and Smelling*, E. G. Carterette and M. P. Friedman (Eds), Academic Press, New York, pp. 245–55.

Beets, M. G. J. and Theimer, E. T. (1970). Odour similarity between structurally unrelated odorants. In: *Taste and Smell in Vertebrates*, G. Wolstenholme and J. Knight (Eds), Churchill, London, pp. 313–21.

Beickert, P. (1951). Halbseitenrhythmus der vegetativen innervation, *Arch. Ohr.-Nask.-u. Kehlk. Heilk*, **157**, 404–11.

Beidler, L. M. (1957). Facts and theory on the mechanism of taste and odor perception. In: *Chemistry of Natural Food Flavors*, J. H. Mitchell, N. J. Leinen, E. M. Mrak and S. D. Bailey (Eds). Chicago Quartermaster Food and Container Institute, Chicago, pp. 7–47.

Beidler, L. M. and Tucker, D. (1956). Olfactory and trigeminal responses to odors. *Fed. Proc.*, **15**, 14.

Belmont, M. R. and Ambarek, A. (1985). Long-wavelength measurement of infrared properties of biological materials and their relation to olfaction. *IEEE Trans. Elect. Insul.*, **EI-20**, 475–9.

Berg, H. W., Pangborn, R. M., Roessler, E. B. and Webb, A. D. (1963). Influence of hunger on olfactory acuity. *Nature*, **197**, 108.

Berglund, B. and Lindvall, T. (1982). Olfaction. In: *The Nose: Upper Airway Physiology and the Atmospheric Environment*. D. F. Proctor and I. Andersen (Eds), Elsevier Biomedical Press, Amsterdam, pp. 279–305.

Blank, D. L. (1974). Mechanism underlying the analysis of odorant quality at the level of the olfactory mucosa. II. Receptor selective sensitivity. *Ann. NY Acad. Sci.*, **237**, 91–101.

Bloom, G. (1954). Studies on the olfactory epithelium of the frog and toad with the aid of light and electron microscopy. *Z. Zellforsch. Mikroscop. Anat. Abt. Histochem.*, **41**, 89–100.

Boelens, H. (1974). Relationship between the chemical structure of compounds and their olfactive properties. *Cosmet. Perfum.*, **89**, 1–7.

Bojsen-Møller, F. (1964). Topography of the nasal glands in rats and some other mammals. *Anat. Rec.*, **150**, 11–24.

Bojsen-Møller, F. (1975). Demonstration of terminalis, olfactory, trigeminal and perivascular nerves in the rat nasal septum. *J. Comp. Neurol.*, **159**, 245–56.

Boulet, M., Daval, G. and Leveteau, J. (1978). Qualitative and quantitative odour discrimination by mitral cells as compared to anterior olfactory nucleus cells. *Brain Res.*, **142**, 123–34.

Bradley, R. M. (1973). Electrophysiological investigations of intravascular taste using perfused rat tongue. *Am. J. Physiol.*, **224**, 300–4.

Breipohl, W., Moulton, D. G., Ummels, M. and Matulionis, D. (1982). Spatial pattern of sensory cell terminals in the olfactory sac of the tiger salamander. I. A scanning electron microscopy study. *J. Anat.*, **134**, 757–69.

Broadwell, R. D. (1975). Olfactory relationships of the telencephalon and diencephalon in the rabbit. *J. Comp. Neurol.*, **163**, 329–46.

Bronshtein, A. A. and Minor, A. V. (1977). The regeneration of olfactory flagella and restoration of electroolfactogram after treatment of the olfactory mucosa with Triton X-100. *Tsitologiya*, **19**, 33–9.

Brookover, C. (1914). The nervus terminalis in adult man. *J. Comp. Neurol.*, **24**, 131–5.

Brookover, C. (1917). The peripheral distribution of the nervus terminalis in an infant. *J. Comp. Neurol.*, **28**, 349–60.

Bruch, R. C. and Kalinoski, D. L. (1987). Interaction of GTP-binding regulatory proteins with chemosensory receptors. *J. Biol. Chem.*, **262**, 2401–4.

Burdach, K. J., Kroeze, J. and Koster, E. P. (1984). Nasal, retronasal, and gustatory perception: An experimental comparison. *Percept. Psychophys.*, **36**, 205–8.

Cabanac, M. (1971). Physiological role of pleasure. *Science*, **173**, 1103–7.

Cagan, R. H. and Zeigler, W. N. (1978). Biochemical studies of olfaction. I. Binding specificity of radioactivity labeled stimuli to an isolated olfactory preparation from Rainbow trout (*Salmo gairdneri*). *Proc. Nat. Acad. Sci.*, **75**, 4679–83.

Cain, D. P. (1975). Effects of insulin injection on responses of olfactory bulb and amygdala single units to odors. *Brain Res.*, **99**, 69–83.

Cain, W. S. (1970). Odor intensity after adaptation and cross adaptation. *Percept. Psychophys.*, **7**, 271–5.
Cain, W. S. (1974). Contribution of the trigeminal nerve to perceived odor magnitude. *Ann. NY Acad. Sci.*, **237**, 28–34.
Cain, W. S. (1975). Odor intensity: mixtures and masking. *Chem. Senses Flav.*, **1**, 339–52.
Cain, W. S. (1976). Olfaction and the common chemical sense: some psychophysical contrasts. *Sen. Proc.*, **1**, 57–67.
Cain, W. S. and Murphy, C. L. (1980). Interaction between chemoreceptive modalities of odour and irritation. *Nature*, **284**, 255–7.
Cartas-Heredia, L., Guevara-Aguilar, R. and Aguilar-Baturoni, H. U. (1978). Oestrogenic influences on the electrical activity of the olfactory pathway. *Brain Res. Bull.*, **3**, 623–30.
Chaput, M. and Holley, A. (1979). Spontaneous activity of olfactory bulb neurons in awake rabbits, with some observations on the effects of pentobarbital anesthesia *J. Physiol. (Paris)*, **75**, 939–48.
Chaput, M. and Holley, A. (1985). Responses of olfactory bulb neurons to repeated odor simulations in awake freely-breathing rabbits. *Physiol. Behav.*, **34**, 249–58.
Chaput, M. and Panhuber, H. (1982). Effects of long duration odor exposure on the unit activity of olfactory bulb cells in awake rabbits. *Brain Res.*, **250**, 41–52.
Chen, Z. and Lancet, D. (1984). Membrane proteins unique to vertebrate olfactory cilia: candidates for sensory receptor molecules. *Proc. Nat. Acad. Sci.*, **81**, 1859–63.
Cherry, R. J., Dodd, G. H. and Chapman, D. (1970). Small molecule-lipid membrane interactions and the puncturing theory of olfaction. *Biochim. Biophys. Acta*, **211**, 409–16.
Cohen, T. and Gitman, L. (1959). Oral complaints and taste perception in the aged. *J. Gerontol.*, **14**, 294–8.
Cole, P. and Haight, J. S. (1986). Posture and the nasal cycle. *Ann. Otol. Rhinol. Laryngol.*, **95**: 233–7.
Cole, P., Forsyth, R. and Haight, J. S. (1983). Effects of cold air and exercise on nasal patency. *Ann. Otol. Rhinol. Laryngol.*, **92**, 196–8.
Costanzo, R. M. and O'Connell, R. J. (1978). Spatially organized projections of hamster olfactory nerves. *Brain Res.*, **139**, 327–32.
Costanzo, R. M. and O'Connell, R. J. (1980). Receptive fields of second-order neurons in the olfactory bulb of the hamster. *J. Gen. Physiol.*, **76**, 53–68.
Davies, J. T. (1962). The mechanism of olfaction. *Symp. Soc. Exptl. Biol.*, **16**, 170–9.
Davies, J. T. (1965). A theory of the quality of odours. *J. Theor. Biol.*, **8**, 1–7.
Davies, J. T. (1969). The 'penetration and puncturing' theory of odor: types and intensity of odors. *J. Colloid. Interface Sci.*, **29**, 296–304.
Davies, J. T. (1971). Olfactory theories. In: *Handbook of Sensory Physiology IV*, L. M. Beidler (Ed.), Springer-Verlag, New York, pp. 322–50.
Davies, J. T. and Taylor, F. H. (1954). A model system for the olfactory membrane. *Nature*, **174**, 693–4.
Davis, B. J., Macrides, F., Youngs, W. M., Schneider, S. P. and Rosene, D. L. (1978). Efferents and centrifugal afferents of the main and accessory olfactory bulbs in the hamster. *Brain Res. Bull.*, **3**, 59–72.

Delaleu, J. and Holley, A. (1983). Investigations of the discriminative properties of the frog's olfactory mucosa using a photoactivable odorant. *Neurosci. Lett.*, **37**, 251–6.
Della-Fera, M. A. and Baile, C. A. (1979). Cholecystokinin octapeptide: continuous picomole injections into the cerebral ventricles of sheep suppress feeding. *Science*, **206**, 471–3.
Demski, L. S. and Northcutt, R. G. (1983). The terminal nerve: A new chemosensory system in vertebrates? *Science*, **220**, 435–7.
DeOlmos, J., Hardy, H. and Heimer, L. (1978). The afferent connections of the main and the accessory olfactory bulb formations in the rat: an experimental HRP-study. *J. Comp. Neurol.*, **18**, 213–44.
Devor, M. (1976). Fiber trajectories of olfactory bulb efferents in the hamster. *J. Comp. Neurol.*, **166**, 31 48.
DeVries, H. and Stuvier, M. (1961). The absolute sensitivity of the human sense of smell. In: *Sensory Communication*, W. A. Rosenblith (Ed.), J. Wiley & Sons, New York, pp. 159–67.
Dishoeck, H. A. E. Van and Versteeg, N. (1957). On the problem of hematogenic olfaction. *Arch. Otolaryngol.*, **47**, 396–401.
Dodd, G. H. (1971). Studies on olfactory recptor mechanisms. *Biochem. J.*, **123**, 31–2.
Doty, R. L. (1975). Intranasal detection of chemical vapors by humans. *Physiol. Behav.*, **14**, 855–9.
Doty, R. L., Brugger, W. E., Jurs, P. C., Orndorff, M. A., Snyder, P. F. and Lowry, L. D. (1978). Intranasal trigeminal stimulation from odorous volatiles: psychometric responses from anosmic and normal humans. *Physiol. Behav.*, **20**, 175–87.
Doty, R. L., Huggins, G. R., Snyder, P. J. and Lowry, L. D. (1981). Endocrine, cardiovascular, and psychological correlates of olfactory sensitivity changes during the human menstrual cycle. *JCPP*, **95**, 45–60.
Doty, R. L., Gregor, T. P. and Settle, R. G. (1986). Influence of intertrial interval and sniff-bottle volume on phenylethyl alcohol odor detection thresholds. *Chem. Senses*, **11**, 259–64.
Døving, K. B. (1964). Studies of the relation between the frog's electro-olfactogram (EOG) and single unit activity in the olfactory bulb. *Acta Physiol. Scand.*, **60**, 150–63.
Døving, K. B. (1966). An electrophysiological study of odour similarities of homologous substances. *J. Physiol.*, **186**, 97–109.
Dubois-Dauphin, M., Tribollet, E. and Dreifuss, J. J. (1980). Relations somatotopiques entre la mugueuse olfactive et la bulbe olfactif chez le triton. *Brain Res.*, **219**, 269–87.
Duchamp, A., Revial, M. F., Holley, A. and MacLeod, P. (1974). Odor discrimination by frog olfactory receptors. *Chem. Senses*, **1**, 213–33.
Dupont, A., Mérand, Y., Savard, P., Leblanc, J. and Dockray, G. J. (1982). Evidence that cholecystokinin is a neutrotransmitter of olfaction in nucleus olfactorius anterior. *Brain Res.*, **250**, 386–90.
Eccles, R. (1977). Cyclic changes in human nasal resistance to air flow. *J. Physiol.*, **272**, 75–91.
Ekman, G. (1958). Two generalized scaling methods. *J. Psychol.*, **45**, 287–95.

Ekman, G., Berglund, B., Berglund, U. and Lindvall, T. (1967). Perceived intensity of odor as a function of time and adaptation. *Scand. J. Psychol.*, **8**, 177–86.
Engen, T. (1971). Olfactory psychophysics. In: *Handbook of Sensory Physiology, Chemical Sense I*, L. M. Beidler (Ed.), Springer-Verlag, New York, pp. 216–44.
Engen, T. (1986). Perception of odor and irritation. *Envir. Internat.*, **12**, 177–87.
Enns, M. P. and Hornung, D. E. (1985). Contributions of smell and taste to overall intensity. *Chem. Senses*, **10**, 357–66.
Erickson, J. R. and Caprio, J. (1984). The spatial distribution of ciliated and microvillous olfactory receptor neurons in the channel catfish is not matched by a differential specificity to amino acid and bile salt stimuli. *Chem. Senses*, **9**, 127–41.
Eskenazi, B., Cain, W. S. and Friend, K. (1986). Exploration of olfactory aptitude. *Bull. Psychon. Soc.*, **24**, 203–6.
Falasco, J. D., Smith, G. P. and Gibbs, J. (1979). Cholecystokinin suppresses sham feeding in the rhesus monkey. *Physiol. Behav.*, **23**, 887–90.
Fesenko, E. E., Novoselov, V. I., Mjasoedov, N. F. and Sidorov, G. V. (1978). Molecular mechanisms of olfactory reception III. Binding of camphor by rat and frog mucosa components. *Stud. Biophys.*, **73**, 71–84.
Fesenko, E. E., Novoselov, V. I. and Krapivinskaya, L. D. (1979). Molecular mechanisms of olfactory reception IV. Some biochemical characteristics of the camphor receptor from rat olfactory epithelium. *Biochim. Biophys. Acta*, **587**, 424–33.
Friedman, L. and Miller, J. G. (1971). Odor incongruity and chirality. *Science*, **172**, 1044–6.
Gennings, J. N., Gower, D. B. and Bannister, L. H. (1977). Studies on the receptors to 5-alpha-androst-16-en-3-one and 5-alpha-androst-16-en-3 alpha-ol in sow nasal mucosa. *Biochim. Biophys. Acta*, **496**, 547–56.
Gesteland, R. C., Lettvin, J. Y., Pitts, W. H. and Rojas, A. (1963). Odor specificities of the frog's olfactory receptors. In: *Olfaction and Taste I*, Y. Zotterman (Ed.), Pergamon Press, New York, pp. 19–34.
Gesteland, R. C., Lettvin, J. Y. and Pitts, W. H. (1965). Chemical transmission in the nose of the frog. *J. Physiol.*, **181**, 525–9.
Getchell, M. L. and Gesteland, R. C. (1972). The chemistry of olfactory reception: stimulus specific protection from sulfhydryl reagent inhibition. *Proc. Nat. Acad. Sci.*, **69**, 1494–8.
Getchell, M. L., Zielinski, B., DeSimone, J. L. and Getchell, T. V. (1987). Odorant stimulation of secretory and neural processes in the salamander olfactory mucosa. *J. Comp. Physiol. A*, **160**, 155–68.
Getchell, T. V. (1969). The interaction of the peripheral olfactory nerve with non-odorous stimuli. In: *Olfaction and Taste III*, C. Pfaffman (Ed.), Rockefeller University Press, New York, pp. 117–24.
Getchell, T. V. (1986). Functional properties of vertebrate olfactory neurons. *Physiol. Rev.*, **66**, 772–818.
Getchell, T. V. and Getchell, M. L. (1975). Signal detecting mechanisms in the olfactory epithelium: molecular discrimination. *Ann. NY Acad. Sci.*, **237**, 62–75.
Getchell, T. V. and Shepherd, G. M. (1975a). Synaptic actions on mitral and tufted cells elicited by olfactory nerve volleys in the rabbit. *J. Physiol.*, **251**, 497–522.

Getchell, T. V. and Shepherd, G. M. (1975b). Short-axon cells in the olfactory bulb: dendrodendritic synaptic interactions. *J. Physiol.*, **251**, 523–48.
Getchell, T. V. and Shepherd, G. M. (1978). Adaptive properties of olfactory receptors analyzed with odour pulses of varying durations. *J. Physiol. Lond.*, **282**, 541–60.
Getchell, T. V., Heck, G. L., DeSimone, J. A. and Price, S. (1980). The location of olfactory receptor sites: inferences from latency measurements. *Biophys. J.*, **29**, 397–412.
Gibbs, J., Young, R. C. and Smith, G. P. (1973). Cholecystokinin-decreased food intake in rats. *JCPP*, **84**, 488–95.
Gibbs, J., Falasco, J. P. and McHugh, P. R. (1976). Cholecystokinin-decreased food intake in Rhesus monkeys. *Am. J. Physiol.*, **230**, 15–8.
Gillan, D. J. (1983). Taste-taste, odor-odor, and taste-odor mixtures: Greater suppression within than between modalities. *Percept. Psychophys.*, **33**, 183–5.
Goetzl, F. R. and Stone, F. (1947). Diurnal variations in acuity of olfaction and food intake. *Gastroenterology*, **9**, 444–53.
Goetzl, F. R., Abel, M. S. and Ahokas, A. J. (1950). Occurrence in normal individuals of diurnal variations in olfactory acuity. *J. Appl. Physiol.*, **2**, 553–62.
Graziadei, P. P. C. (1971a). The olfactory mucosa of vertebrates. In: *Handbook of Sensory Physiology, Chemical Senses I*. L. M. Beidler (Ed.), Springer-Verlag, New York, pp. 27–58.
Graziadei, P. P. C. (1971b). Topological relations between olfactory neurons. *Z. Zellforsch. Mikroskop. Anat.*, **118**, 449–66.
Graziadei, P. P. C. and Metcalf, J. F. (1971). Autoradiographic and ultrastructural observations on the frog's olfactory mucosa. *Z. Zellforsch. Mikroskop. Anat.*, **116**, 305–18.
Graziadei, P. P. C. and Monti-Graziadei, G. A. (1979). Neurogenesis and neuron regeneration in the olfactory system of mammals. I. Morphological aspects of differentiation and structured organization of the olfactory sensory neurons. *J. Neurocytol.*, **8**, 1–18.
Graziadei, M., Karlan, M. S., Bernstein, J. J. and Graziadei, P. (1980). Reinnervation of the olfactory bulb after section of the olfactory nerve in monkey (*Saimiri scuireus*). *Brain Res.*, **189**, 343–54.
Güttich, H. (1964). Intravenös verabreichte Riechstoffe: gustatorisches Riechen. *HNO*, **13**, 42–5.
Haberly, L. B. (1985). Neuronal circuitry in olfactory cortex: Anatomy and functional implications. *Chem. Senses*, **10**, 219–38.
Haberly, L. B. and Price, J. L. (1977). The axonal projection patterns of the mitral and tufted cells of the olfactory bulb in the rat. *Brain Res.*, **129**, 152–7.
Haberly, L. B. and Price, J. L. (1978). Association and commissural fiber systems of the olfactory cortex of the rat. *J. Comp. Neurol.*, **178**, 711–740.
Hammer, F. J. (1951). The relation of odor, taste and flicker-fusion thresholds to food intake. *JCPP*, **44**, 103–11.
Hasegawa, M. and Kern, E. (1977). The human nasal cycle. *Mayo Clin. Proc.*, **52**, 28–48.
Havrankova, J., Roth, J. and Brownstein, M. (1978). Insulin receptors are widely distributed in the central nervous system of the rat. *Nature*, **272**, 827–9.
Heetderks, D. L. (1927). Observations on the reaction of normal nasal mucous membrane. *Am. J. Med. Sci.*, **174**, 231–44.

Heldman, J. and Lancet, D. (1986). Cyclic AMP-dependent protein phosphorylation in chemosensory neurons: Identification of cyclic nucleotide-regulated phosphoproteins in olfactory cilia. *J. Neurochem.*, **47**, 1527–33.
Henkin, R. I. (1974). Sensory changes during the menstrual cycle. In: *Biorhythms and Human Reproduction*. M. Ferin, F. Halberg, R. M. Richart, and L. V. Wiele (Eds), J. Wiley & Sons, New York, pp. 277–85.
Higashino, S. Takeuchi, H. and Amoore, J. (1969). Mechanism of olfactory discrimination in the olfactory bulb of the bullfrog. In: *Olfaction and Taste III*. C. Pfaffman (Ed.), Rockefeller Press, New York, pp. 192–211.
Holley, A. and Døving, K. B. (1977). Receptor sensitivity, acceptor distribution, convergence and neural coding in the olfactory system. In: *Olfaction and Taste VI*. J. LeMagnen and P. MacLeod (Eds), Information Retrieval Ltd. London, pp. 113–23.
Holley, A., Duchamp, A., Revial, M. F. and Juge, A. (1974). Qualitative and quantitative discrimination in the frog's olfactory receptors: analysis from electrophysiological data. *Ann. NY Acad. Sci.*, **237**, 102–14.
Hornung, D. E. and Enns, M. P. (1984). The independence and integration of olfaction and taste. *Chem. Senses*, **9**, 97–106.
Hornung, D. E., Leopold, D., Youngentob, S., Sheehe, P., Gagne, G., Thomas, F. D. and Mozell, M. M. (1987). Airflow patterns in a human nasal model. *Arch. Otolaryngol. Head Neck Surg.*, **113**, 169–72.
Hyman, A., Metzer, T. and Calderone, L. (1979). The contribution of olfaction to taste discrimination. *Bull. Psychonom. Soc.*, **13**, 359–62.
Innes, D. L. and Michal, E. K. (1970). Effects of progesterone and estrogen on the electrical activity of the limbic system. *J. Exp. Zool.*, **175**, 487–92.
Jackson, R. T. (1965). Effect of sympathetic stimulation on nasal blood shunt. *Exp. Neurol.*, **3**, 318–23.
Jackson, R. T. (1970). Pharmacologic responsiveness of the nasal mucosa. *Ann. Otol. Rhinol. Laryngol.*, **79**, 461–7.
Jafek, B. W. (1983). Ultrastructure of human nasal mucosa. *Laryngoscope*, **93**, 1576–99.
Jahr, C. E. and Nicoll, R. A. (1982). An intracellular analysis of dendrodentritic inhibition in the turtle *in vitro* olfactory bulb. *J. Physiol.*, **326**, 213–34.
James, J. E. and Daly, M. (1969). Nasal reflexes. *Proc. R. Soc. Med.*, **62**, 1287–93.
Janowitz, H. D. and Grossman, M. I. (1949). Gusto-olfactory thresholds in relation to appetite and hunger sensations. *J. Appl. Physiol.*, **2**, 217–22.
Jennes, L. (1986). The olfactory gonadotropin-releasing hormone immunoreactive system in mouse. *Brain Res.*, **386**, 351–63.
Jennes, L. and Stumpf, W. E. (1980). LHRH-systems in the brain of the golden hamster. *Cell Tiss. Res.*, **209**, 239–56.
Johnson, A., Josephson, R. and Hawke, M. (1985). Clinical and histological evidence for the presence of the vomeronasal organ in adult humans. *J. Otolaryngol.*, **14**, 71–9.
Jones, F. N. and Woskow, M. H. (1964). On the intensity of odor mixtures. *Ann. NY Acad. Sci.*, **116**, 484–94.
Kafka, W. A. (1971). Specificity of odor-molecule interaction in single cells. In: *Gustation and Olfaction*. G. Ohloff and A. F. Thomas (Eds), Academic Press, New York, pp. 61–72.

Kashiwayangi, M. and Kurihara, K. (1985). Evidence for non-receptor odor discrimination using neuroblastoma cells as a model for olfactory cells. *Brain Res.*, **359**, 97–103.
Kauer, J. S. (1974). Response patterns of amphibian olfactory bulb neurones to odour stimulation. *J. Physiol.*, **243**, 695–715.
Kauer, J. S. (1977). Odor processing in the salamander olfactory bulb. In: *Olfaction and Taste VI*. J. Le Magnen and P. MacLeod (Eds), Information Retrieval Ltd, London, pp. 125–33.
Kauer, J. S. (1981). Olfactory receptor cell staining using horseradish peroxidase. *Anat. Rec.*, **200**, 331–6.
Kauer, J. S. and Shepherd, G. M. (1977). Analysis of the onset phase of olfactory bulb unit responses to odour pulses in the salamander. *J. Physiol.*, **272**, 495–516.
Keverne, E. B., Murphy, C. L., Silver, W. L., Wysocki, C. J. and Meridith, M. (1986). Non-olfactory chemoreceptors of the nose: Recent advances in understanding the vomeronasal and trigeminal systems. *Chem. Senses*, **11**, 119–33.
Kissileff, H. R., Pi-Sunyer, F. X., Thornton, J. T. and Smith, G. P. (1981). C-terminal octapeptide of cholecystokinin decreases food intake in man. *Am. J. Clin. Nutr.*, **34**, 154–60.
Klopping, H. L. (1971). Olfactory theories and the odors of small molecules. *J. Agric. Food Chem.*, **19**, 999–1004.
Koelega, H. S. and Köster, E. P. (1974). Some experiments on sex differences in odor perception. *Ann. NY Acad. Sci.*, **237**, 234–46.
Koster, E. P. (1972). Recovery of the olfactory sensitivity after adaptation and cross-adaptation with xylene isomers. In: *Olfaction and Taste IV*, D. Schneider (Ed.), Wissenschaffliche, Stuttgart, pp. 149–55.
Koyama, N. and Kurihara, K. (1972). Effect of odorants on lipid monolayers from bovine olfactory epithelium. *Nature*, **236**, 402–4.
Kubie, J. and Moulton, D. G. (1979). Regional patterning of response to odors in the salamander olfactory mucosa. *Soc. Neurosci. Abst.*, **5**, 129.
Kubie, J., Mackay-Sim, A. and Moulton, D. G. (1980). Inherent spatial patterning of responses to odorants in the salamander olfactory epithelium. In: *Olfaction and Taste VII*. H. van der Starre (Ed.), Information Retrieval Ltd, London, pp. 163–6.
Laing, D. G. (1982). Characterization of human behavior during odour perception. *Perception*, **11**, 221–30.
Laing, D. G. (1983). Optimum odour perception is achieved by sniffing naturally. *Perception*, **12**, 99–117.
Laing, D. G. and Willcox, M. E. (1983). Perception of components in binary odor mixtures. *Chem. Senses*, **7**, 249–64.
Laing, D. G., Panhuber, H., Willcox, M. E. and Pittman, E. P. (1984). Quality and intensity of binary odor mixtures. *Physiol. Behav.*, **33**, 309–19.
Lancet, D. (1986). Vertebrate olfactory reception. *Ann. Rev. Neurosci*, **9**, 329–55.
Land, L. J. (1973). Localized projection of olfactory nerves to rabbit olfactory bulb. *Brain Res.*, **63**, 153–66.
Land, L. J. and Shepherd, G. M. (1974). Autoradiographic analysis of olfactory receptor projections in the rabbit. *Brain Res.*, **70**, 506–10.
Lawless, H. (1987). An olfactory analogy to release from mixture suppression in taste. *Bull. Psychonom. Soc.*, **25**, 266–8.

Lawless, H., Rozin, P. and Shenker, J. (1985). Effects of oral capsaicin on gustatory, olfactory and irritant sensations and flavor identification in humans who regularly or rarely consume chili pepper. *Chem. Senses*, **10**, 579–89.

LeGros Clark, W. E. (1951). The projection of the olfactory epithelium on the olfactory bulb in the rabbit. *J. Neurol. Neurosurg. Psychiat.*, **14**, 1–10.

Leitereg, T. J., Guadagni, D. G., Harris, J., Mon, T. R. and Teranishi, R. (1971). Evidence for the difference between the odours of the optical isomers (+)- and (−)-carvone. *Nature*, **230**, 455–56.

LeMagnen, J. (1944/1945). Etude des facteurs dynamiques de l'excitation olfactive. *Ann. Psychol.*, **45–46**, 77–89.

LeMagnen, J. (1952). Les phenomenes olfacto-sexuels chez l'homme. *Arch. Sci. Physiol.*, **6**, 125–60.

Lettvin, J. and Gesteland, R. C. (1965). Speculations on smell. *Cold Spring Harbor Symp.*, **30**, 217–25.

Lovell, M. A., Jafek, B. W., Moran, D. T. and Rowley, J. C. (1982). Biopsy of human olfactory mucosa: an instrument and a technique. *Arch. Otolaryngol.*, **108**, 247–9.

Mackay-Sim, A. and Kubie, J. (1981). The salamander nose: a model system for the study of spatial coding of odorant quality. *Chem. Senses*, **6**, 249–57.

Mackay-Sim, A. and Patel, U. (1984). Regional differences in cell density and cell genesis in the olfactory epithelium of the salamander *Ambystoma tigrinum*. *Exp. Brain Res.*, **57**, 99–106.

Mackay-Sim, A. and Shaman, P. (1983). Topographic coding of odorant quality is maintained at different concentrations in the salamander olfactory epithelium. *Brain Res.*, **297**, 207–17.

Mackay-Sim, A., Shaman, P. and Moulton, D. G. (1982). Topographic coding of odorant quality: odorant specific patterns of epithelial responsivity in the salamander. *J. Neurophysiol.*, **48**, 584–96.

MacLeod, N., Reinhardt, W. and Ellendorff, F. (1979). Olfactory bulb neurons of the pig respond to an identified steroidal pheromone and testosterone. *Brain Res.*, **164**, 323–7.

McNulty, P. B. (1974). Intensity-time curves for flavored oil-in-water emulsion. *J. Food Sci.*, **39**, 55–7.

Macrides, F. (1977). Dynamic aspects of central olfactory processing. In: *Chemical Signals in Vertebrates*. D. Muller-Schwarze and M. M. Mozell (Eds), Plenum Press, New York, pp. 499–514.

Macrides, F. and Davis, B. J. (1983). The olfactory bulb. In: *Chemical Neuroanatomy*. P. Emson (Ed.), Raven Press, New York, pp. 391–426.

Mair, R. G. (1982a). Response properties of rat olfactory bulb neurones. *J. Physiol.*, **326**, 341–59.

Mair, R. G. (1982b). Adaptation of rat olfactory bulb neurons. *J. Physiol.*, **326**, 361–9.

Mair, R. G., Bouffard, J. A., Engen, T. and Morton, T. H. (1978). Olfactory sensitivity during the menstrual cycle. *Sen. Proc.*, **2**, 90–8.

Mancia, M., Baumgarten, R. and Green, J. D. (1962). Response patterns of olfactory bulb neurons. *Arch. Ital. Biol.*, **100**, 449–62.

Marco, J., Morera, H. and Gimenez, J. (1956). Olfaction in laryngectomized patients. *Acta Otolaryngol.*, **46**, 114–26.

Marshall, D. A. and Maruniak, J. A. (1986). Masera's organ responds to odorants. *Brain Res.*, **366**, 329–32.
Maruniak, J. A., Silver, W. L. and Moulton, D. G. (1983*a*). Olfactory receptors respond to bloodborne odorants. *Brain Res.*, **265**, 312–16.
Maruniak, J., Mason, J. R. and Kostelc, J. (1983*b*). Conditioned aversions to an intravascular odorant. *Physiol. Behav.*, **30**, 617–20.
Mason, J. R., Leong, F. C., Plaxco, K. W. and Morton, T. H. (1985). Two-step covalent modification of proteins. Selective labelling of Schiff base forming sites and selective blockade of the sense of smell *in vivo*. *J. Am. Chem. Soc.*, **107**, 6075–84.
Matthews, D. (1972*a*). Response patterns of single neurons in the tortoise olfactory epithelium and olfactory bulb. *J. Gen. Physiol.*, **60**, 166–80.
Matthews, D. (1972*b*). Response patterns of single units in the olfactory bulb of the rat to odor. *Brain Res.*, **47**, 389–400.
Matthews, D. F. and Tucker, D. (1966). Single unit activity in the tortoise olfactory mucosa. *Fed. Proc.*, **25**, 329.
Menco, B., Dodd, G. H., Davey, M. and Bannister, L. H. (1976). Presence of membrane particles in freeze-etched bovine olfactory cilia. *Nature*, **263**, 597–9.
Menevse, A., Dodd, G. and Poynder, T. M. (1977). Evidence for the specific involvement of cyclic AMP in the olfactory transduction mechanism. *Biochem. Biophys. Res. Commun.*, **77**, 671–7.
Menevse, A., Dodd, G. and Poynder, T. M. (1978). A chemical-modification approach to the olfactory code. Studies with a thiol-specific reagent. *Biochem. J.*, **176**, 845–54.
Meredith, M. M. (1983). Sensory physiology of pheromonal communication. In: *Pheromones and Reproduction in Mammals*. J. G. Vandenbergh (Ed.), Academic Press, New York, pp. 200–52.
Meredith, M. (1986). Patterned response to odor in mammalian olfactory bulb: the influence of intensity. *J. Neurophysiol.*, **56**, 572–97.
Minor, A. V. (1971). Electro-olfactogram. Its origin and mechanism of generation. *Fiziol. Zh. SSSR*, **57**, 1115–22.
Minor, A. V. and Vasileva, V. S. (1980). Electrophysiological investigation of the sex pheromone of the boar *Sus scrofa* (in Russian). *Zh. Evol. Biokhim. Fiziol.*, **26**, 616–19.
Minor, A. V., Vaileva, V. S. and Zinkevich, E. P. (1980). Electric responses of the olfactory linings of pigs on the sex pheromone of the boar: androst-16-en-3-one and its analogues (in Russian). *Dokl. Akad. Nauk SSSR*, **254**, 1494–6.
Moncrieff, R. W. (1946). *The Chemical Senses*. J. Wiley & Sons, New York, p. 76.
Moncrieff, R. W. (1955). A technique for comparing the threshold concentrations for olfactory, trigeminal and ocular irritations. *Quart. J. Exp. Psychol.*, **7**, 128–32.
Moran, D. T., Rowley, J. C., Jafek, B. W. and Lovel, M. A. (1982*a*). The fine structure of the olfactory mucosa in man. *J. Neurocytol.*, **11**, 721–46.
Moran, D. T., Rowley, J. C. and Jafek, B. W. (1982*b*). Electron microscopy of human olfactory epithelium reveals a new cell type: the microvillar cell. *Brain Res.*, **253**, 39–46.
Morley, J. E. (1982). The ascent of cholecystokinin (CCK)—from gut to brain. *Life Sci.*, **30**, 479–93.

Moulton, D. G. (1967). Spatio-temporal patterning of response in the olfactory system. In: *Olfaction and Taste II*. T. Hayashi (Ed.), Pergamon Press, New York, pp. 109–6.
Moulton, D. G. (1971). The olfactory pigment. In: *Handbook of Sensory Physiology, Chemical Senses I*. L. M. Beidler (Ed.), Springer-Verlag, New York, pp. 59–74.
Moulton, D. G. (1975). Cell renewal in the olfactory epithelium. In: *Olfaction and Taste V*. D. A. Denton and J. P. Coughlan (Eds), Academic Press, New York, pp. 439–41.
Moulton, D. G. (1976). Spatial patterning of response to odors in the peripheral olfactory system. *Physiol. Rev.*, **56**, 578–93.
Moulton, D. G. and Beidler, L. M. (1967). Structure and function in the peripheral olfactory system. *Physiol. Rev.*, **47**, 1–52.
Moulton, D. G., Celebi, G. and Fink, R. P. (1970). Olfaction in mammals—two aspects: proliferation of cells in the olfactory epithelium and sensitivity to odours. In: *Taste and Smell in Vertebrates*. G. E. Wolstenholme and J. Knight (Eds), Churchill, London, pp. 227–50.
Mower, G. D., Mair, R. G. and Engen, T. (1977). Influence of internal factors on the perceived intensity and pleasantness of gustatory and olfactory stimuli. In: *The Chemical Senses and Nutrition*. M. R. Kare and O. Maller (Eds), Academic Press, New York, pp. 103–21.
Mozell, M. M. (1962). Olfactory mucosal and neural responses in the frog. *Am. J. Physiol.*, **203**, 353–8.
Mozell, M. M. (1964a). Evidence for sorption as a mechanism of the olfactory analysis of vapors. *Nature*, **203**, 1181–2.
Mozell, M. M. (1964b). Olfactory discrimination: electrophysiological spatio-temporal basis. *Science*, **143**, 1336–7.
Mozell, M. M. (1966). The spatiotemporal analysis of odorants at the level of the olfactory receptor sheet. *J. Gen. Physiol.*, **50**, 25–41.
Mozell, M. M. (1970). Evidence for a chromatographic model of olfaction. *J. Gen. Physiol.*, **56**, 46–63.
Mozell, M. M. (1971). Spatial and temporal patterning. In: *Handbook of Sensory Physiology, Chemical Senses I*. L. M. Beidler (Ed.), Springer-Verlag, New York, pp. 205–15.
Mozell, M. M. and Hornung, D. E. (1981). Imposed and inherent olfactory mucosal activity patterns: an experimental design. *Chem. Senses*, **6**, 267–76.
Murphy, C. and Cain, W. S. (1980). Taste and olfaction: independence vs. interaction. *Physiol. Behav.*, **24**, 601–5.
Murphy, C., Cain, W. S. and Bartoshuk, L. M. (1977). Mutual action of taste and olfaction. *Sens. Proc.*, **1**, 204–11.
Mustaparta, H. (1971). Spatial distribution of receptor responses to stimulation with different odours. *Acta Physiol. Scand.*, **82**, 154–66.
Nakamura, T. and Gold, G. H. (1987). A cyclic nucleotide-gated conductance in olfactory receptor cilia. *Nature*, **325**, 442–4.
Nakashima, M., Mori, K. and Takagi, S. (1978). Centrifugal influence on olfactory bulb activity in the rabbit. *Brain Res.*, **154**, 301–16.
Nathan, M. H. and Moulton, D. G. (1981). 2-Deoxyglucose analysis of odorant-related activity in the salamander olfactory epithelium. *Chem. Senses*, **6**, 259–66.

Nicoll, R. A. (1972). Olfactory nerves and their excitatory action in the olfactory bulb. *Exp. Brain Res.*, **14**, 185–97.
O'Connell, R. J. and Mozell, M. M. (1969). Quantitative stimulation of frog olfactory receptors. *J. Neurophysiol.*, **32**, 51–63.
Ohloff, G. (1980). Stereochemistry-activity relationships in human odor sensation: 'The Triaxial Rule', In: *Olfaction and Taste VII.* H. van der Starre (Ed.), Information Retrieval Ltd, London, pp. 3–11.
Ohloff, G. (1986). Chemistry of odor stimuli. *Experentia*, **42**, 271–9.
Ohloff, G. and Giersch, W. (1980). Stereochemistry-activity relationships in olfaction. Odorants containing a proton donor/proton acceptor unit. *Helv. Chim. Acta*, **63**, 76–94.
Ohloff, G., Vial, C., Demole, E., Enggist, P., Giersch, W., Jegou, E., Caruso, A., Polonsky, J. and Lederer, E. (1986). Conformation-odor relationships in norlabdane oxides. *Helv. Chim. Acta*, **69**, 163–73.
Ohno, I., Ohyama, M., Hanamure, Y. and Ogawa, K. (1981). Comparative anatomy of olfactory epithelium. *Biomed. Res.*, **2**, 455–8.
Oshima, K. and Gorbman, A. (1969). Effect of estradiol on NaCl-evoked olfactory bulbar potentials in goldfish: dose-response relationships. *Gen. Comp. Endocr.*, **13**, 92–7.
Ottoson, D. (1956). Analysis of the electrical activity of the olfactory epithelium. *Acta. Physiol. Scand.*, **35**, Suppl. 122.
Ottoson, D. (1973). Generator potentials. In: *Transduction Mechanisms in Chemoreception.* T. M. Poynder (Ed.), Information Retrieval Ltd, London, pp. 231–9.
Overbosch, P. (1986). A theoretical model for perceived intensity in human taste and smell as a function of time. *Chem. Senses*, **11**, 315–29.
Overbosch, P., Van den Enden, J. C. and Keur, B. M. (1986). An improved method for measuring perceived intensity/time relationships in human taste and smell. *Chem. Senses*, **11**, 331–8.
Pace, U. and Lancet, D. (1986). Olfactory GTP-binding protein: signal transducing polypeptide of vertebrate chemosensory neurons. *Proc. Nat. Acad. Sci.*, **83**, 4947–51.
Pace, U., Hanski, E., Salomon, Y. and Lancet, D. (1985). Odorant-sensitive adenylate cyclase may mediate olfactory reception. *Nature*, **316**, 255–8.
Pager, J. (1977). The regulatory food intake behavior: some olfactory central correlates. In: *Olfaction and Taste VI.* J. LeMagnen and P. MacLeod (Eds), Information Retrieval Ltd, London, pp. 135–42.
Pauling, L. (1946). Analogies between antibodies and simpler chemical substances. *Chem. Eng. News*, **24**, 1064–5.
Pearson, A. A. (1941). The development of the nervus terminalis in man. *J. Comp. Neurol.*, **75**, 39–66.
Pearson, A. A. (1942). The development of the olfactory nerve, nervus terminalis and the vomeronasal nerve in man. *Ann. Otol. Rhinol. Laryngol.*, **51**, 317–33.
Pfaff, D. W. and Gregory, E. (1971). Olfactory coding in olfactory bulb and medial forebrain bundle of normal and castrated male rats. *J. Neurophysiol.*, **34**, 208–16.
Phillips, H. S., Hostetter, G., Kerdelhue, B. and Kozlowski, G. P. (1980). Immunocytochemical localization of LHRH in the central olfactory pathways of the hamster. *Brain Res.*, **193**, 574–9.

Phillips, P. D. and Vallowe, H. H. (1975). Cyclic fluctuations in odor detection by female rats and the temporal influences of exogenous steroids on ovariectomized rats. *Proc. Penn. Acad. Sci.*, **49**, 160–4.

Pietras, R. J. and Moulton, D. G. (1974). Hormonal influences on odor detection in rats: changes associated with the estrous cycle, pseudopregnancy, ovariectomy, and administration of testosterone propionate. *Physiol. Behav.*, **12**, 475–91.

Polak, E. H. (1973). Multiple profile-multiple receptor site model for vertebrate olfaction. *J. Theor. Biol.*, **40**, 469–84.

Polak, E., Trotier, D. and Baliquet, E. (1978). Odor similarities in structurally related odorants. *Chem. Senses*, **3**, 369–80.

Potter, H. and Chorover, S. (1976). Response plasticity in hamster olfactory bulb: peripheral and central processes. *Brain Res.*, **116**, 417–29.

Price, J. L. and Powell, T. P. (1970). The mitral and short axon cells of the olfactory bulb. *J. Cell Sci.*, **7**, 631–51.

Price, S. (1978). Anisol binding protein from dog olfactory epithelium. *Chem. Senses*, **3**, 51–5.

Price, S. (1981). Receptor protein in vertebrate olfaction. In: *Biochemistry of Taste and Olfaction*. R. H. Cagan and M. R. Kare (Eds), Academic Press, New York, pp. 69–84.

Price, S. (1984). Mechanisms of stimulation of olfactory neurons: an essay. *Chem. Senses*, **8**, 341–54.

Pryor, G. T., Steinmetz, G. and Stone, H. (1970). Changes in absolute detection and subjective intensity of supra-threshold stimuli during olfactory adaptation and recovery. *Percept. Psychophys.*, **8**, 331–5.

Puchelle, E., Aug, F., Pham, Q. and Bertrand, A. (1981). Comparison of three methods for measuring nasal mucocilliary clearance in man. *Acta Otolaryngol.*, **91**, 297–303.

Randebrock, R. E. (1971). Molecular theory of odor with the alpha-helix as potential perceptor. In: *Gustation and Olfaction*. G. Ohloff and A. F. Thomas (Eds), Academic Press, New York, pp. 111–25.

Rehfeld, J. F. (1978). Immunochemical studies on cholecystokinin. *J. Biol. Chem.*, **263**, 4022–30.

Rehn, T. (1978). Perceived odor intensity as a function of air flow through the nose. *Sen. Proc.*, **2**, 198–205.

Revial, M. F., Sicard, G., Duchamp, A. and Holley, A. (1982). New studies on odour discrimination in the frog's olfactory receptor cells. I. Experimental results. *Chem. Senses*, **7**, 175–90.

Rhein, L. D. and Cagan, R. H. (1980). Biochemical studies of olfaction: isolation, characterization, and odorant binding activity of cilia from rainbow trout olfactory rosettes. *Proc. Nat. Acad. Sci.*, **77**, 4412–6.

Rittner, M. and Schmidt, U. (1982). The influence of the sexual cycle on the olfactory sensitivity of wild female mice (*Mus musculus domesticus*). *Z. Säugetierkunde*, **47**, 47–50.

Russell, G. F. and Hills, J. I. (1971). Odor differences between enantiomeric isomers. *Science*, **172**, 1043–4.

Rozin, P. (1982). 'Taste-smell confusions' and the duality of the olfactory sense. *Percept. Psychophys.*, **31**, 397–401.

Saini, K. D. and Breipohl, W. (1977). Frequency of mast cells in olfactory mucosa of Rhesus monkeys. *Cell Tiss. Res.*, **178**, 61–72.

Saito, A., Sankaran, H., Goldfine, I. D. and Williams, J. A. (1980). Cholecystokinin receptors in the brain: characterization and distribution. *Science*, **208**, 115–56.
Saito, A., Williams, J. A. and Goldfine, I. D. (1981). Alterations in brain cholecystokinin receptors after fasting. *Nature*, **289**, 599–600.
Samanen, D. W. and Forbes, W. B. (1984). Replication and differentiation of olfactory receptor neurons following axotomy in the adult hamster: A morphometric analysis of postnatal neurogenesis. *J. Comp. Neurol.*, **225**, 201–11.
Scalia, F. and Winans, S. S. (1975). The differential projections of the olfactory bulb and accessory bulb in mammals. *J. Comp. Neurol.*, **161**, 31–56.
Schafer, R. and Brower, K. R. (1975). Psychophysical recognition of functional groups on odorant molecules. In: *Olfaction and Taste V*. D. A. Denton and J. P. Coughlan (Eds), Academic Press, New York, pp. 313–6.
Schafer, R., Fracek, S., Criswell, D. and Brower, K. R. (1984). Protection of olfactory responses from inhibition by ethyl bromacetate, diethylamine and other chemically active odorants by certain esters and other compounds. *Chem. Senses*, **9**, 55–72.
Schemper, T., Voss, S. and Cain, W. S. (1981). Odor identification in young and elderly persons: sensory and cognitive limitations. *J. Gerontol.*, **36**, 446–52.
Schiffman, S. (1977). Food recognition by the elderly. *J. Gerontol.*, **32**, 586–592.
Schiffman, S. and Leffingwell, J. C. (1981). Perception of odors of simple pyrazines by young and elderly subjects; a multidimensional analysis. *Pharmacol. Biochem. Behav.*, **14**, 787–98.
Schiffman, S. and Pasternak, M. (1979). Decreased discrimination of food odors in the elderly. *J. Gerontol.*, **34**, 73–9.
Schmidt, C. and Schmidt, U. (1980). Changes of olfactory sensitivity during the estrous cycle in female laboratory mice. *Chem. Senses*, **5**, 359–365.
Schmidt, R. F. (1978). *Fundamentals of Sensory Physiology*. Springer-Verlag, New York, pp. 33–4.
Schneider, D. (1971). Specialized odor receptors of insects. In: *Gustation and Olfaction*. G. Ohloff and A. F. Thomas (Eds), Academic Press, New York, pp. 45–60.
Schneider, R. A. and Wolf, S. (1955). Olfactory perception thresolds for citral utilizing a new type olfactorium. *J. Appl. Physiol.*, **8**, 337–42.
Schneider, R. A., Costiloe, J. P., Howard, R. P. and Wolf, S. (1958). Olfactory perception thresholds in hypogonadal women: changes accompanying administration of androgen and estrogen. *J. Clin. Endocrinol Metab.*, **18**, 379–82.
Schneider, S. P. and Macrides, F. (1978). Laminar distributions of interneurons in the main olfactory bulb of the adult hamster. *Brain Res. Bull.*, **3**, 73–82.
Schwanzel-Fukuda, M. and Silverman, A. J. (1980). The nervus terminalis of the guinea pig: a new luteinizing hormone-releasing hormone (LHRH) neuronal system. *J. Comp. Neurol.*, **191**, 213–25.
Seroogy, K. B., Brecha, N. and Gall, C. (1985). Distribution of cholecystokinin-like immunoreactivity in the rat main olfactory bulb. *J. Comp. Neurol.*, **239**, 373–83.
Sharp, F. R., Kauer, J. S. and Shepherd, G. M. (1975). Local sites of activity-related glucose metabolism in rat olfactory bulb during olfactory stimulation. *Brain Res.*, **98**, 596–600.

Shevrygin, B. V. (1973). The movement of air to the olfactory region of the nose in man. *Fiziol. Zh. SSSR*, **19**, 247–9.
Shibuya, T. (1969). Activities of single olfactory receptor cells. In: *Olfaction and Taste III*. C. Pfaffmann (Ed.), Rockefeller Press, New York, pp. 109–16.
Shibuya, T. S. and Tucker, D. (1967). Single unit responses of olfactory receptors in vultures. In: *Olfaction and Taste II*. T. Hayashi (Ed.), Pergamon Press, Oxford, pp. 219–33.
Shipley, M. T. and Adamek, G. D. (1984). The connections of the mouse olfactory bulb: A study using orthograde and retrograde transport of wheat germ agglutinin conjugated to horseradish peroxidase. *Brain Res. Bull.*, **12**, 669–88.
Shirley, S., Polak, E. H. and Dodd, G. (1983). Chemical modification studies on rat olfactory mucosa using a thiol-specific reagent and enzymatic iodination. *Eur. J. Biochem.*, **132**, 485–94.
Sicard, G. (1986). Electrophysiological recordings from olfactory receptor cells in adult mice. *Brain Res.*, **397**, 405–8.
Sicard, G. and Holley, A. (1984). Receptor cell responses to odorants: similarities and differences among odorants. *Brain Res.*, **292**, 283–296.
Silver, W. L. and Maruniak, J. A. (1981). Trigeminal chemoreception in the nasal and oral cavities. *Chem. Senses*, **6**, 295–307.
Silver, W. L. and Moulton, D. G. (1982). Chemosensitivity of rat nasal trigeminal receptors. *Physiol. Behav.*, **28**, 927–31.
Silver, W. L., Mason, J. R., Adams, M. A. and Smeraski, C. (1986). Nasal trigeminal chemoreception: Responses to n-aliphatic alcohols. *Brain Res.*, **376**, 221–9.
Simon, H., Drettner, B. and Jung, B. (1977). Messung des Schleimhauttransportes in menschlichen Nase mit 51 Cr markierten Harzkügelchen. *Acta Otolaryngol. (Stockholm)*, **83**, 378–86.
Sklar, P. B., Anholt, R. and Snyder, S. H. (1987). The odorant-sensitive adenylate cyclase of olfactory receptor cells: Differential stimulation by distinct classes of odorants. *J. Biol. Chem.*, **261**, 15538–43.
Skeen, L. C. (1977). Odor-induced patterns of deoxyglucose consumption in the olfactory bulb of the tree shrew, *Tupaia glis*. *Brain Res.*, **124**, 147–53.
Skeen, L. C. and Hall, W. C. (1977). Efferent projections of the main and the accessory olfactory bulb in the tree shrew (*Tupaia glis*). *J. Comp. Neurol.*, **172**, 1–36.
Smith, G. (1942). Age incidence of atrophy of olfactory nerves in man. *J. Comp. Neurol.*, **77**, 589–94.
Sperber, G. O. (1977). Coacervate-like membrane structures and olfactory nerves in man. *J. Comp. Neurol.*, **77**, 589–94.
Steinmetz, G., Pryor, G. T. and Stone, H. (1970). Olfactory adaptation and recovery in man as measured by psychophysical techniques. *Percept. Psychophys.*, **8**, 327–30.
Stevens, D. A. and Lawless, H. T. (1981). Age-related changes in flavor perception. *Appetite*, **2**, 127–36.
Stevens, J. C. and Cain, W. S. (1985). Age related deficiency in the perceived strength of six odorants. *Chem. Senses*, **10**, 517–29.
Stevens, J. C. and Cain, W. S. (1986a). Aging and the perception of nasal irritation. *Physiol. Behav.*, **37**, 323–8.

Stevens, J. C. and Cain, W. S. (1986b). Smelling via the mouth: Effects of aging. *Percept. Psychophys.*, **40**, 142–6.
Stevens, J. C. and Cain, W. S. (1987). Old-age deficits in the sense of smell as gauged by thresholds, magnitude matching, and odor identification. *Psychol. Aging*, **2**, 36–42.
Stevens, J. C., Plantinga, A. and Cain, W. S. (1982). Reduction of odor and nasal pungency associated with aging. *Neurobiol. Aging*, **31**, 125–32.
Stevens, J. C., Bartoshuk, L. M. and Cain, W. S. (1984). Chemical senses and aging: taste versus smell. *Chem. Senses*, **9**, 167–79.
Stoksted, P. (1952). The physiologic cycle of the nose under normal and pathologic conditions. *Acta Otolaryngol. (Stockholm)*, **42**, 175–9.
Stoksted, P. and Thomsen, K. (1953). Changes in the nasal cycle under stellate ganglion block. *Acta Otolaryngol. (Stockholm)*, Suppl. **109**, 176–92.
Stone, H. and Rebert, C. S. (1970). Observations on trigeminal-olfactory interactions. *Brain Res.*, **21**, 138–42.
Stone, H., Williams, B. and Carregal, E. J. (1968). The role of the trigeminal nerve in olfaction. *Exp. Neurol.*, **21**, 11–19.
Takagi, S. F., Wyse, G. A., Kitamura, H. and Ito, K. (1968). The roles of sodium and potassium ions in the generation of the electro-olfactogram. *J. Gen. Physiol.*, **51**, 522–78.
Taylor, M. (1961). An experimental study of the influence of the endocrine system on the nasal respiratory mucosa. *J. Laryngol.*, **75**, 972–7.
Teatini, G. P. and Pincini, G. (1961). Estimulacion olfactoria por via hematica. *Acta. Oto-Rino-Laringol. Ibero-Amer.*, **12**, 417–31.
Teghtsoonian, R., Teghtsoonian, M., Berglund, B. and Berglund, U. (1978). Invariance of odor strength with sniff vigor: an olfactory analogue to size constancy. *J. Exp. Psych. Hum. Percept. Perf.*, **4**, 144–52.
Terasawa, E. and S. (1967). Electrical activity during the estrous cycle of the rat: cyclic changes in limbic structures. *Endocrinology*, **85**, 207–16.
Thommesen, G. (1982). Specificity and distribution of receptor cells in the olfactory mucosa of char (*Salmo alpinus* L.). *Acta Physiol. Scand.*, **115**, 47–56.
Thommesen, G. and Døving, K. B. (1977). Spatial distribution of the EOG in the rat: a variation with odour quality. *Acta Physiol. Scand.*, **99**, 270–80.
Thornhill, R. A. (1970). Cell division in the olfactory epithelium of the lamprey, *Lampreta fluviatilis*. *Z. Zellforsch. Mikroscop. Anat.*, **109**, 147–57.
Tucker, D. (1961). Physiology of olfaction. *Am. Perfumer*, **76**, 48–53.
Tucker, D. (1963). Physical variables in the olfactory stimulation process. *J. Gen. Physiol.*, **46**, 453–89.
Tucker, D. (1971). Nonolfactory responses from the nasal cavity: Jacobson's organ and the trigeminal system. In: *Handbook of Sensory Physiology, Chemical Senses I*. L. M. Beidler (Ed.), Springer-Verlag, New York, pp. 151–81.
Tucker, D. and Beidler, L. M. (1956). Autonomic nervous system influence on olfactory receptors. *Am. J. Physiol.*, **187**, 637.
Ulrich, C. E., Haddock, M. P. and Alarie, Y. (1972). Airborne chemical irritants. Role of the trigeminal nerve. *Arch. Environ. Health*, **24**, 37–42.
Van As, W., Kauer, J. S., Menco, B. P. and Koster, E. P. (1985). Quantitative aspects of the electro-olfactogram in the tiger salamander. *Chem. Senses*, **10**, 1–21.
Van Toller, S. and Dodd, G. H. (1987). Prebyosmia and olfactory compensation for the elderly. *Brit. J. Clin. Pract.*, **41**, 725–8.

Vanderhaeghen, J. J., Lostra, F., DeMey, J. and Gilles, C. (1980). Immunohistochemical localization of cholecystokinin- and gastrin-like peptides in the brain and hypopysis of the rat. *Proc. Nat. Acad. Sci.*, **77**, 1190–4.
Van Drongelen, W., Holley, A. and Døving, K. B. (1978). Convergence in the olfactory system: quantitative aspects of odour sensitivity. *J. Theor. Biol.*, **71**, 39–48.
Vierling, J. S. and Rock, J. (1967). Variations in olfactory sensitivity to Exaltolide during the menstrual cycle. *J. Appl. Physiol.*, **22**, 311–15.
Vinnikov, Y. A. (1974). *Sensory Reception*. Springer-Verlag, Berlin.
Vinnikov, Y. A., Pyatkina, G. A., Shakhmatova, Y. I. and Natochin, Y. V. (1979). The structure and ionic composition of the olfactory and respiratory mucus in sturgeons and the oxidative hypothesis of olfaction (in Russian). *Dokl. Akad. Nauk SSSR*, **245**, 750–3.
Vodyanoy, V. and Murphy, R. B. (1983). Single channel fluctuations in bimolecular lipid membranes induced by rat olfactory epithelial homogenates. *Science*, **220**, 717–9.
Vodyanoy, V. and Vodyanoy, I. (1987). ATP and GTP are essential for olfactory response. *Neurosci. Lett.*, **73**, 253–8.
Voirol, E. and Daget, N. (1986). Comparative study of nasal and retronasal olfactory perception. *Lebensm.-Wiss. u. Technol.*, **19**, 316–9.
Walker, J. C., Tucker, D. and Smith, J. C. (1979). Odor sensitivity mediated by trigeminal nerve in the pigeon. *Chem. Senses*, **4**, 107–16.
Walsh, R. R. (1956). Single cell spike activity in the olfactory bulb. *Am. J. Physiol.*, **186**, 255–63.
Witkin, J. and Silverman, A. (1983). Luteinizing hormone-releasing hormone (LHRH) in rat olfactory systems. *J. Comp. Neurol.*, **218**, 426–32.
Wright, R. H. (1954). Odour and molecular vibration. I. Quantum and thermodynamic considerations. *J. Appl. Chem.* **4**, 611–15.
Wright, R. H. (1974). Predicting olfactory quality from far infrared spectra. *Ann. NY Acad. Sci.*, **237**, 129–36.
Wright, R. H. (1977). Odor and molecular vibration: neural coding of olfactory information. *J. Theor. Biol.*, **64**, 473–502.
Wright, R. H. (1982). Odour and molecular volume. *Chem. Senses*, **7**, 211–13.
Wright, R. H. and Burgess, R. E. (1969). Musk odor and far infrared vibration frequencies. *Nature*, **224**, 1033–5.
Wright, R. H. and Burgess, R. E. (1971). Molecular mechanisms of olfactory discrimination and sensitivity. In: *Gustation and Olfaction*. G. Ohloff and A. F. Thomas (Eds), Academic Press, New York, pp. 61–72.
Wright, R. H. and Burgess, R. E. (1975). Molecular coding of olfactory specificity. *Can. J. Zool.*, **53**, 1247–53.
Wysocki, C. J. (1979). Neurobehavioral evidence for the involvement of the vomeronasal system in mammalian reproduction. *Neurosci. Biobehav. Rev.*, **3**, 301–41.
Wysocki, C. J., Beauchamp, G. K., Reidinger, R. R. and Wellington, J. L. (1985). Access of large and nonvolatile molecules to the vomeronasal organ of mammals during social and feeding behaviors. *J. Chem. Ecol.*, **11**, 1147–59.
Yamamoto, C., Yamamoto, T. and Iwama, K. (1963). The inhibitory system in the olfactory bulb studied by intracellular recording. *J. Neurophysiol.* **26**, 403–15.

Youngentob, S. L., Stern, N. M., Mozell, M. M., Leopold, D. A. and Hornung, D. E. (1986). Effect of airway resistance on perceived odor intensity. *Am. J. Otolaryngol.*, **7**, 187–93.

Youngs, W., Schneider, S. and Macrides, F. (1976). Some response properties of deep short axon cells in the olfactory bulb of the Syrian golden hamster. *Neurosci. Abstr.*, **2**, 167.

Zinkevich. E. P. and Minor, A. V. (1969). On the olfactory mechanism. *Proceedings First All-Union Conference on the Structure and the Function of the Olfactory Analyzer of Animals and Man and their Modelling.* Nauka, Moscow, p. 42.

Zinkevich, E. P. and Treboganov, A. D. (1973). Investigation of the chemical conversions of an odoriferous substance on an olfactory lining. In: *Mechanisms of the Work of the Receptor Elements of the Sense Organs* (In Russian). Nauka, Leningrad, pp. 108–12.

Chapter 3

TEXTURE PERCEPTION AND MEASUREMENT

J. G. Brennan

*Department of Food Science and Technology,
University of Reading, UK*

1. INTRODUCTION

Texture of foods and its measurement continue to interest researchers worldwide. Recognition of the complex nature of this sensory attribute has led to increased use of profiling techniques, both sensory and instrumental, in texture studies. In relating sensory and non-sensory data the use of sophisticated statistical techniques, such as principal component analysis, is now quite common in such work. There is also much interest in relating sensory textural attributes to well-defined physical properties of foods. There is a renewed awareness of the need for more exchange of information and cooperative research between oral researchers and food texture scientists. The use of electromyography as a tool in texture studies may provide more opportunities for such cooperation.

1.1. Terminology and Classification of Terms

Many definitions of the term 'texture' have been proposed. One which I favour was put forward by the British Standards Institution as follows (Anon, 1975):

> *Texture* (noun): The attribute of a substance resulting from a combination of physical properties and perceived by the senses of touch (including kinaesthesis and mouthfeel), sight and hearing. Physical properties may include size, shape, number, nature and conformation of constituent structural elements.

Thus texture is clearly defined as a sensory attribute and so only measurable directly by sensory means. Three senses touch, sight and hearing may be involved in sensory assessment of texture.

Many attempts have been made to identify, define and classify specific textural terms. Szczesniak (1963) published a classification of textural terms which is still widely used today in its original or modified form. Three categories of textural characteristics were proposed as follows:

1. *mechanical characteristics*—relating to the reaction of food to stress,
2. *geometrical characteristics*—relating to the size, shape and orientation of the particles within the food,
3. *other characteristics*—relating to the perception of the moisture and fat contents of the food.

The mechanical characteristics were subdivided into:

a. *primary* parameters of hardness, cohesiveness, viscosity, elasticity and adhesiveness;
b. *secondary* parameters of brittleness, chewiness and gumminess.

Definitions for these terms were proposed.

Sherman (1969) proposed an alternative classification to that of Szczesniak in which the only criterion was whether a characteristic was a fundamental property or whether it was derived by a combination of two or more attributes in unknown proportions. Characteristics which Szczesniak classed as geometrical (particle size, shape, size distribution, air content, air cell size, size distribution and shape) became primary characteristics in the Sherman scheme. Rheological properties viscosity, elasticity and adhesion were classed as secondary characteristics. All other attributes were placed in the tertiary category. The latter was further subdivided according to the type of mechanical process involved, i.e. mastication, disintegration after mastication and non-masticatory mechanical treatment of the sample prior to sensory assessment in the mouth. Some of these tertiary attributes were as follows:

Mastication	hard, soft
	brittle, plastic, crisp, rubbery, spongy,
	smooth, coarse, powdery, lumpy, pasty,
	creamy, watery, soggy, sticky, tacky
Disintegration	greasy, gummy, stringy,
	melt down properties
Non-masticatory mechanical properties	appearance
	sampling and slicing characteristics,
	spreading, creaming characteristics,
	pourability.

A glossary of textural terms was proposed by Jowitt (1974). These were arranged into four main categories as follows:
1. General—structure, texture consistency.
2. Terms relating to the behaviour of materials under stress, e.g. firm, soft, sticky, crisp, thick.
3. Terms relating to the structure of the material,
   (i) relating to particle size and shape, e.g. smooth, powdery, gritty, mealy
   (ii) relating to shape and arrangement of structural elements, e.g. flaky, fibrous, spongy
4. Terms relating to mouthfeel characteristics, e.g., juicy, greasy, creamy.

Definitions were proposed for all terms.

Szczesniak (1979) reported a study of the mouthfeel characteristics of beverages. A list of 33 different beverages was issued to 103 untrained people and they were asked to list all of the popular terms that came into their minds and to think of as many descriptive words as possible referring to how the individual beverages felt in the mouth. About 136 terms were generated. Of these 75 were mentioned three or more times. The terms

TABLE 1
CLASSIFICATION OF SENSORY MOUTHFEEL TERMS FOR BEVERAGES
From Szczesniak (1979)

	Category	Total response (%)	Typical words
I	Viscosity-related terms	30·7	thin, thick, viscous
II	Feel on soft tissue surfaces	17·6	smooth, pulpy, creamy
III	Carbonation-related terms	11·2	bubbly, tingly, foamy
IV	Body-related terms	10·2	heavy, watery, light
V	Chemical effect	7·3	astringent, burning, sharp
VI	Coating of oral cavity	4·5	mouthcoating, clinging, fatty, oily
VII	Resistant to tongue movement	3·6	slimy, syrupy, pasty, sticky
VIII	Afterfeel—mouth	2·2	clean, drying, lingering, cleansing
IX	Afterfeel—physiological	3·7	refreshing, warming, thirst-quenching, filling
X	Temperature-related	4·4	cold, hot
XI	Wetness-related	1·3	wet, dry

generated were classified into 11 categories as shown in Table 1. The author discussed the terms in each category and concluded that the classification, although useful, may be an over-simplification.

The authors quoted above and many others have pointed out the importance of defining terms and adopting some form of classification of such terms when undertaking work involving the assessment of the texture of foods. However, as yet no one classification has been adopted internationally.

2. PERCEPTION OF TEXTURAL AND MOUTHFEEL CHARACTERISTICS

While the three senses of touch, sight and hearing may be involved in the perception of textural and mouthfeel characteristics of foods, in the majority of cases the sense of touch plays the most important role.

2.1. Texture Perception in the Mouth

Very often sensory assessments of texture are made on the basis of the sensations perceived when the food sample is manipulated in the mouth, i.e. when it is bitten, masticated and swallowed. During such manipulation there is a reciprocal interaction between the texture of the food and the buccal work acting to change the texture to a state suitable for swallowing (Pierson and Le Magnen, 1970). The sense organs involved in mastication and hence texture perception may be grouped into (1) those in the superficial structure of the mouth (2) those around the roots of the teeth and (3) those in the muscles and tendons (Oldfield, 1960).

The hard palate has an almost intact cellular surface which is very sensitive to touch. Coarseness (and other geometrical characteristics) of food are sensed by receptors in this surface. The soft structures of the mouth are provided with a network of free nerve endings and also a variety of organised terminations both encapsulated and unencapsulated. Although the exact function of these organs is, as yet, not clear it is likely that the free nerve endings respond to touch and light pressure and probably chemical and thermal stimuli. Some of the more organised structures are likely to respond to greater pressures, associated with distortion of the tissue by stretching. The tongue moves the food in the mouth to the correct position for chewing. It may press soft foods against the hard palate thus exerting pressure and sensing the response of the food to such pressure.

The teeth play an important role in the sensory evaluation of texture. Branches of the dental nerve terminate in the periodontal membrane which surrounds the tooth in the jaw. These are sensitive to minute pressures but can tolerate large total pressures. There is considerable evidence that the teeth exhibit a degree of movement in both the horizontal and vertical directions. This movement occurs in two stages. A high initial rate of movement at low loads is followed by a lower rate at high loads. This movement is due to elastic deformation of the sockets and it stops when pain is felt. It causes the periodontal membrane to be compressed and this is the stimulus detected when assessing the textural characteristics of foods. Much work has been reported concerned with the measurement of tooth movement and the loads exerted on the teeth during mastication. Some of this work has been discussed by Yeatman and Drake (1983) and Williams (1975).

There is little evidence to suggest that the sense organs located in the muscles and tendons play an important direct role in texture perception. However, these muscles and tendons control jaw movements during mastication thus influencing load distribution and tooth contact. In this way they play an indirect role in texture perception in that they influence the sensations perceived by the receptors in other parts of the mouth. Williams (1975) reviewed early publications dealing with the masticatory process.

Boyar and Kilcast (1986a) published a more recent review of the physiology of mastication. An analysis of the jaw movements indicated that the mandible (lower jaw) moves through five stages during each masticatory stroke: (1) closing stage from the lowest position to when it contacts the food; (2) the contact phase from the first contact with the food to the position of maximum elevation; (3) the squeeze phase: a pause in the position of maximum elevation; (4) the separation phase from the point when the teeth start to separate until they leave the crushed bolus of food; and (5) the opening phase from the end of the separation phase to the lowest position of the mandible.

The tongue, cheeks and lips play a part in mastication; they control and direct the food to the surfaces of the teeth. The tongue also plays an important selective role in deciding whether particles of food are sufficiently small to be swallowed as well as functioning in the first stages of swallowing.

Boyar and Kilcast (1986a) also discussed the different methods used to study the masticatory process including the analysis of sounds produced by the masticatory mechanism, the use of cinematography, X-rays, synchro-transmitters, light emitting diodes and electromyography. This last method involves the study of muscle activity during mastication. Electrodes are attached to the skin and these enable recording of the small electrical potentials created by the action currents flowing in the underlying muscles. A record of this type is known as an electromyogram (EMG). The record obtained cannot always be attributed with certainty to a particular muscle. They stated that for a complete understanding of the motion of the mandible both the muscle activity and the movement pattern need to be recorded simultaneously. Considering the masticatory cycle, during the rise of the mandible action potentials develop in the temporal and masseter muscles at the start of movement; EMG activity increases towards the termination of movement. At the start of the occlusal phase (teeth come together) a short 'silent' period appears in the EMG. After this 'silent' period EMG patterns with the largest amplitude occur. Approximately three-quarters of the total EMG activity occurs during the closing phase and the remainder during the occlusal phase. During the lowering of the mandible no action potentials are recorded from the elevator muscles.

The motor pause or (silent period) has been studied by other workers and was discussed in this review by Boyar and Kilcast (1986a). More motor pauses were produced by hard, breakable boli than by soft, plastic ones. The frequency of occurence of these pauses was significantly higher in the first part of the chewing cycle than in the last part for all foods tested. Motor pauses also occurred when loads were released at the breakage of brittle boli and when contact was made with hard objects, particles or teeth.

These authors also discussed the forces developed during biting and chewing. Maximum bite forces of up to 50 kg have been reported. A sensation of pressure on the joints, muscles, tendons, gums and teeth indicates that any further increase in bite force may cause pain. Under normal conditions the bite force exerted is well below that which would cause pain. The level of the forces developed during ordinary chewing is controlled by a neuromuscular mechanism activated by nervous impulses via sensory pathways from the different parts of the masticatory system. Force measurements have shown that during chewing the force increases approximately linearly with time until it reaches a maximum. Beyond this there is a non-linear phase. These force patterns vary between subjects and different foods. The maximum force during chewing is produced when tooth contact occurs.

In the design of instrumental methods of texture measurement very little account has been taken of the published work on the masticatory process. For example it has been shown that teeth move more slowly when biting and chewing hard foods compared with soft foods. As far as is known this fact has been ignored by texture scientists. Since texture perception is clearly related to oral sensitivity some screening of panelists involved in texture work on the basis of oral sensitivity should be undertaken.

In a research note Boyar and Kilcast (1986b) report on the use of EMG to study the breakdown patterns of different gels. The integrated EMG data from the masseter muscle of one subject chewing a carrigeen gel and a gelatine gel revealed differences in the manner in which the gels broke down. They both required similar forces to initiate breakdown. In the case of the gelatin gel these forces fell off rapidly as the gel dispersed in the saliva and was swallowed. Oral forces remained at a higher level for much longer in the case of the carrigeen gel. These patterns confirm the behaviour of the two gels as assessed subjectively. Preliminary observations suggest that the relative EMG outputs from different foods remained constant from subject to subject. It was also observed that relative outputs from different foods remained constant when tested on different occasions. The authors outlined a programme of work using EMG to study textural characteristics of foods to be undertaken at the Food Research Association, Leatherhead, UK.

Tornberg et al. (1985) reported on a study in which the chewing patterns of a range of cooked meats and meat products were examined with the use of ten strain gauges attached to a prosthetic appliance used by one subject. From the UV recorder traces obtained from the gauges, the maximum force or load exerted by the teeth was noted. The total chewing times and the total number of chewing cycles were also noted and from these data the time for a chewing cycle was calculated. The individual peak heights and the front slopes of the peaks were also measured. The samples were also tested using a Warner–Bratzler shearing device, see below, and the peak shear forces recorded. The test subject scored the samples on a three-point scale for resistance to initial chewing, elasticity, chewiness and ease of disintegration. A sum of these scores for each sample was denoted as 'total toughness impression'. The subject also assessed amount of residue and juiciness.

One observation made was that deformation rates in the mouth were relatively high, up to 200–400 cm min^{-1} for whole meat. The significance of these high rates compared with the much lower rates normally used in instrumental testing was discussed. Multiple linear regression analysis was performed between the sensory scores, some of the measurements from the recorded chewing patterns and the Warner–Bratzler maximum loads. The number of chewing cycles correlated best with all combinations of the sensory attributes. The Warner–Bratzler values did not correlate significantly with any combination of the sensory attributes. A highly significant correlation with sensory attributes was achieved by the product of the loading rate and the number of chews.

Heath and Lucas (1987) discussed the need for more collaborative work between oral researchers and food texture scientists. They pointed out that a proper understanding of the mechanical nature of food is lacking in oral research whereas the parameters concerning dynamics of tooth–food–tooth and tongue–food–palate contact are not well understood by food texture scientists. They commended recent efforts to bring both areas of research together, including those of Boyar and Kilcast (1986a,b) and Tornberg et al. (1985). They suggested areas for collaboration including: the fracture characteristics of foods which could be studied by accurate jaw movement profiles coupled with force transducers—this type of work could provide better information for the design of mechanical tests and might better distinguish between first bite and the cyclical pattern of loading on a molar; swallowing—how does the decision to swallow depend on specific textural attributes?; the use of both a psychophysical and physiological approach to the study of what people can actually sense during biting and chewing; a study of the importance of the first moments that the food is in the mouth (before and during the first bite). These authors also recommended that participants in texture panels should be assessed for oral status by a dentist.

The mechanisms of perception outlined above are mainly involved in the assessment of the texture of solid and semi-solid foods. Other mechanisms may be used when liquid foods are being consumed and their mouthfeel characteristics are being assessed. Shama and Sherman (1973) have shown that the physical stimulus in the sensory perception of the viscosity of low viscosity liquids, < 70 centipoise, is the shear rate at a constant shear stress, 100 dynes cm^{-2}. Cutler et al. (1983) reported a study in which perceived in-mouth thickness of a range of fluid foods and model systems were assessed by a trained sensory panel using ratio scaling and the results were correlated with those obtained by objective measurements of viscosity. For Newtonian samples a good log–log relationship was obtained between thickness and viscosity. In the case of non-Newtonian samples it was concluded that the oral perception of viscosity was obtained over a range of shear rates. In the case of extremely shear-thinning materials the perceived thickness is dominated by their high viscosity at low rates of shear.

A good correlation between perceived thickness and perceived stickiness was obtained with most of the samples. However with some samples (chocolate spread, tomato ketchup, chicken soup, milk and high concentrations of xanthan gum) this correlation was poor. The presence of structural particles in these materials may account for the poor correlation.

Szczesniak (1979) postulated that for beverages the rate of flow under a given force is the sensory measure of viscosity and this may be experienced as the velocity of liquid movement in the mouth or the velocity of flow into the mouth when sipped. She further states that viscosity is the most striking and clearly recognisable mouthfeel parameter of beverages. Other relevant categories of mouthfeel terms in Table 1 are VI *coating of the oral cavity* and VII *resistance to tongue movement*.

Kokini *et al.* (1977) assumed that liquid perception in the mouth takes place largely between the tongue and the roof of the mouth. They suggested that most sensory textural terms associated with liquids could be placed into three groups represented by the terms 'thickness', 'smoothness' and 'slipperiness'. The first of these is closely associated with viscous force, the second with frictional force and the third with a combined force involving viscous and frictional components.

2.2. Non-Oral Methods

Information about the textural characteristics of foods may be obtained prior to mastication. The visual appearance of the sample may provide some cues, e.g. the colour of fruit may indicate its state of ripeness and hence its firmness and the limp appearance of a leafy vegetable would suggest a lack of crispness. When a sample of a solid food is manipulated in the hand by squeezing, impressing or bending or is cut with a knife or penetrated with a fork cues as to its firmness, toughness, crispness or fibrousness may be revealed. The behaviour of liquid or semi-liquid foods when shaken, poured, or spread may yield information about their viscosity, smoothness or stickiness. The importance of non-oral cues to texture was recognised by Sherman (1969) in his classification scheme.

Non-oral techniques have been used in many studies of texture, e.g. to assess the bending properties of biscuits (Brennan *et al*, 1975a; Williams, 1975), the firmness of pectin gels (Gross *et al.*, 1980) and the hardness of beef patties (Anderson and Lundgren, 1981). In a study of the stickiness of spreads and jams (Mohamed and Brennan, 1979; Mohamed, 1981*b*) both oral and non-oral methods were used. The latter procedure involved placing a small sample of the test material on top of the index finger, bringing it into contact with the thumb and then separating the two digits. The results from the non-oral tests were slightly, but not significantly, more reproducible than those obtained from the oral method. In another study (Kisseoglou and Sherman, 1983) a panel of students compared the pourability and spreadability of two commercial brands of salad dressing and a carboxymethyl cellulose (CMC)-sugar solution. The panel's judgement of pourability was based on the rate at which the samples flowed down the side of the container and spreadability was based on the area of plate which was eventually covered by each sample. Correlation of panel responses with viscometric data for the three samples enabled the shear stress and shear rate ranges associated with sensory evaluation of these two properties to be identified. Pokorńy *et al.* (1984) reported work on the texture of margarines. The hardness of a range of soft and semi-soft margarines was judged by cutting with a knife and by spreading on bread. Spreadability of the samples was also assessed by the spreading on bread. The results thus obtained were compared with rheometric measurements made on the samples. It was concluded that the sensory analysis could not be replaced satisfactorily by rheometry. Peleg (1980) published a discussion paper on the sensitivity of the fingers, tongue and jaws as mechanical testing instruments. The main implications from his analysis were that there ought to be a difference in the mechanical sensititivity of the jaws and fingers, that our general sensitivity vanishes when we test materials that are much harder than our sensing tissues and that the human senses are more sensitive in the lower portion of the hardness scale.

2.3. The Role of Sound in Texture Perception

It has long been assumed that the sounds emitted when certain foods are bitten and chewed are to some extent a reflection of the texture of these foods. This is most likely in the case of dry crisp (biscuits, potato crisps) and wet crisp (apples, celery) foods. In fact an advertisement for one well known brand of potato crisps uses the slogan 'Britain's noisiest crisps'.

Drake (1963*a,b*) recorded the sounds produced when food samples were chewed and also crushed between wooden blocks. Differences in sound amplitude were obtained between two samples of meat, one tough and one tender. A positive correlation was found between the recorded sound amplitude and the loudness of the sound perceived by panelists when a range of crispy/brittle foods were tested. Vickers (1979) and Vickers and Bourne (1976) reported on studies in which a number of dry and wet crisp foods were bitten and the sounds produced were recorded. Crisp foods seemed to be characterised by a very uneven amplitude-time pattern. This was explained in terms of the breakdown of dry and wet cellular structures. In a later publication (Vickers and Wasserman, 1979) a study was reported in which 18 different foods were crushed by hand using rubber coated pliers and the sounds produced were recorded. After analysis of these sounds it was concluded that there were two basic sensory criteria for distinguishing foods sounds. One is loudness and the other uneveness and discontinuity. In another study (Vickers and Christensen, 1980) 20 subjects judged the crispness, loudness and firmness of 16 food samples by both biting and chewing using magnitude estimation. They also scored the foods for 13 textural characteristics using category scales. Crispness appeared to be very closely related to loudness and less to firmness. 'Loud', 'snap' and 'crackly' were the three sensory terms most closely related to crispness. In a later paper (Christensen and Vickers, 1981) results were presented from an experiment in which foods were assessed for crispness with and without the biting and chewing sounds blocked by loud masking noises. Subjects had no difficulty in judging crispness in the presence of the blocking noise. An explanation proposed was that vibratory stimuli form the basis for crispness determination. Such vibrations could produce both auditory and tactile sensations. In another study (Vickers, 1984) 30 subjects judged crackliness and hardness of 11 foods under three test conditions: (1) normal, (2) with biting and chewing sounds blocked and (3) by listening to recorded biting and chewing sounds. Compared with hardness in which tactile cues are more useful than auditory cues, either oral or auditory stimuli can be used to make crackliness judgements. The number of sharp noise bursts produced when a food is bitten and chewed provides a relatively good indication of sensory crackliness. Two other papers deal with crispness and crunchiness in foods (Vickers, 1984, p. 157; Vickers, 1985). It was found that foods that were more crisp than crunchy always produced higher-pitched sounds as compared with foods that were more crunchy than crisp. Crispness scores were generally higher for the bite sounds than for the chew sounds. The crisper sounds were typically both higher in pitch and louder than the crunchier sounds.

Mohamed (1981a) and colleagues (Jowitt and Mohamed, 1980; Mohamed et al., 1982; Brennan, 1983) subjected samples of dry and wet crisp foods to deformation under a constant rate of loading. Force–time and deformation–time traces were obtained. The experiments were carried out in an anechoic chamber and the sounds emitted during the tests were recorded. These were analysed to produce a quantity known as the equivalent continuous sound level (L_{eq}). This represents the level of sound which, if it persisted, would result in the same total of sound energy reaching the microphone. The food samples were also assessed by a sensory panel for crispness and hardness by magnitude estimation. The sound patterns produced by both types of food were similar, i.e. a number of tones occurred simultaneously and there were no dominant overtones or harmonics. Good correlations were obtained between sensory crispness and the L_{eq} values calculated from the sound recordings. When the results obtained from the mechanical tests and sound recordings were combined together by multiple linear regression a number of well-fitting equations were derived for the prediction of sensory crispness. The inclusion of the L_{eq} values improved the accuracy of the predictions as compared with expressions involving only the results of the mechanical tests.

3. PROCEDURES USED IN THE SENSORY ASSESSMENT OF TEXTURE

Both large, untrained, *consumer* panels and smaller, *trained* panels are used in the sensory assessment of texture. However, the use of consumer panels has found only limited application in this specific area of study. They are more usually used to assess overall quality in which texture may be an important component. Large panels have been used in the collection and classification of textural terms (Yoshikawa *et al.*, 1970; Szczesniak, 1979). A large panel was used in a study of consumer preference and properties of raw and cooked, milled rice (Del Mundo and Juliano, 1981). Texture was one of the parameters studied. An untrained panel was used to assess the crispness and acceptability of four snack foods, conditioned to different values of water activity (Katz and Labuza, 1981). For all four products crispness and acceptability decreased with increasing levels of water activity. Consumer panels each numbering 180 participants in each of three cities were recently used to assess the influence of marbling on the palatability of beef loin steaks (Savell *et al.*, 1987).

Cardello *et al.* (1982a) studied the relationships between judgements of the perceived texture of foods by trained and consumer panelists. The general conclusions drawn from the experiments were that, through experience, trained texture profile panels develop a broader perceptual range of texture, but that regression equations can be developed to relate these data to consumer data.

Smaller panels consisting of 3–30 selected members are widely used in texture studies. Panelists are educated in the terminology and techniques of sensory assessment. The criteria of selection, degree and type of training varies according to the type of test being carried out and the information being sought. The following types of test are carried out by such panels.

3.1. Discriminatory Tests

This term is used to cover threshold and difference testing. *Threshold* tests are not widely used in texture studies. Mohamed (1981*b*) used a threshold test to determine the lowest concentration of sugar, in a sugar syrup, below which stickiness was no longer perceived. A threshold value for objectionable fibrousness was identified in a study of asparagus, rhubarb, celery, spinach and french beans (Anzaldua Morales, 1982; Brennan, 1983). *Difference* testing may involve the use of paired comparison, triangle or duo-trio tests. Panelists involved in such tests usually receive some training. Examples of the application of difference testing in texture work are in the evaluation of the firmness and crumbliness of cheese (Culioli and Sherman, 1976; Carter and Sherman, 1978), the tenderness of chicken meat (Moraes *et al.*, 1981) and the sound produced when samples of crisp foods were crushed (Vickers and Wasserman, 1979).

3.2. Descriptive Tests

This term is used here to denote the techniques of ranking and scaling which describe and/or quantify the difference between samples. *Ranking* is a procedure seldom used in texture work. It is sometimes used in the early stages of a study to select test samples, e.g. it was used by Williams (1975) to select biscuit types for a correlation study. *Scaling* involves indicating the intensity of an attribute on a scale. The scale may be a numerical one, the length ranging between 0–3 and 0–100, and one or more points may be anchored by a word, description or reference sample. Graphic scales may be provided. These may be structured or unstructured. Both types are very widely used in texture studies. Some examples are as follows: structured scales—for apples (Brennan *et al.*, 1977), beverages (Pangborn *et al.*, 1978), jams and spreads (Mohamed and Brennan, 1979), crisp foods (Vickers and Christensen, 1980), asparagus and rhubarb (Anzaldua Morales, 1982); unstructured scales—biscuits (Williams, 1975), textured proteins (Breene, 1978), beef patties (Voisey and Larmond, 1977), chewing gum (Bogarty *et al.*, 1980) and ice cream (Moore and Shoemaker, 1981). The term 'scoring' is often used to represent scaling without the use of graphic scales and 'rating' where graphic scales are used.

There are two major categories of scaling. In the case of interval or category scaling it is assumed that each scale represents a continuum divided into equal parts by the scale points. This in turn assumes that the relationship between perceived sensory intensity and objectively measured physical intensity can be described by a linear or semi-logarithmic function. In the case of ratio or proportional scaling this relationship is assumed to be described by a power function of the type:

$$S = kI^n$$

where S is the sensory intensity, I the physical intensity and k and n are constants. Magnitude estimation (ME) is a method of ratio scaling. Samples are assigned scores, with reference to a particular textural characteristic, in proportion to the score assigned to a reference sample.

In recent years there has been an increase in the use of ratio scaling, including ME, in texture work. Williams (1975) and Brennan et al. (1975a) used ratio scaling to judge the hardness of biscuits, Christensen (1979) the viscosity of aqueous solutions thickened with carboxymethyl cellulose, Mohamed and colleagues (Jowitt and Mohamed, 1980; Mohamed, 1981a; Mohamed et al., 1982; Brennan, 1983) the crispness of biscuits and apples, Katz and Labuza (1981) the crispness and acceptability of snack foods, Cutler et al. (1983) in-mouth thickness of fluid foods.

Anzaldua Morales (1982) and Brennan (1983) used both interval and ratio scaling to assess the fibrousness and firmness of asparagus and rhubarb. The latter technique yielded more reproducible results which correlated better with instrumental data. Cardello et al. (1982b) rescaled the food items in the six standard, interval scales of the General Foods texture profile, see below, using ratio scaling. The interval scale data were concave downward relative to the ratio scale data. This was said to demonstrate that different results may be obtained with different scaling methods.

The principles of interval and ratio scaling have been discussed many times in the literature by Moskowitz and colleagues (Moskowitz and Kapsalis, 1975, 1976; Moskowitz and Chandler, 1977; Moskowitz, 1977, 1981; Kapsalis and Moskowitz, 1978) and are covered in Chapter 6 of this book.

Multiple scales, either interval or ratio, may be used to obtain a more complete description of the textural characteristics of foods. In sensory profiling a number of scales are used together with qualitative descriptions and are applied in an organised way to build up a 'picture' of the textural features of a food. Brandt et al. (1963) adopted the classification of textural terms proposed by Szczesniak (1963), discussed earlier in this chapter. Standard rating scales for many of the *mechanical parameters* were developed (Civille and Szczesniak, 1973). Each scale was of the interval type and each point was identified by a reference food. Each reference food was identified by its descriptive name, brand or type, manufacturer's name, sample size and temperature of serving. Panelists were instructed on how to assess each mechanical characteristic. One version of these instructions is given in Table 2.

The *geometrical characteristics* were evaluated and described in terms of type and amount present (Brandt et al., 1963). In some cases the amount

TABLE 2
MODIFIED DEFINITIONS OF MECHANICAL PROPERTIES IN THE GENERAL FOODS TEXTURE PROFILE

Mechanical characteristics

Primary properties

Hardness	The force required to compress a substance between the molar teeth (for solids) or between the tongue and palate (for semi-solids) to a given deformation or to penetration.
Cohesiveness	The extent to which a material can be deformed before it ruptures.
Viscosity	The force required to draw (slurp) a liquid from a spoon over the tongue.
Springiness (elasticity)	The amount of recovery from a deforming force: the rate at which a deformed material returns to its undeformed condition after the deforming force is removed.
Adhesiveness	The force required to remove material that adheres to the mouth (generally the palate) during the normal eating process.

Secondary properties

Fracturability (brittleness)	The force with which a sample crumbles, cracks or shatters; the horizontal force with which the fragments move away from the point where the vertical force is applied. Fracturability is the result of a high degree of hardness and low degree of adhesiveness.
Chewiness	The length of time or the number of chews required to masticate a solid food to a state pending for swallowing. Chewiness is a product of hardness, cohesiveness and springiness.
Gumminess	A denseness that persists throughout mastication, the energy required to disintegrate a semi-solid food to a state ready for swallowing. Gumminess is a product of a low degree of hardness and a high degree of cohesiveness.

was indicated on a scale (Brandt *et al.*, 1963; Civille and Szczesniak, 1973).

The other characteristics were also assessed. In most cases this was done qualitatively but again, occasionally scales were used (Brandt *et al.*, 1963; Civille and Szczesniak, 1973). The order in which the various characteristics were to be assessed was specified. The procedure was divided into three stages: *initial* (first bite), *masticatory* and *residual*. The characteristics to be evaluated in each stage were specified.

Panelists received extensive training in the terminology and methodology of profiling. Initially they used the foods that comprised the standard rating scales and later they gained experience in preparing profiles for a range of other foods. Useful advice on the training of profile panels has been published (Brandt et al., 1963; Civille and Szczesniak, 1973; Bourne et al., 1975). Experience gained in the first decade or so in which texture profiling was used has been discussed (Szczesniak, 1975; Civille and Liska, 1975). Some modifications to the above General Foods Corporation scheme have been introduced. The terms 'springiness' and 'fracturability' have replaced 'elasticity' and 'brittleness' respectively. Surface properties perceived by the soft tissues of the mouth, before the first bite are now included. A 14-point intensity rating scale has been developed to quantify the mechanical characteristics. Munoz (1986) further modified some of the standard rating scales, introduced new reference materials and developed new rating scales for: wetness, adhesiveness to lips, roughness, self adhesiveness, springiness, cohesiveness of mass, moisture absorption, adhesiveness to teeth and manual adhesiveness.

In recent years sensory texture profile analysis has been widely used. Some researchers used the General Foods profile scheme or modifications thereof. Others developed their own schemes. Some examples of such work follow. Borderies et al. (1983) used both sensory and instrumental profile analysis on fish fillets and minced fish. No significant correlation was found between sensory texture profiles and the results of instrumental analyses on the fillets. However, good sensory–instrumental correlations were found in the case of the minced fish. This was attributed to the greater homogeneity of the fish minces. Dransfield et al. (1984a) developed a sensory texture profile for cooked meats. Thirteen assessors judged 66 meats for nine textural attributes. A statistical rotational fitting technique reduced these nine attributes to five.

In another study (Dransfield et al., 1984b) researchers from the same institute developed a sensory texture profile for UK beefburgers. Initially, 24 texture descriptors were agreed by the panel. After discussion these were reduced to five (Tables 3 and 4). These five descriptors together with overall texture were used to study 25 retail brands of frozen and chilled beefburgers. As a result the descriptors 'rubberiness' and 'ease of fragmentation' were combined into one descriptor. In a later paper (Jones et al., 1985) a study is reported in which the weights, dimensions, chemical composition, compressive strength and expressible fluid of the same 25 brands of beefburger were measured before and after cooking. All these results were compared with the sensory data reported in Dransfield et al. (1984b).

TABLE 3
DEFINITION OF ATTRIBUTES OF BEEFBURGER TEXTURE
From Dransfield et al. (1984b)

After tasting and by discussion and agreement, 11 assessors grouped 24 texture descriptors under defined attribute headings placed in temporal order. For each descriptor, its frequency of use (maximum = 66) and the number of assessors using it (maximum = 11) is given in parentheses.

1. Rubberiness (initial sensations on biting and chewing)
    rubbery   'degrees of rubberiness'   soft/hard (4/3)
    spongy (3/2)   resistant (2/1)

2. Fragmentation (nature and ease of initial breakdown on chewing)
    crumbly (17/8)   friable (6/1)
    cohesive (4/2)   loose/tight (2/1)
    'mealy' (3/2)

3. Comminution (nature and size of particles making up the mix)
    lumpy (8/3)   compact (1/1)
    coarse/fine (4/3)   smooth (1/1)
    granular (2/1)

4. Character of the particles (nature and breakdown by chewing of the particles of the mix)
    tough/tender (34/10)   'doughy' (2/2)   'breadlike' (1/1)
    'chewy' (8/3)   gristly (3/1)
    rubbery (6/5)   elastic (2/1)
    'pasty' (3/3)   stringy (1/1)

5. Moistness (the sensations of wetness and juiciness)
    dry/moist (7/4)   juicy (4/1)

Beefburgers with a more rubbery, cohesive and coarser, tougher texture had higher compressive strength, higher defatted meat content and a more acceptable texture. The major difference in texture between the brands was a combination of the sensory properties rubberiness and tenderness of particles. The second principal sensory component was composed mainly of moistness and comminution attributes and was related to the chemical assays and expressible fluid.

Garruti and Bourne (1985) used a profile based on the General Foods model to study the texture of cooked red kidney beans. Onayemi (1985) developed sensory profiles for yam and cassava products, Berry (1986) did likewise for restructured beef steaks which varied in meat particle size. The profile analysis could distinguish differences between the samples. Onayemi

TABLE 4
TEXTURE PROFILE FOR BEEFBURGERS
From Dransfield et al. (1984b)

BEEFBURGER—TEXTURE PROFILE FORM _____
                                                        Date _____

Name _____

You are given samples of beefburgers; please ignore colour and flavour and judge for texture only, making the following judgments in the sequence indicated.

Mark the scales with a cross at the appropriate point.

1) *Rubberiness*
   Non-rubbery_____Rubbery

2) *The ease of fragmentation*
   Cohesive_____Readily separating

3) *The degree of comminution* (the ease with which the particles separate is irrelevant)
   Coarse_____Fine

4) *The character of the particles*
   Extremely tough_____Extremely tender

5) *Moistness*
   Dry_____Wet (juicy)

6) *Summing-up* judgment of *overall* texture
   Very poor_____Very good

7) Any other description of outstanding character of these samples

| Elastic | Greasy | Spongy | Pasty | Other adjectives |

et al. (1987) used sensory texture profiling to look at three different varieties of yam. Samples were taken from different parts of the tuber and cooked for different periods of time. The chemical composition and yield force of the samples were also measured. The sensory profile used in this study is shown in Table 5. The possibility of using sensory profiling as a basis for selecting yams for some speciality uses was suggested.

Syarief *et al.* (1985) subjected sensory profile data for butter, cheeses, fish gels, frankfurters, peanut butters, raw potatoes and baked sweet potatoes

TABLE 5
TEXTURE PROFILE OF COOKED STORED YAM
From Onayemi *et al.* (1987)

Stage		White yam	Yellow yam	Water yam
I	Surface properties to finger-feel and action with fork			
	Dryness on surface	1	1	1
	Moistness	x	2	2
	Firmness	1	1·5	2
	Brittleness	1	1	1
II	First bite			
	Hardness/firmness	1·5	2·5	2·0
	Fracturability	1·0	1·0	1·5
	Brittleness	3·0	2·0	2·0
III	Mastication (mouthfeeling)			
	Mealiness or easy breakdown	3·0	2·0	1·0
	Waxy and lack of disintegration firm and close	1·0	2·5	1·5
	Smoothness	1·0	1·5	1·5
IV	Residual after mastication and rolling three times with tongue and palate			
	Fracturability	3·5	1·5	1·5
	Graininess and lack of disintegration	x	2·0	2·0
	Difficulty of swallowing	1·0	1·0	1·0

Evaluation based on a 5-point intensity scale for dryness with x = just detectable (dry), 1 = dry, 2 = slightly dry, 3 = moderately dry, 4 = very dry for the middle section of the stored cooked yam. Other characteristics, moistness, firmness and hardness were evaluated by the same scale.

to principal component analysis to study the interdependence and underlying dimensions of the sensory characteristics. The results suggested that the character notes used by the panel, for each product, were not independent but formed a pattern of interdependency. They concluded that the texture profile analysis for each product could be simplified. This is in line with the findings reported by Dransfield *et al.* (1984*a,b*).

3.3. General Comments on Procedures for Sensory Assessment of Texture

The type of panel, the degree of training and the type of test used in sensory assessment of texture will vary according to the information required and the purpose for which it is required. Abbot (1973) published a useful guide to the selection of panels and tests.

The general organisation of panel work for texture assessment does not differ in any major way from that required to assess other sensory properties. The following points are particularly relevant to texture work. The size and shape of the sample presented to panelists should be standardised where possible. The form in which uncooked food is presented should be standardised, i.e. whether it is peeled, sliced or ground. In the case of cooked food, the cooking conditions and the temperature of serving should be controlled. Some foods are sensitive to humidity changes and should be conditioned to and tested in an atmosphere of specified humidity. Implements or containers used by panelists, e.g. knives, forks, spoons, cups should be standardised as far as possible.

Texture assessment can be influenced by differences in other sensory characteristics. Christensen (1983) discussed the effects of colour on aroma, flavour and texture. Urbanyi (1983) reported a study which revealed that the consistency of food samples influenced the evaluation of taste, particularly with foods the consistency of which is an important parameter. Whether the reverse effect occurs was not mentioned. Izutsu and Wani (1985) published a review paper on food texture and taste. In it it was suggested that texture is the most important sensory property in foods which have relatively low flavour intensity. They proposed the introduction of a new concept, physical taste, for making effective evaluations of the palatability of foods. If desirable, differences in other sensory properties can be masked when texture is being assessed. Differences in appearance can be masked by blindfolding the panelists or the use of coloured lighting. The influence of flavour differences can be reduced by the addition of foreign flavours to the samples or administered to the panelists. Washing the mouth out with salt solution and sucking a peppermint were used in the study of the texture of asparagus and rhubarb respectively (Anzaldua

Morales, 1982). The sounds emitted during mastication were blocked by a masking noise when testing crisp foods (Christensen and Vickers, 1981). In many cases however such procedures may distract the panelists and make them less discriminating. Good panel training is the most effective way of minimising the above effects.

4. NON-SENSORY METHODS

Apart from the use of sensory techniques there are three other general approaches to the study of the textural properties of foods. The most common involves the use of instruments to evaluate the physical properties of the samples. Another approach is to examine the structure of the foods and a third is to determine their chemical composition, in particular that of their structural components.

4.1. Instrumental Methods

The most common principle employed in instrumental texture measurement is to cause a probe to come into contact with the sample. The sample is deformed and the extent of the deformation and/or the resistance offered by the sample is noted and used as an index of the texture of the food. There are very many instruments described in the literature and detailed discussion of them is outside the scope of this chapter. The reader is referred to a number of reviews and critical discussions of such instruments published elsewhere (Mohsenin, 1970; Szczesniak, 1973; p. 71; Voisey, 1976; Voisey and deMan, 1976; Brennan, 1980; Bourne, 1982; Prentice, 1984).

The type of test performed with these instruments may be purely empirical in nature, i.e. measure some poorly defined mechanical property of the sample in instrumental units. Some instruments are designed to simulate to some extent the process of mastication and so give results which may relate better to sensory assessments of texture. Other instruments measure well-defined mechanical/physical properties of the sample such as modulus of elasticity or viscosity.

4.1.1. Empirical Methods

These usually involve the sample being subjected to a complex pattern of forces. However, often one type of action predominates and this provides a basis for classifying such tests.

In a *puncture* or *penetration* the probe is made to penetrate into the sample and the force necessary to achieve a certain depth of penetration or the penetration depth achieved in a specified time under defined conditions is measured and used as an index of hardness, firmness or toughness of the food. Examples of commercial instruments featuring puncture are: fruit pressure testers, the Bloom Gelometer, the Boucher Jelly Tester, cone penetrometers and multi-probe instruments such as the Christel Texturometer, the Lynch-Mitchell Maturometer and the Armour Tenderometer.

In a *shear* test one or a number of blades is made to shear through a food sample under specified conditions. The maximum force required and/or the work necessary to achieve this is taken as a measure of toughness, firmness or fibrousness of the food. Instruments in this category include: the Warner-Bratzler Shear, the Kramer Shear Cell and the FMC Pea Tenderometer.

In a *compression* test the extent of the compression achieved under a specified load in a given time or the load required to achieve a specified degree of compression is measured and used as a texture indicator. The Baker Compressometer, the Ball Compressor and the Firm-o-meter or Firmness Meter fall into this category.

In an *extrusion* test the food sample is forced through one or more orifices and the maximum force developed or the average force registered or the work done during the test or the amount of material extruded in a specified time is measured and used as an index of firmness, toughness, consistency or spreadability. Cells featuring forward extrusion through perforated plates or wire meshes are available. In back extrusion the food is expressed back between a loose fitting plunger and the inner surface of the cell.

Cutting devices consist of wires or blades which are made to cut through the food sample and the maximum force developed or the time necessary to cut through a standard sized sample is measured. Such measurements are taken as representing fibrousness or firmness of the samples under test. Rotating blades are also available, e.g. the rotating knife tenderometer and the FMBRA Biscuit texture meter.

Flow and mixing devices are used to assess the viscosity/consistency of liquid and semi-liquid foods. The Adams Consistometer and the Bostwick Consistometer are simple instruments which measure the extent to which samples spread or flow under defined conditions. Other instruments measure the resistance offered by the sample to a mixing element turning in it, e.g. the Farinograph and the Amylograph.

4.1.2. Imitative Instruments

These only simulate the complex process of mastication to a very limited degree, see discussion of texture perception. They usually feature tooth-shaped probes and/or an action which copies the movement of the jaws but only in a vertical plane. The General Foods Texturometer, details of which were first published in 1963 (Friedman *et al.*, 1963), is probably the most widely used imitative instrument. The principle of this instrument is shown in Fig. 1. The food sample is located on a plate attached to a beam. As the eccentric wheel is rotated, the plunger descends onto the sample, deforming it. Strain gauges attached to the beam detect its movement and this is recorded as a force–time trace. The usual procedure is to subject the sample to two successive cycles of the plunger and typical force–time curves thus obtained are shown in Fig. 2. The interpretation of these curves, recommended by the original authors and based on the classification of terms and the sensory profile method discussed earlier (Brandt *et al.*, 1963), is shown in Table 6. This is known as an instrumental texture profile. The applications for this instrument up to 1975 were reviewed by Breene (1975). Other relevant publications are also listed (Tanaka, 1975; Szczesniak and Hall, 1975; Szczesniak, 1975, p. 139; Brennan *et al.*, 1975*b*; Saio and Watanabe, 1978; Saio, 1978; Furukawa *et al.*, 1979; Okabe, 1979; Juliano *et al.*, 1981; Ram and Nigam, 1983). Other imitative instruments used in texture work are the Volodkevich Bite Tenderometer and the Masticometer, which are discussed elsewhere (Brennan, 1980). Other instruments

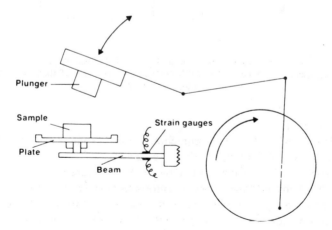

FIG. 1. Principle of the General Foods texturometer. From Brennan (1980).

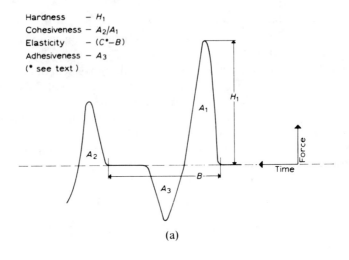
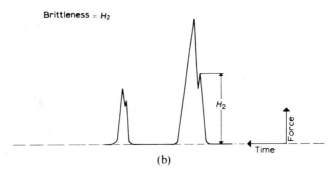

FIG. 2. Typical curves from the General Foods Texturometer. (a) non-brittle food, (b) brittle food. See Table 6 for guide to the interpretation of these curves. From Brennan (1980).

have been used to obtain instrumental texture profiles, in particular the Instron Universal Testing Machine, see below.

A more *fundamental* approach to the study of texture involves measuring well-defined physical properties of the food sample and attempting to relate these to its sensory textural characteristics. The potential advantages of this approach include the ability to express results in well-defined units and to quantify the effects of varying sample and probe geometry and other test conditions. The difficulties encountered, however, are many and stem from

TABLE 6
INTERPRETATION OF TRACES FROM THE GENERAL FOODS TEXTUROMETER

Hardness	Height of the first peak (H_1) normalised to a one-volt input (sample size specified)
Cohesiveness	Ratio of the area under the second peak to that under the first peak (A_2/A_1)
Springiness (elasticity)	The difference between the distance B, measured from the initial sample contact to contact on the second cycle, and the distance C, the same measurement made on a completely inelastic standard material such as clay (sample size specified)
Adhesiveness	The area, in arbitrary instrumental units (A_3), of the negative peak formed when the plunger is pulled from the sample (sample size and voltage input is specified)
Fracturability (brittleness)	This is characterised by the multi-peak shape of the first cycle trace and is measured as the height of the first significant break in the peak (H_2), normalised to a one-volt input (sample size specified)
Chewiness	Hardness × cohesiveness × elasticity
Gumminess	Hardness × cohesiveness × 100

the heterogeneous nature of most foods and the fact that few exhibit true elastic, viscous or plastic behaviour. Many physical properties of foods change with time and vary according to the conditions of storage. In spite of these difficulties much work of this kind has been undertaken.

At one end of the rheological spectrum some rigid foods may exhibit a degree of elastic behaviour. Compressive, tensile and bend tests have been used to obtain values of the modulus of elasticity and breaking strength of such materials. Where little true elastic behaviour is observed, less well-defined measurements have been taken from force–deformation curves and used as indices of physical properties (see Fig. 3). Various machines have been used to carry out such tests. One which has been used frequently is the Instron Universal Testing Machine. This is a multi-purpose instrument which features a crosshead which is driven vertically at a range of constant speeds. It may also be programmed to cycle over a fixed distance or load range. A series of load cells is available which enables the load capacity to be varied. Force–time traces are obtained on a strip chart. Another multi-purpose instrument, the Stevens Compression Response Analyser, has been used in texture studies recently (Brennan, 1984). See Fig. 4.

At the other end of the rheological spectrum are liquids which exhibit viscous flow. They may exhibit well-recognised flow patterns, i.e.

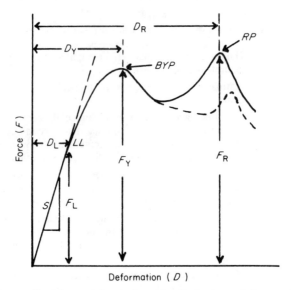

FIG. 3. Generalised force–deformation trace. S, elastic modulus may be calculated from this slope. BYP, bioyield point. RP, rupture point. LL, linear limit. F_L, D_L, force and deformation at linear limit. F_Y, D_Y, force and deformation at bioyield point. F_R, D_R, force and deformation at rupture point. From Brennan (1980).

Newtonian, pseudo-plastic (shear thinning) or dilatant (shear thickening). A large number of viscometers, with varying degrees of sophistication, is available for characterising the flow properties of liquid foods. Some success has been achieved in relating these properties to textural and mouthfeel characteristics of the foods.

In between these two extremes of the rheological spectrum are products which exhibit viscoelastic, viscoplastic and plasto-viscoelastic behaviour. Such behaviour may be represented by mechanical models featuring elastic, viscous and plastic elements. For more solid foods such models may be constructed from the results of carefully executed experiments using instruments similar to the Instron, mentioned above. For liquids, similar data may be obtained from the more sophisticated viscometers. Again, some success has been achieved in relating such models to the textural and mouthfeel characteristics of some foods. Examples of this type of work have been discussed elsewhere (Brennan, 1980; Prentice, 1984). It is an interesting avenue of research which in the long term should enable work on texture to be based on a more scientific foundation.

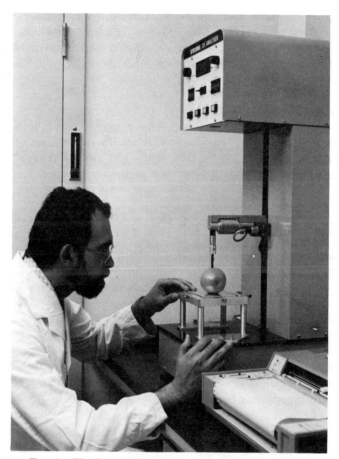

Fig. 4. The Stevens Compression Response Analyser.

Vibrational techniques have also been applied to texture studies. Samples of food have been subjected to low frequency vibration and their behaviour observed and recorded. Information on damping characteristics and modulus of elasticity can be obtained. Examples of such work are reported in the literature (Finney, 1972; Webb *et al.*, 1975; Petrell *et al.*, 1980); Yamamoto *et al.*, 1980; Hayes and Chingon, 1982).

4.2. Structural Studies

The definition of the term 'texture' quoted earlier implies that the physical properties and hence the textural characteristics of foods are a reflection of their structures. Thus studying the structure of foods is another way of compiling information on their physical/textural properties. General discussions of the relationship between structure and the physical/textural characteristics of foods have been published (Sherman, 1973; Stanley and Tung, 1976).

4.3. Chemical Analysis

The determination of the chemical composition of foods is another means whereby a more complete explanation of their physical/textural properties can be obtained. Szczesniak (1973, p. 109) and Jowitt (1977) have discussed the role of chemical analysis in texture studies.

5. RELATING SENSORY AND NON-SENSORY DATA

As stated before texture is a sensory attribute and can only be measured directly by sensory means. Thus it is necessary to relate data obtained by non-sensory means to sensory data to establish the accuracy of the former as a measurement of texture.

Methods of relating sensory and non-sensory data have been discussed by Moskowitz (1981) and Szczesniak (1987) and are further discussed in Chapters 9 and 10 in this book. A whole range of statistical techniques has been used over the years to relate sensory and non-sensory texture data. Simple regression analysis to relate one sensory attribute to one physical or chemical property has been very widely used and there are very many reports of this in the literature. An acceptance of the fact that texture is a complex attribute has led to the widespread use of multiple linear regression. The usual procedure is to denote a textural characteristic as the dependent variable and mechanical or chemical parameters as the independent variables. In this way regression equations may be developed to relate the textural characteristic to two or more mechanical measurements on a linear, semi-log or log-log basis. Such equations may be used to predict the value of the sensory property from the results obtained from instrumental testing and/or chemical analysis. Reports of studies of this kind abound in the literature. Examples from the references already cited in this chapter are: Mohamed (1981) in which the sensory stickiness of syrups, jams and spreads was related to their viscosity, surface tension and results obtained from plate and extrusion tests; Mohamed (1981a), Mohamed *et al.* (1982) and Brennan (1983) in which the sensory crispness of wet and dry crisp foods was related to results obtained from compression, puncture and 'bite' tests and the sounds produced during these tests; Brennan, (1983) and Anzaldua Morales (1982) in which sensory fibrousness of asparagus and rhubarb was related to results obtained from the Volodkevich Bite Tenderometer, a back extrusion cell and chemically-determined parameters of alcohol insoluble solids and the ratio of protopectin to pectin.

Multivariate techniques have grown in popularity with researchers concerned with texture studies in recent years. Some examples from references cited above are: Vickers and Wasserman (1979) where they were used in a study of food sounds; Vickers and Christensen (1980) where they were used to determine which attributes of wet and dry crisp foods related best to crispness; Cardello *et al.* (1982a) where they were applied to comparing the performance of trained and untrained panels; Dransfield *et al.* (1984) where they were used to study the relationships between the sensory characteristics of cooked meats; Dransfield *et al.* (1984b) where they were applied in the development of a texture profile for beefburgers; Jones *et al.* (1985) where they were used to correlate mechanical properties, composition and sensory attributes of beefburgers; Syarief *et al.* (1985) where they were used to study the interpendency of sensory textural attributes.

REFERENCES

Abbott, T. A. (1973). Sensory assessment of textural attributes of foods. In: *Texture Measurement of Foods*, A. Kramer and A. S. Szczesniak (Eds), D. Reidel Publishing Co., Dordrecht, Holland, p. 17.

Anderson, Y. and Lundgren, B. (1981). *J. Texture Studies*, **12**, 217.

Anon (1975). Glossary of terms relating to the Sensory Analysis of Food. B.S. 5098: 1975. British Standards Institution, London.

Anzaldua Morales, A. (1982). The texture of fibrous fruits and vegetables. PhD thesis (unpublished), University of Reading, England.

Berry, B. W. (1986). *J. Sens. Studies*, **1**, 15.

Bogaty, H., Twitty, M. and Talmage, J. M. (1980). *J. Texture Studies*, **10**, 231.

Borderias, A. J., Lamua, M. and Tejeda, M. (1983). *J. Food Technol.*, **18**, 85.

Bourne, M. C. (1982). *Food Texture and Viscosity, Concept and Measurement*, Academic Press, London.

Bourne, M. C., Sandoval, A. M. R., Villalobos, M. C. and Buckle, T. S. (1975). *J. Texture Studies*, **6**, 43.

Boyar, M. M. and Kilcast, D. (1986a). *J. Texture Studies*, **17**, 221.

Boyar, M. M. and Kilcast, D. (1986b). *J. Food Sci.*, **51**, 859.

Brandt, M. A., Skinner, E. Z. and Coleman, J. A. (1963). *J. Food Sci.*, **28**, 404.

Breene, W. M. (1975). *J. Texture Studies*, **6**, 53.

Breene, W. M. (1978). *J. Texture Studies*, **9**, 77.
Brennan, J. G. (1980). Food texture Measurement. In: *Developments in Food Analysis Techniques*, R. D. King (Ed.), Applied Science Publishers, London, p. 1.
Brennan, J. G. (1983). Measurement of particular textural characteristics of some fruits and vegetables. In: *Sensory Quality in Food and Beverages*, A. A. Williams and R. K. Atkin (Eds), Ellis Horwood, Chichester, p. 173.
Brennan, J. G. (1984). *Proc. Inst. Food Sci. Technol.*, **17**, 219.
Brennan, J. G., Jowitt, R. and Williams, A. (1975a). Sensory and instrumental measurement of crispness and brittleness in biscuits. *Proceedings Fourth International Congress of Food Science and Technology, Madrid 1974*, Instituto de Agroquimica y Tecnologia de Alimentos, Valencia, Spain, Vol. 2, p. 130.
Brennan, J. G., Jowitt, R. and Williams, A. (1975b). *J. Texture Studies*, **6**, 83.
Brennan, J. G., Jowitt, R. and Mohamed, A. M. A. (1977). *Ann. Appl. Biol.*, **87**, 121.
Cardello, A. V., Mellar, C., Kapsalis, J. G., Segars, R. A., Sawyer, F. M., Murphy, C. and Moskowitz, H. R. (1982a). *J. Food Sci.*, **47**, 1186.
Cardello, A. V., Mates, A. and Sweeney, J. (1982b). *J. Food Sci.*, **47**, 1738.
Carter, E. J. V. and Sherman, P. (1978). *J. Texture Studies*, **9**, 311.
Christensen, C. M. (1979). *J. Texture Studies*, **10**, 153.
Christensen, C. M. (1983). *J. Food Sci.*, **48**, 787.
Christensen, C. M. and Vickers, Z. M. (1981). *J. Food Sci.*, **46**, 574.
Civille, G. V. and Liska, I. H. (1975). *J. Texture Studies*, **6**, 19.
Civille, G. V. and Szczesniak, A. S. (1973). *J. Texture Studies*, **4**, 204.
Culioli, J. and Sherman, P. (1976). *J. Texture Studies*, **7**, 353.
Cutler, A. N., Morris, E. R. and Taylor, I. J. (1983). *J. Texture Studies*, **14**, 377.
Del Mundo, A. M. and Juliano, B. O. (1981). *J. Texture Studies*, **12**, 107.
Drake, B. (1963a). *J. Food Sci.*, **28**, 233.
Drake, B. (1963b). *J. Food Sci.*, **30**, 556.
Dransfield, E., Francombe, M. A. and Whelehan, O. P. (1984a). *J. Texture Studies*, **15**, 33.
Dransfield, E., Jones, R. C. D. and Robinson, J. M. (1984b). *J. Texture Studies*, **15**, 337.
Finney, E. E. Jr. (1972). *J. Texture Studies*, **3**, 263.
Friedman, H. H., Whitney, J. E. and Szczesniak, A. S. (1963). *J. Food Sci.*, **28**, 390.
Furukawa, T., Dhta, S. and Yamamoto, A. (1979). *J. Texture Studies*, **10**, 333.
Garruti, R. D. S. and Bourne, M. C. (1985). *J. Food Sci.*, **50**, 1067.
Gross, M. O., Rao, V. N. M. and Smit, C. J. B. (1980). *J. Texture Studies*, **11**, 271.
Hayes, C. F. and Chingon, H. T. G. (1982). *J. Texture Studies*, **13**, 397.
Heath, M. R. and Lucas, P. W. (1987). *J. Texture Studies*, **18**, 111.
Izutsu, T. and Wani, H. (1985). *J. Texture Studies*, **16**, 1.
Jones, R. C., Dransfield, E. and Robinson, J. M. (1985). *J. Texture Studies*, **16**, 241.
Jowitt, R. (1974). *J. Texture Studies*, **5**, 351.
Jowitt, R. (1977). Correlation between measurable properties (chemical/physical) and texture. *7th European Symposium on Food Product and Process Selection in the Food Industry*, European Federation of Chemical Engineering, Eindhoven, p. 163.
Jowitt, R. and Mohamed, A. A. (1980). An improved instrument for studying crispness in foods. In: *Food Process Engineering, Vol. 1*, P. Linko, Y. Malkki, J. Olkku and J. Larinkari (Eds), Applied Science Publishers, London, p. 292.

Juliano, B. O., Perez, C. M., Barber, S., Blakeney, A. B., Iwasaki, T., Shibuya, N., Keneaster, K. K., Chung, S., Laignnelet, B., Launay, B., Del Mundo, A. M., Suzuki, H., Shiki, J., Tsuti, S., Tokoyama, J., Tatsumi, K. and Webb, B. D. (1981). *J. Texture Studies*, **12**, 17.
Kapsalis, J. G. and Moskowitz, H. R. (1978). *J. Texture Studies*, **9**, 371.
Katz, E. E. and Labuza, T. P. (1981). *J. Food Sci.*, **36**, 403.
Kiosseoglou, V. D. and Sherman, P. (1983). *J. Texture Studies*, **14**, 277.
Kokini, J. L., Kadene, J. B. and Cussler, F. L. (1977). *J. Texture Studies*, **8**, 195.
Mohamed, A. M. A. (1981a). Sensory and instrumental measurement of food texture with particular reference to stickiness. PhD thesis (unpublished), University of Reading, England.
Mohamed, A. M. A. (1981b). Sensory and instrumental measurement of food texture with particular reference to stickiness. PhD thesis (unpublished), University of Reading, England.
Mohamed, A. A. (1981a). Sensory and instrumental measurement of food instrumental and sensory techniques. *Proceedings 2nd International Congress of Engineering and Food, Helsinki*, p. 285.
Mohamed, A. A., Jowitt, R. and Brennan, J. G. (1982). *Food Engineering*, **1**, 55; 123.
Mohsenin, N. N. (1970). *Physical Properties of Plant and Animal Materials, Vol. I*, Gordon and Breach, London.
Moore, I. J. and Shoemaker, C. F. (1981). *J. Food Sci.*, **46**, 399.
Moraes, M. A. C., Schneider, I. S. and Forster, R. J. (1981). *J. Texture Studies*, **12**, 63.
Moskowitz, H. R. (1977). *J. Food Quality*, **3**, 195.
Moskowitz, H. R. (1981). *J. Food Quality*, **4**, 15.
Moskowitz, H. R. and Chandler, J. W. (1977). New uses of magnitude estimation. In: *Sensory Properties of Food*, G. G. Birch, J. G. Brennan and K. J. Parker (Eds), Applied Science Publishers, London, p. 189.
Moskowitz, H. R. and Kapsalis, J. G. (1975). Toward a general theory of texture psychophysics. *Proceedings Fourth International Congress of Food Science and Technology, Madrid 1974*, Instituto de Agroquimica y Tecnologia de Alimentos, Valencia, Spain, Vol. 2, p. 102.
Moskowitz, H. R. and Kapsalis, J. G. (1976). Psychophysical relations in texture. In: *Rheology and Texture in Food Quality*, J. M. de Man, P. W. Voisey, V. F. Raspek and D. W. Stanley (Eds), AVI Publishing Co., Westport, Connecticut, p. 554.
Munoz, A. M. (1986). *J. Sens. Studies*, **1**, 55.
Okabe, M. (1979). *J. Texture Studies*, **10**, 131.
Oldfield, R. C. (1960). Perception in the mouth. In: *Texture in Foods, Monograph No. 7*, Soc. Chem. Ind., London, p. 3.
Onayemi, O. (1985). *J. Texture Studies*, **16**, 263.
Onayemi, O., Babaloza, R. O. and Badanga, H. (1987). *J. Texture Studies*, **18**, 17.
Pangborn, R. M., Gibbs, Z. M. and Tassen, C. (1978). *J. Texture Studies*, **9**, 415.
Peleg, M. (1980). *J. Texture Studies*, **10**, 245.
Petrell, R. J., Mohsenin, N. N. and Wallner, S. (1980). *J. Texture Studies*, **10**, 217.
Pierson, A. and Le Magnen, J. (1970). *J. Texture Studies*, **1**, 327.
Pokorny, J., Davídek, J. and Dobiášová, S. (1984). *J. Texture Studies*, **15**, 295.
Prentice, J. H. (1984). *Measurements in the Rheology of Foodstuffs*, Elsevier Applied Science Publishers, London.

Ram, B. P. and Nigam, S. N. (1983). *J. Texture Studies*, **14**, 245.
Saio, K. (1978). *J. Texture Studies*, **9**, 159.
Saio, K. and Watanabe, T. (1978). *J. Texture Studies*, **9**, 135.
Savell, J. W., Branson, R. E., Cross, H. R., Stiffler, D. M., Wise, J. W. and Griff, G. C. (1987). *J. Food Sci.*, **52**, 517.
Shama, F. and Sherman, P. (1973). *J. Texture Studies*, **4**, 111; 254.
Sherman, P. (1969). *J. Food Sci.*, **34**, 458.
Sherman, P. (1973). Structure and textural properties of foods. In: *Texture Measurement of Foods*, K. Kramer and A. S. Szczesniak (Eds), D. Reidel Publishing Co., Dordrecht, Holland, p. 52.
Stanley, D. W. and Tung, M. A. (1976). Microstructure of food and its relation to texture. In: *Rheology and Texture in Food Quality*, J. M. de Man, P. W. Voisey, V. F. Rasper and D. W. Stanley (Eds), AVI Publishing Co., Westport, Connecticut, p. 28.
Syarief, H., Hamann, D. D., Giesbrecht, F. G., Young, C. T. and Munroe, R. J. (1985). *J. Texture Studies*, **16**, 29.
Szczesniak, A. S. (1963). *J. Food Sci.*, **28**, 385.
Szczesniak, A. S. (1973). Instrumental methods of texture measurement. In: *Texture Measurements of Foods*, A. Kramer and A. S. Szczesniak (Eds), D. Reidel Publishing Co., Dordrecht, Holland, pp. 71; 109.
Szczesniak, A. S. (1975). *J. Texture Studies*, **6**, 5.
Szczesniak, A. S. (1975). *J. Texture Studies*, **6**, 139.
Szczesniak, A. S. (1979). Classification of mouthfeel characteristics of beverages. In: *Food Texture and Rheology*, P. Sherman (Ed.), Academic Press, London, p. 1.
Szczesniak, A. S. (1987). *J. Texture Studies*, **18**, 1.
Szczesniak, A. S. and Hall, B. J. (1975). *J. Texture Studies*, **6**, 117.
Szczesniak, A. S., Brandt, M. A. and Friedman, H. H. (1963). *J. Food Sci.*, **28**, 397.
Tanaka, M. (1975). *J. Texture Studies*, **6**, 101.
Tornberg, E., Fjelkner-Modig, S., Ruderus, H., Glantz, P., Randow, K. and Stafford, D. (1985). *J. Food Sci.*, **50**, 1059.
Urbanyi, Gy. (1983). *Acta Alimentaria*, **12**, 109.
Vickers, Z. (1979). Crispness and crunchiness of food. In: *Food Texture and Rheology*, P. Sherman (Ed.), Academic Press, London, p. 33.
Vickers, Z. M. (1984). *J. Texture Studies*, **15**, 49; 157.
Vickers, Z. M. (1985). *J. Texture Studies*, **16**, 85.
Vickers, Z. and Bourne, M. C. (1976). *J. Food Sci.*, **41**, 1158.
Vickers, Z. M. and Christensen, C. M. (1980). *J. Texture Studies*, **11**, 291.
Vickers, Z. M. and Wasserman, S. S. (1979). *J. Texture Studies*, **10**, 319.
Voisey, P. W. (1976). Instrumental measurement of food texture. In: *Rheology and Texture in Food Quality*, J. M. de Man, P. W. Voisey, V. F. Rasper and D. W. Stanley (Eds), AVI Publishing Co., Westport, Connecticut, p. 79.
Voisey, P. W. and deMan, J. M. (1976). Application of instruments for measuring food texture. In: *Rheology and Texture in Food Quality*, J. M. de Man, P. W. Voisy, V. F. Rasper and D. W. Stanley (Eds), AVI Publishing Co., Westport, Connecticut, p. 142.
Voisey, P. W. and Larmond, E. (1977). *J. Texture Studies*, **8**, 93.
Webb, N. B., Rao, V. N. M., Civille, G. V. and Hamman, D. D. (1975). *J. Texture Studies*, **6**, 329.

Williams, A. (1975). Sensory and instrumental measurement of food texture, PhD thesis (unpublished), University of Reading, England.
Yamamoto, H., Iwamoto, M. and Haginums, S. (1980). *J. Texture Studies*, **11**, 117.
Yeatman, J. N. and Drake, B. K. (1973). Physiological aspects of texture perception. In: *Texture Measurements of Foods*, A. Kramer and A. S. Szczesniak (Eds), D. Reidel Publishing Co., Dordrecht, Holland, p. 10.
Yoshikawa, S., Nishimaru, S., Tashiro, T. and Yoshida, M. (1970). *J. Texture Studies*, **1**, 437; 443.

Chapter 4

COLOUR VISION AND APPEARANCE MEASUREMENT

DOUGLAS B. MACDOUGALL

*Agricultural and Food Research Council,
Institute of Food Research, Bristol, UK*

1. INTRODUCTION

Philosophising on vision and colour perception can be traced back to the time of the Ancient Greeks who thought the eye had a lantern-like function. In the Middle Ages Alhazan, in the 10th century, experimented with the pinhole camera and Leonardo, in the 15th, was convinced that the eye contained the image of the scene. The modern era for the study of light and vision can be regarded as starting in the 17th century (Boynton, 1979). Kepler reasoned that, because the eye has a crystalline lens, the receptive layer for the image must be at the back of the eye—the retina—and an additional process would be required for appreciation of the sensation. Newton's experiments on refraction of light led him to rationalise that the rainbow did not possess colour but it was the spectrum's rays that contained the power to produce the sensation in the observer's experience (Wright, 1967). The stimulus for perception of colour does not reside in the object but in the light reflected from it. The first printed record of a logical colour arrangement was in the 17th century, but the concept of organising colours into a sequence can be traced to Aristotle who placed the major hues between black and white. By the 13th century this linear arrangement had become a colour solid (Parkhurst and Feller, 1982). Modern colour systems developed from Newton's seven rainbow colours. The widely used uniform lightness and chroma colour space of Munsell is based on five hues; the DIN system has a hue circuit of 24 steps; the increasingly popular Swedish Natural Colour System is based on the six psychologically unique perceptions of black, white, red, green, yellow and blue (Hård, 1970) and the OSA system is a regular rhombohedral lattice (Wyszecki and Stiles, 1982).

Vision has the advantage over the other senses in that an observer's appreciation of an object's appearance can be recorded pictorially. The artist's portrayal of a scene may be regarded as an attempt to convey his sensory appreciation to others. This includes both the colour of objects in the scene and the effect of light. Comparison of classic Dutch and Italian schools of painting reveals regional differences between cool and warm daylight on the landscape colours. Pictorial representation can only be partly true; the pigment layers in a painting, or in a coloured photograph, will produce a colour mismatch under different lights, and for those who are colour-blind, the appearance will not be the same. From the early experience of mixing pigments to the classic experiments of Maxwell, Young, Helmholtz and others in mixing coloured lights in the last century (MacAdam, 1970), it became clear that, for observers with normal colour vision, colour space must be tridimensional with at least three photosensitive pigments in the retina to account for it. By the late 1920s, the eye's sensitivity to light relative to wavelength was established. With the construction of colorimeters (colour mixing spectrophotometers), the so-called 'standard observer' was defined (Wright, 1980) and the CIE (Commission Internationale de L'Eclairage) system of colour measurement adopted. This technique of tristimulus colorimetry, defined more than 50 years ago (Wright, 1981, 1982) and subsequently used by the textile, paint, paper, ceramic and other industries, is increasingly being used in the food industry as a quality control tool (Francis and Clydesdale, 1975), especially since the development of automatic colour-measuring equipment. The system has improved considerably since 1931, particularly in improved visual spacing, and one of its major industrial applications is now the computer formulation of pigment mixtures and prediction of colour appearance (Best, 1981).

In the following section on colour vision the tristimulus concept, the basis of the CIE system, will be related to the visual mechanism. This is now recognised as an opponent response processes giving rise to four unique hues—red, yellow, green and blue.

2. COLOUR VISION

For convenience, vision and appearance will be distinguished thus: vision can be regarded as the process of seeing whereas appearance is the recognition and assessment of the properties—surface structure, opacity, colour, etc.—associated with the object seen. Vision is a sequence of events starting with the external stimulus, the radiant flux incident on the eye, and proceeding through the reception of light on the retina to its interpretation after transmission through the visual pathway (Rodieck, 1979) to the cortex of the brain (Zeki, 1980). Vision is the psychological response to the objective stimulus generated by the physical nature of the object viewed.

The visual apparatus consists of the eye, that is a light-detector wavelength analyser, and a neural pathway to the brain in which the visual cortex provides a kind of interpretation map of the retinal image (Boynton, 1979). The apparatus is so constructed that the sensation perceived can be thought of as being projected back out into the world from which it originated. This leads to the erroneous interpretation of imputing to the physical object the sensation it generated. The coloured sensation exists in the observer's mind and not in the physical entity. Because vision is binocular the image perceived is spatial with recognisable boundaries. Stationary or mobile object positions are instantly located and the constancy of colour appearance maintained over large ranges in light intensity and quality by the accommodation and adaptation of the eye. The detection and translation of the external stimuli into coloured sensations begins in the 200 million cells in the retina.

2.1. The Eye

The human eye is virtually spherical with a diameter of approximately 2 cm. Its mobility is controlled by six muscles to give a near circular field of view of 100°. The exterior of the eyeball consists of three membranes. The sclera, the tough outer opaque white membrane, is continuous posteriorly with the sheath of the optic nerve and anteriorly with the transparent cornea. The middle layer, the pigmented choroid, contains the blood capillary network. The iris diaphragm, which controls the amount of light entering the eye, and the ciliary body, which suspends the lens, arise out of the anterior of the choroid layer. The inner membrane, the light-detecting retina, lines the inside of the posterior of the eye (Brown, 1965).

2.2. The Retina

Light from the external world passes sequentially through, and is partly absorbed by, the cornea, the aqueous humor which occupies the space between the cornea and the lens, the lens, and the vitreous humor which fills the rest of the eye. The image is focused onto a depression in the retina approximately 1·5 mm in diameter, the fovea, the region about the visual pole where vision is most acute. The fovea is located in a 2–3 mm diameter yellow pigmented area, the macula lutea, but the central area of the foveal pit, the foveola, is non-pigmented and free of blood vessels. There are two types of receptor cells, the rods and cones. The photoptic 'colour' detecting cones are exclusively and most densely packed in the centre of the fovea. The rods, the receptors for low intensity scotoptic colourless vision, increase in density to 20° from the fovea and then decrease towards the periphery of the eye.

The retina is a multilayered structure which was divided into ten layers by Polyak (1941) on the basis of its nerve cells and fibres. Stell (1972) reduced this to three functional layers: the photoreceptor cells; the intermediate neurons containing bipolar, horizontal and amacrine cells; and the ganglion cells interconnected by synaptic layers. The photoreceptors are located in multifolded disk-shaped membranous structures in the outer segment of the rods and cones. A photon passing through the disks has a high probability of being absorbed. This depends on its wavelength and which photopigment is involved. Photoreceptor absorption is somewhat similar to a transducer where light energy is converted into electrical signals to convey information to the brain (Berridge and Irvine, 1984). Photon absorption causes a conformational change in the photopigment molecule, rhodopsin, which proceeds through a sequence of steps until the pigment bleaches (Wald, 1968).

Regeneration of rhodopsin after bleaching takes place either in the eye or via vitamin A in the liver. Photon capture and onset of conformational change in rhodopsin decreases the permeability of the cell membrane to sodium ions and creates hyperpolarisation of the electrical potential across the membrane (Normann and Werblin, 1974). It is still questionable which biological pathway produces sufficient amplification of the photon capture event to achieve subsequent change in membrane conductance to initiate the neural signal (Berridge and Irvine, 1984; Brown *et al.*, 1984; Fein *et al.*, 1984). An internal transmitter system is required such as inositol triphosphate which has been shown to be involved in the control of the transduction mechanism of the excitatary cascade that opens up to 1000 ionic channels from a single photon. Only rhodopsin, the rod pigment, has been extracted and its absorption spectrum characterised; it absorbs maximally at 505 nm. However, microspectrodensitometry has fairly conclusively established the existence of three cone pigments (Bowmaker and Dartnall, 1980) which had been named by Rushton (1965) as cyanolabe, chlorolabe and erythrolabe. A more efficient technique than retinal densitometry for determination of the spectral curves of the three pigments is to use colour blind observers deficient in one pigment (Smith and Pokorny, 1975). The absorption maxima are approximately 450, 530 and 560 nm (Hurvich, 1981; Jacobs, 1981).

2.3. Trichromacy and Colour Matching

Cone vision is trichromatic. The colour of any light can be matched by a suitable mixture of red, green and blue monochromatic primary lights. Part of the match over the visible spectrum is negative; that is for certain wavelengths one of the matching primaries must be added to a portion of the spectrum being matched to elicit agreement. This trichromatic principle is what would be expected if the visual effect of the stimulus was produced by absorption of three photoreceptors. It follows that a linear relationship should exist between empirical colour-matching functions and the spectral absorption of the pigments (Stiles, 1978). The shape of the colour-matching function curves will depend on the choice of the primaries but their general shape is similar. Elucidation of the foveal spectral sensitivity curves, initiated by Stiles and Crawford (1933), resulted in the determination of the so-called π mechanisms for retinal vision using a two-colour threshold technique in which one adapting wavelength was used to bleach two of the photoreceptors and the threshold versus background radiance of the test wavelength assessed. For a full discussion of the π mechanisms, chromatic adaptation, increment thresholds and visual response functions the reader is referred to the appropriate sections in Boynton (1979) and Wyszecki and Stiles (1982) and the collected papers of Stiles (1978). Although no unique set of spectral sensitivity curves for human cone vision has been established (Hunt, 1982), probably the best set is that of Estévez (1982) based on the 1955 data of Stiles and Burch (1959) which is in excellent agreement with the original data of Guild (1931) and Wright (1928/29) used along with the CIE 1924 V_λ curve to establish to 2° standard observer. The V_λ curve, however, has been found to be inaccurate in the blue-violet part of the spectrum (Wright, 1981) but it is unlikely that an improved curve will be incorporated into colour measurement systems. Because of practical inconvenience (Wyszecki and Stiles, 1982) it has not been included in the revised CIE (1986) recommendations for colorimetry.

2.4. Opponent Mechanisms

Although colour vision is initiated in the photopigment receptors in the retina, and can be described by three trichromatic variables, there is ample evidence (Jameson and Hurvich, 1968) that colour vision depends on interactions at subsequent stages in the visual system between responses derived from the three types of cones (De Valois and De Valois, 1975). The post-receptor process involves an opponent colour mechanism giving rise to four unique hues, red (R), green (G), blue (B), yellow (Y) and a lightness or white (W) to black (Bl) response. Electrophysiological experiments on the lateral geniculate nucleus at the termination of the optic nerve before the signals are transmitted to the visual cortex indicate that the opponent interactions probably take place in the retina, since the ganglion cells in the retina are quantitatively similar (Abramov, 1981). Hurvich's (1981) model of photopigment absorption and opponent neural process links the cone absorptions, α, β, γ, with maxima at 450, 530 and 560 nm respectively, to three opponent processes G:R, B:Y and W:Bl in which G is activated by β, R by $\alpha + \gamma$, B by α, Y by $\beta + \gamma$, and W:Bl by $\alpha + \beta + \gamma$. Hunt's (1982) model for predicting colour appearance has five channels in which there are three zones—a linear, a non-linear power function and an interpretative zone. The first transformation is to compress the dynamic range of the R, G and B signals, represented in this model as square-root functions. The second stage (Hunt and Pointer, 1985) is the formation of an achromatic signal, weighted for the relative abundance of the cones, and three opponent colour difference signals consistent with Hurvich's (1981) scheme for the generation of response functions from cone absorption. Hunt speculates that the role of some of the post-cone cells, for example the horizontal cells, might be to add signals linearly, and others, for example the bipolar cells, to subtract after the power function stage. The model provides a colour order system with good prediction of the unique hues, red, green, blue and yellow, and of constant hues and saturation compared with the Munsell and Swedish Natural Colour systems. The achromatic signal provides perception of fine detail and combinations of the opponent and summation chromatic power signals provide the perceptions of hue, colourfulness and saturation, and combinations of achromatic and chromatic provide perceptions of brightness, lightness and chroma. The model also includes a chromatic adaptation transformation which means that tristimulus values can be converted to appearance specification relative to daylight and tungsten illumination.

2.5. Chromatic Adaptation

Chromatic adaptation is the modification of visual response to the perception of a colour brought about by conditioning the visual system to the chromatic nature of the surroundings, or pre-exposing the observer to light of a different spectral power distribution. Despite much experimentation on the phenomenon few general principles have emerged (Wyszecki and Stiles, 1982). Recent studies have been concerned with predicting the changes that occur under different illuminants and at different light levels. Bartleson (1979*a*) used direct scaling to estimate the magnitude of the changes throughout the entire colour solid. Lightness varied with luminance and colourfulness with colour temperature. This agrees with

Hunt's (1977) summation of previous work: that visual contrast of greys is affected by luminance and colourfulness increases as illumination level is increased. The concept of colourfulness as opposed to saturation or chroma was used by Pointer (1980, 1982) in constructing grids of constant hue based on the Munsell system for subjective magnitude estimation from which other grids for other illuminants could be derived.

Bartleson (1979b) used the fundamental response curves of the cones to construct a method of prediction which agrees well with other methods, and Nayatani et al. (1982) have constructed a model based on a von Kries transformation followed by a non-linear transformation which predicts successfully the perceptual effects of change in lamp, for example between CIE standard illuminants A and D_{65}, and the changes in colourfulness with luminance level. A more comprehensive model for predicting colour appearance under a variety of adapting conditions has been proposed by Nayatani et al. (1986) using logarithmic functions. Hunt's (1987a) most up-to-date model incorporates physiologically likely hyperbolic functions which provide colour predictions for both related and unrelated colours, at any level of illumination, whether photopic, mesopic, or scotopic, for a wide range of stimulus intensities, and for backgrounds of different colours and reflectances. The model applies to the realistic situation in which the eye's fixation wanders and the field size involves both rods and cones.

The magnitude of the effect of chromatic adaptation is not always realised because of the limitation of the observer's memory for individual colours and the phenomenon of colour constancy. Visual adaptation serves to keep the eye in balance (Boynton, 1979); the experience that any 'white' background under a 'coloured' lamp tends to look white means a coloured object will tend to appear the same as it does under a less coloured lamp that elicits the same conditioned visual response to white from the background.

The partly predictable effect of light quality on chromatic adaptation illustrates the difficulty in separating the concept of vision from that of appearance. The mechanism of seeing is modulated by the light from the scene while the characteristics of appearance are modified by the light incident upon the object.

3. APPEARANCE MEASUREMENT

Colour is probably the most important appearance characteristic of foods, especially if some other aspect of quality is related to the colour, for example the ripening of fruit or the colour changes which occur with deterioration and spoilage. Appearance is more than colour. It is the combination of the visually perceived information contained in the light reflected, transmitted or scattered by the object. Translucency and opacity may be as important in describing appearance as is colour and the directional reflectance from the surface. Light scatter from the structural matrix and absorbance from pigment dissolved or suspended in it interactively affect both translucency/opacity and colour. Relatively small changes in scatter in some foods, for example fresh meat (MacDougall, 1982), may produce larger colour changes than are attributable to the normal range of pigment concentration. Because of the factors governing the visual process, for example memory, colour constancy, contrast and adaptation, and colour blindness, the measurement of appearance characteristics, particularly colour, has to be regarded as a two-stage procedure. The first stage is physical (or chemical) and the second is psychological. The physical factors are the size, shape and uniformity of the object and the pigmentation and structure that attenuate and redirect the light from the object. The conversion of the physical information to the psychological stage for colour specification is accomplished by the psychophysical step of translating reflectance or transmittance to tristimulus values and then to an appropriate colour space.

Foods have an infinite variety of appearance characteristics. Their surfaces may be wet, dry, glistening, matt, irregular or flat although few natural foods are flat. Their subsurface may be transparent, hazy, translucent or opaque and the pigmentation uniform or patchy. This has the consequence that colour measuring procedures may have to be somewhat different from those recommended for measuring flat opaque surface colours, paint and paper, etc., for which colour measuring instruments are principally designed. Compromises in sample preparation coupled with limitations of optical geometry may lead to difficult decisions having to be made in presenting the sample to the instrument; for example the question arises whether to exclude or include the specular component of gloss or matt surfaces and there is always uncertainty about the edge effects of translucent materials on their reflectance (Atkins and Billmeyer, 1966; Hunter, 1975; MacDougall, 1983). Careful standardisation of instruments is essential. For interlaboratory product control, procedural agreement must be related to the context of the variables associated with the product (Kent, 1987), for example, the particle size of powdered foods and the concentration of translucent liquids. Instrument presentation geometry affects the measured reflectance of translucent foods and is an important factor in the quality grading of tomato paste (Brimelow, 1987). In a study on evaporated milk, MacDougall (1987) showed that straight-line relationships exist between the logarithm of the ratio of absorption to scatter, aperture area and concentration. Psychophysical data must be related to the visual assessment of appearance, otherwise, as will be shown in the example of orange juice, sensible interpretation may be impossible.

3.1. Colour Specification

The factors required to specify colour in the CIE system are the spectral power distribution of the illuminant, the reflection or transmission factor of the object and the colour matching functions $\bar{x}(\lambda)$, $\bar{y}(\lambda)$, $\bar{z}(\lambda)$ of the standard observer (Tarrant, 1981). The mathematical procedure is given in any standard text on colour, for example Wright (1980), Judd and Wyszecki (1975) and Hunt (1987b). The system is based on the trichromatic principle, but instead of using 'real' primaries—red (R), green (G) and blue (B)—which may require negative matching, the system is based on imaginary primaries X, Y and Z. The transformation from R, G, B to X, Y, Z for a 2° field of view was accomplished in 1931 in such a way that all values of x, y, z are positive and y is identical to V (the photoptic observer's spectral sensitivity function). Primary Y therefore contains the entire lightness stimuli as a function of wavelength, but Y is not linearly related to visual lightness. The Y-value for object colours is known as luminous reflectance or transmittance. The familiar 1931 CIE x, y chromaticity diagram, bounded by the spectrum locus from 380–770 nm, is the basic colour space to which all others are referred. The chromaticity coordinates are $x = X/(X+Y+Z)$ and $y = Y/((X+Y+Z)$. Hence, every colour can be uniquely located in 1931 CIE space by its Y-value and chromaticity provided the illuminant and the observer's visual field are defined.

Until recently, the most frequently used reference light was CIE illuminant C which was intended to represent average daylight with a correlated colour temperature of 6774 K but is now superseded by D_{65} (6500 K) which includes part of the near ultra-violet in its spectral power distribution. The 2° field, based on the colour-matching functions of the foveal cones, is applicable strictly only to small objects. In 1964 the CIE introduced a supplementary standard observer for a 10° field. This large field observer's colour-matching functions obey the trichromatic generalisation but have limitations that do not apply to the 2° foveal condition. The large field includes areas of the retina that contain rods. The 10° functions relate to visual matching in high intensity light to obviate intrusion by the rods which could upset the prediction. At low levels of illumination the scotoptic mechanism contributes to the match and the system becomes tetrachromatic (Trezona, 1973). The current trend is to use D_{65} and the 10° observer as reference. The CIE have just revised their recommended procedures for colorimetry (CIE, 1986). The information is also published in the newly issued ASTM Standards (1987) and Hunt (1987b) along with weighting factors for a variety of practical illuminants including different types of fluorescent lamps.

3.2. Uniform Colour Space

The 1931 CIE Y, x, y system has one serious fault; it is far from being equally visually spaced. Although it specifies colours with straight line additivity of colour stimuli, its non-uniformity has distortions similar to the Mercator projection of the world (Billmeyer and Saltzman, 1981). Lines of constant hue are curves, constant chroma which should be circular are oval and equal visual distances increase several-fold in size from purple–red towards green (MacAdam, 1942; Stiles, 1946). Improved uniform spacing has used both linear and non-linear transformations of Y, x, y space but it is unlikely that an ideal uniform space is attainable (Billmeyer and Saltzman, 1981). For a review of progress from the 1931 to the 1976 recommended CIE spaces the reader is referred to the article by Wyszecki (1981) in the publication of the golden jubilee of the CIE. The traditional way of constructing three-dimensional colour scales has been to choose an approximately uniform chromaticity diagram and lightness scale, with scaling factors to make perceptual chromaticity differences equal to lightness differences. The three near-uniform colour spaces of practical importance are the Hunter (1958) L, a, b opponent colour space which has been incorporated into most colour meters and the 1976 CIELUV and CIELAB spaces (Robertson, 1977). The 1976 CIE spaces were an attempt to reduce the many scales in use to two (CIE, 1978), but since then studies on their reliability for industrial acceptability judgements in comparison with newer scales with improved performance developed for the textile industry indicate that they have serious disadvantages for routine colour matching (McDonald, 1980, 1985; McLaren, 1981).

The lightness coordinate for both 1976 CIE spaces is the same but there is no simple relationship between the chromaticness diagrams because of the concepts in their construction. The 1976 CIE L^*, a^*, b^* colour space, known as CIELAB, is a non-linear cube root transformation of the 1931 tristimulus values to approximate the Munsell system. The formulae are

$$L^* = 116(Y/Y_0)^{1/3} - 16$$
$$a^* = 500[(X/X_0)^{1/3} - (Y/Y_0)^{1/3}]$$
$$b^* = 200[(Y/Y_0)^{1/3} - (Z/Z_0)^{1/3}]$$

where X_0, Y_0, Z_0 refer to the nominally white object colour stimulus. CIELAB is the colour space most used for general application.

The 1976 CIE L^*, u^*, v^* colour space has the same cube-root relationship of L^* to Y which is virtually equal to the Munsell lightness scale value V. It has a linearly transformed chromaticity diagram similar in concept to that of the 1931 space in which additive mixtures of two components lie on a straight line. A weighting of L^* is incorporated to make the space near uniform. The formulae are

$$L^* = 116(Y/Y_0)^{1/3} - 16$$
$$u^* = 13L^*(u' - u'_0)$$
$$v^* = 13L^*(v' - v'_0)$$

where

$$u' = 4X/(X + 15Y + 3Z)$$
$$v' = 9Y/(X + 15Y + 3Z)$$

and u'_0 and v'_0 are the values of u' and v' with X_0, Y_0 and Z_0 (Robertson, 1977).

3.3. Terminology

For a full discussion on the inter-relationships of the CIE (1978) recommended colour terms, their complete definitions, and illustrations of their application, the reader is referred to Hunt (1978). Colour terms can be divided into the subjective and objective. The subjective, the perceived or psychosensorial are brightness, lightness, hue, saturation, chroma and colourfulness. Colourfulness is a recently introduced term and is that aspect of visual sensation according to which an area appears to exhibit more or less chromatic colour. Although hue is easily understood as that attribute described in colour names (red, green, purple, etc.), the difference between saturation and chroma is less easily comprehended. Saturation is colourfulness judged in proportion to its brightness whereas chroma is colourfulness relative to the brightness of its surroundings. A similar difference exists between lightness and brightness; lightness is relative brightness. Lightness is unaffected by illumination level because it is the proportion of light reflected whereas the sensation of brightness increases with increase in illumination.

The objective terms refer to the stimulus and are divided into the psychophysical and psychometric. Psychophysical quantities are evaluated from spectral power distributions, reflectance or transmittance and observer response, and provide the basis for the psychometric qualities, which correspond more nearly to what we perceive. There is a psychometric term for each perceptual one. For CIELUV space the more important terms are

$$\text{psychometric lightness} = L^*$$
$$\text{psychometric hue} = h^*_{uv} = \tan^{-1}(v^*/u^*)$$
$$\text{psychometric chroma} = C^*_{uv} = (u^{*2} + v^{*2})^{1/2}$$
$$\text{psychometric saturation} = s^*_{uv} = C^*_{uv}/L^*$$

CIELAB has similar properties except that no simple correlate of saturation exists.

Colour differences, ΔE^*, can be expressed either as the coordinates of colour space or in terms of the correlates of lightness, chroma and hue. For CIELAB

$$\Delta E^* = [(\Delta L^*)^2 + (\Delta a^*)^2 + (\Delta b^*)^2]^{1/2}$$

and

$$\Delta E^* = [(\Delta L^*)^2 + (\Delta C^*)^2 + (\Delta H^*)^2]^{1/2}$$

where ΔH^* is used rather than Δh^* because the latter is angular. For small colour differences away from the L^* axis, if h^* is expressed in degrees, then

$$\Delta H^* = C^* \Delta h^* (\pi/180)$$

3.4. Absorption and Scatter

The reflectance of opaque or translucent materials depends on attenuation by pigment absorption, scatter by structural elements in the object produced by differences in refractive index at boundaries, and external and internal reflection at the air/object surface (MacDougall, 1986). Although few foods are sufficiently flat for their surface characteristics to be measured, components of surface reflection, i.e. the degree of glossiness or mattness, should be accounted for in the measurement of colour. Most colour meters are designed to eliminate the specular component but most spectrophotometers have the option of including or excluding the specular component for total or diffuse reflectance. For techniques of measuring surface characteristics the reader is referred to Hunter (1975) and the ASTM (1987) Standards.

The method used most for separating the relative contributions that absorption and scatter have on reflectance is the two-variable procedure of Kubelka and Munk (Kubelka, 1948; Allen, 1978) which is fully illustrated in Judd and Wyszecki (1975). Advanced turbid media theory, for example the many-flux model of Mudgett and Richards (1972), explains some of the limitations inherent in the Kubelka–Munk theory, for example the effect of the angular distribution of the radiation with changing depth, but the many-flux theory has not yet found widespread industrial application. The Kubelka-Munk theory relates reflectivity (R_∞), the reflectance at infinite thickness, to the absorption and scatter coefficients (K and S) by

$$K/S = (1 - R_\infty)^2/2R_\infty$$

where K and S are calculated from the reflectance of thin layers on white and black backgrounds, or from the reflectivity (R_∞), and a thin layer on a white background; K and S thus calculated are approximations. More accurate values require incorporation of correction factors to account for the internal reflection at the air/object boundary (Saunderson, 1942; Allen, 1978), but such corrections are probably unnecessary for specifying food opacity.

Surface colour calculated from the reflectivity spectrum, and the angular distribution of the specular and diffuse components are usually sufficient information to describe appearance if the object is highly scattering and opaque, for example paint where virtually all the reflectance is returned from the upper 1 mm. Specification of colour by itself is inadequate to describe non-opaque materials, even if they are only slightly translucent. To specify degree of translucency, K and S or the internal transmittance (T_i) is required.

Pigment oxidation in fresh meat, translucence in bacon, dilution of orange juice and dyed milk are particularly good examples of foods that illustrate the effects of illumination and the interaction of absorption and scatter on colour appearance.

3.5. Fresh Meat Oxidation

On exposure to oxygen the purple ferrous haem pigment, myoglobin, forms the bright red covalent complex, oxymyoglobin, which oxidises to the unattractive brown pigment, metmyoglobin (MacDougall, 1982). The level of metmyoglobin required for consumer discrimination for rejection is approximately 20% (Hood and Riordan, 1973). The appearance of meat is greatly affected by the colour-rendering properties of the lamps used for display (Halstead, 1978). Fluorescent lamps recommended by the manufacturers for displaying meat have enhanced red emission that maintains the preferred colour of oxymyoglobin and shifts the early stages of metmyoglobin development from brown towards red. The effect of red enhancement on meat colours has been shown to elicit a greater visual

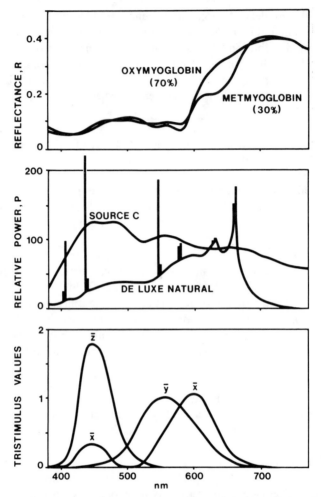

FIG. 1. Components required for psychometric specification of the colour of red (70% oxymyoglobin) and brown (30% metmyoglobin) beef for CIE Source C and a De Luxe Natural fluorescent lamp for the 2° standard observer.

colour change in making brown appear red than in making red appear more red (MacDougall, 1981; MacDougall *et al.*, 1985*a*). For some people red enhancement could make meat appear too red. Broad band tubes with high colour rendering such as 'Artificial Daylight' separate food colours with maximum discrimination and should be used for accurate quality judgements. Red enhancing tubes, with their property to flatter

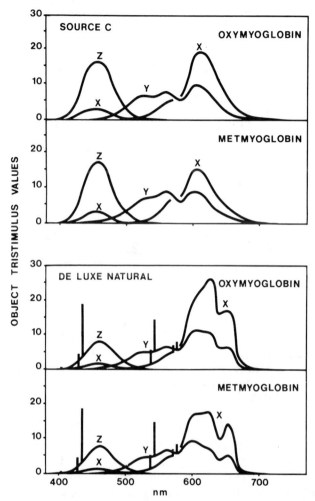

FIG. 2. Intermediate step in specification of colour showing products from Fig. 1 in determining tristimulus values.

appearance, reduce the observer's ability to assess differences in the colourfulness of bright colours. Figures 1, 2 and 3 illustrate the method of calculating psychometric space from red and brown meat, the illuminant and the observer (CIE, 1971, 1978). Figure 2 shows the relative effects of the difference in oxidation of oxymyoglobin compared with the difference between CIE Source C and a commercial red enhancing illuminant. Lamps generative of redwards shifts give large values of C^*, and poorer colour

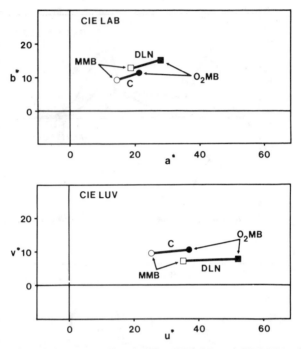

FIG. 3. Psychometric chroma diagrams for CIELAB and CIELUV colour spaces for meat colours (O_2MB—oxymyoglobin, MMB—metmyoglobin) for Source C (C) and a De Luxe Natural (DLN) fluorescent lamp, both of which have their achromatic location, the nominally white object colour stimulus, at the origins.

rendering lamps decrease C^* for brown samples. In the case of De Luxe Natural (DLN) (Thorn EMI Lighting Ltd), one of the lamps recommended for meat display, the visually judged redness of red meat increases by >80% of the difference between the red and brown as viewed under Artificial Daylight (MacDougall, 1981). A similar increase occurred for the brown sample.

Although psychometric spacing is indicative of the increase or decrease in C^* as deduced from either the a^*, b^* or u^*, v^* chroma diagrams (Fig. 3) which have the illuminants commonly located at the origin, the visual hue is not adequately represented. There is a difference in hue angle fundamental to the ways the spaces are constructed; reds have lower values of $h°$ in u^*, v^* than a^*, b^* (Robertson, 1977) and when fluorescent tubes are used as the origin in the diagram, the relative position of different colours may change between the two chroma diagrams (Mori and Fuchida, 1982). This is seen in

Fig. 3 where DLN changes places relative to Source C. Mori and Fuchida (1982) conclude from their study that CIELAB is the best for use in colour rendering specification of commonly used light sources and the terms $(Y/Y_0)^{1/3}$, etc., included in the formula seem to be a practical correction for chromatic adaptation.

3.6. Bacon Opacity

Muscle cut immediately after slaughter is translucent, but during rigor mortis as the pH falls from 7·0 to 5·5 it becomes more light scattering and opaque. Factors contributing to variation in opacity in raw meat are the rate of chilling, the rate of fall in pH and the availability of muscle glycogen for conversion to lactic acid. Excessively rapid glycolysis results in low pH while the carcass is still warm and the meat appears pale, soft and exudative with high values of S (mm^{-1}), the measured Kubelka-Munk scatter coefficient. Inadequate reserves of glycogen may produce insufficient lactic acid for the pH to fall below 6·0 and the transition from translucent to semi-opaque does not happen and the muscle appears dark. MacDougall (1982) has related the lightness, hue and chroma to S (mm^{-1}) and pigment concentration for fresh meat and shown that the range in lightness that results from the variation in S (mm^{-1}) from normal carcass handling can be confused with that from large changes in pigment concentration. A portable fibre optic probe has been developed to measure light scatter in meat without cutting the carcass (MacDougall, 1984).

The main use of the Kubelka–Munk analysis is in the colour formulation industries (Best, 1981), but it has been used successfully in studies on meat (MacDougall, 1970; Birth, 1978). To illustrate how it separates the absorption from the scatter components in meat, two samples of bacon, 'Wiltshire' and 'Sweetcured', differing greatly in structure and lightness, were assessed. The Wiltshire had the typical translucent appearance of the cold cured product and the Sweetcured which has a mild heating step in the process was typically pale (MacDougall et al., 1985b). Reflectance spectra were measured on a Pye Unicam SP8-100 spectrophotometer and are shown in Fig. 4. The information from pigment absorption was all contained in K with no crossover of information to S. The scatter coefficient was a power function of wavelength between 700 and 450 nm. It was not possible to determine K and S directly between 380 and 450 nm because of the intense absorption from the Soret region but estimated values of K and T_i were calculated from the extrapolation of S, and shown in Fig. 5 as dashed lines.

The value of S (mm^{-1}) for the Sweetcured was twice that of the Wiltshire

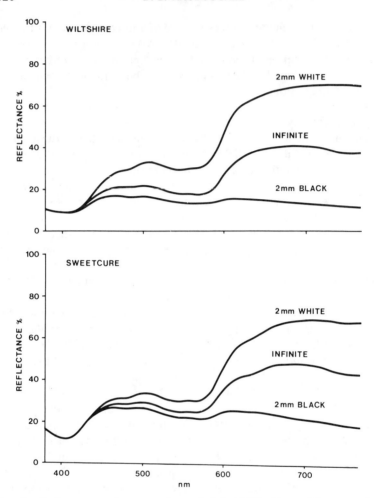

FIG. 4. Reflectance spectra, specular excluded, of Wiltshire and Sweetcure bacon at infinite thickness and at 2 mm thickness on black and white backgrounds.

and the entire visual difference between the samples is explainable by the change in T_i relative to thickness. There was negligible difference in pigment concentration between the two samples. For tristimulus value Y, S (mm^{-1}) and K (mm^{-1}) were 0·16 and 0·26 for the Wiltshire and 0·30 and 0·27 for the Sweetcured. Measured Y_∞ (%) was 22·1 and 29·9 respectively; these values differ from the theoretical Y_∞ values based on the ratio of measured K/S, the discrepancy arising because no Saunderson-type correction for air/object boundary was applied. It is my experience that a simple correction is impossible using the two-flux theory that would be applicable over the entire range of translucency to opacity found in meat. As a working tool for measuring opacity, no correction is necessary provided the measurement technique remains constant—in this case, 2 mm cells lined with 'near' zero reflectance black and with white plastic

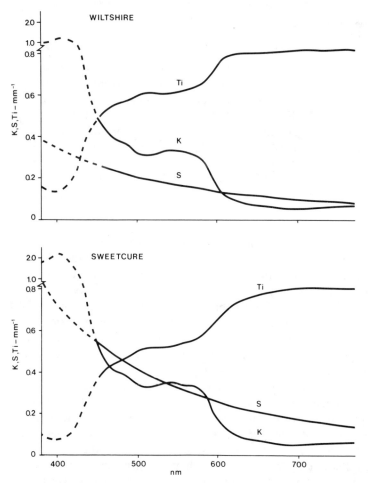

FIG. 5. Kubelka–Munk absorption (K) and scatter (S) coefficients and internal transmittance (T_i) for unit thickness of 1 mm for Wiltshire and Sweetcure bacon. Continuous line, calculated from 2 mm thickness on black and white; dashed line, estimated from extrapolation of S.

($Y_\infty = 0.82$). The values of T_i calculated for different thicknesses of the bacon showed that for a 3 mm thick slice, 30% of the light was travelling in a forward direction for the Wiltshire but this was reduced to 20% for the Sweetcured.

3.7. Orange Juice Scatter

Fruit juice, jam and other highly chromatic translucent gels are difficult to measure because the incident light is dispersed in the sample. The effect of optical geometry on measured colour values for orange juice has been discussed by Huggart et al. (1966), Rummens (1970) and Eagerman (1978). Best results for such translucent materials are usually obtained if the sample port is large relative to the incident beam (Kent, 1987; MacDougall, 1987). The effects of concentration and optical geometry on measured colour and the Kubelka–Munk K, S and T_i value for orange juice were studied by MacDougall (1983) who found a 50% increase in Y_∞ was obtained by increasing the sample port diameter from 2 cm to 5 cm while maintaining the incident beam at 1 cm. Over a range of dilutions from 4-fold concentrate to 0·2, the scatter coefficient, S (mm^{-1}), decreased with dilution for X, Y and Z as the high value of the absorption coefficient, K (mm^{-1}) for Z, decreased to approach the much lower value of K (mm^{-1}) for X and Y. The decrease in K (mm^{-1}) for Z is as anticipated for a blue absorbing yellow–red transmitting pigment. The consequence of the loss of scattering power on dilution is a reduction in Y_∞, and therefore the most dilute juice is instrumentally the darkest, and the most concentrated, the lightest. This, however, is not what the observer perceives.

A set of dilutions was presented to a ten member panel in 100 ml 4·5 cm diameter glass containers filled to a depth of 4 cm. Illumination was by 700 lux Artificial Daylight and the samples, on white discs on a grey background, were viewed by essentially reflected light. Panelists, unaware of the concentrations, were asked to estimate the visual difference between adjacent samples and their scores were adjusted to give a cumulative total colour difference of 1·0 between the most dilute and most concentrated. The results are shown in Fig. 6 along with the Hunter psychometric spacing determined on the SP8-100 spectrophotometer. The panel cumulative mean score was linear with the hue angle $h°$, but lightness L and chroma C were linear only over the more dilute suspensions and the cumulative calculated colour difference ΔE was curvilinear. The observer's estimate of change in colour over the range of very translucent yellow to semi-opaque orange may be more affected by the change in $h°$ and C than L. The panelists unanimously stated that the most dilute suspension was the lightest which is the opposite of that determined instrumentally.

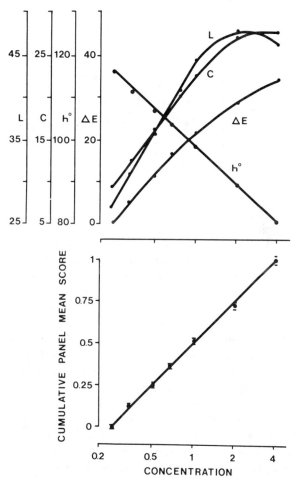

FIG. 6. Relationship of cumulative colour difference (ΔE), lightness (L), hue angle ($h°$), chroma (C) and panel mean score ($\pm$s.e.) of visual difference between adjacent samples to the logarithm of the concentration of orange juice. Reconstituted juice at normal concentration = 1·0. (Reprinted by permission of Ellis Horwood Publishers, Chichester.)

For strongly coloured scattering materials in dilute suspension, instrumental measurement of colour, even supplemented by information on scatter, is inadequate to fully describe appearance. The instrument does not measure what the observer perceives because it measures reflectance over a limited solid angle from a narrow beam of incident illumination whereas the observer sees the light after multidirectional illumination which may cause the material to glow.

124　　　　　　　　　　D. B. MACDOUGALL

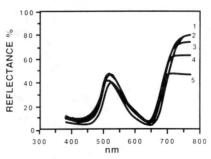

FIG. 7. Reflectance spectra of 'near constant' K/S series of evaporated milk and green dye dilutions; 1 to 5 increase with decrease in concentration of dye.

3.8. Dyed Milk Spacing

The orange juice study demonstrated the special case of changing translucency while maintaining near constant K/S. In practice, this is an approximation because the scattering coefficient is not linear with dilution at low scattering power. Forward scattering increases with separation of the scattering elements in the suspension. An experiment to compare near constant K/S dilution with independent change in K and S was carried out on mixtures of evaporated milk and a green food dye. Two 1:1 series of dyed dilutions were prepared. In one, the dye was diluted proportionally with the milk and in the other the dye increased as the milk decreased. The mid-points in each series were therefore identical and produced a cross-arrangement of scatter to absorption. Reflectance spectra of the near constant K/S set (Fig. 7) are separated at >700 nm where absorption from the dye is minimal, but, where dye absorbs strongly <650 nm, the spectra are virtually superimposed. The values for CIELAB lightness, L^*, chroma, C^*, and hue angle, h^*, in the top right to bottom left diagonal of the cross-arrangement in Fig. 9 show small but progressive changes. On dilution h^*

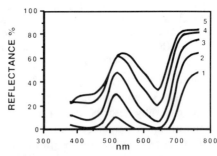

FIG. 8. Reflectance spectra of 'opposite order' K/S series of evaporated milk and green dye dilutions; 1 to 5 increase with decrease in concentration of dye.

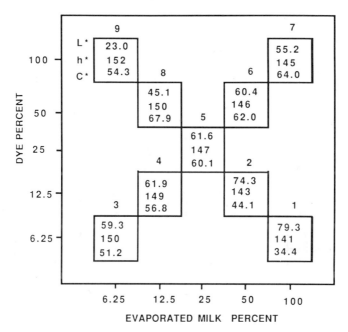

FIG. 9. Cross-arrangement of CIELAB L^*, C^*, h^* for series of dilutions of evaporated milk and green dye. Order of panel selection from lightest to darkest numbered 1-9.

increases and C^* decreases but L^* increases and then decreases. This is similar to that found in tomato paste (MacDougall, 1987). The inverse order of dye concentration relative to milk separates the reflectance spectra in order of dye, the most dilute milk with the most dye giving the darkest suspension (Fig. 8). The L^*, C^*, h^* values in the bottom right to top left diagonal again show progressive change but the order of h^* is reversed and C^* increases and then decreases with dye dilution which is similar to dying textiles (McLaren, 1986). L^* shows the greatest increase with dilution of dye. The absorption and scatter coefficients (Fig. 10) were determined using 2 mm black- and white-backed cells. In both series K and S increased with increased concentrations of dye and milk but the range in K/S for the 'opposite order' set was >25 times that in the 'near constant' set.

The panel were asked to arrange the samples in order from lightest to darkest. The order is recorded in Figs 9 and 10. In both sets the lightest had the least dye irrespective of the order of milk concentration. However the entire 'near constant' set was interposed between the first two and the last two samples of the 'opposite order' set. In this demonstration, a

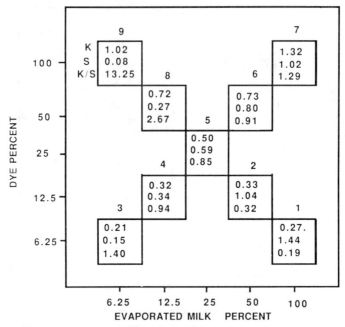

FIG. 10. Cross-arrangement of Kubelka–Munk absorption and scatter coefficients, K and S (mm^{-1}), for series of dilutions of evaporated milk and green dye.

purposefully wide range in milk and dye was selected to demonstrate the massiveness of the effects that small but independent changes in K or S can have on appearance and measured colour values. Although large concomitant changes in K and S may cause confusion in relating translucence to measured colour, small independent changes produce much larger differences in visual appearance. This would explain the large changes in lightness or brightness that occur in the darkening of over-ripe fruit, the loss of brilliance in dark-cutting meat and the lightening that occurs during cooking of fish and meat products from the increase in scatter that accompanies protein denaturation.

ACKNOWLEDGEMENTS

I would like to thank Mr G. Fitton for writing the computer program and Miss H. Poyser and the Institute's Appearance panelists for their assistance in the dyed milk experiment.

REFERENCES

Abramov, I. (1981). Physiological mechanisms of colour vision. In *AIC Color 81*, Deutsche Farbwissenshaftliche Gesellschaft, Berlin.
Allen, E. (1978). Advances in colorant formulation and shading. In: *AIC Color 77*, Adam Hilger, London, pp. 153–79.
ASTM (1987). *Standards on color and appearance measurement*. American Society for Testing and Materials, Philadelphia.
Atkins, J. T. and Billmeyer, F. W. (1966). Edge-loss errors in reflectance and transmittance measurement of translucent materials. *Mater. Res. Stand.*, **6**, 564–9.
Bartleson, C. J. (1979a). Changes in color appearance with variations in chromatic adaptation. *Color Res. Appl.*, **4**, 119–38.
Bartleson, C. J. (1979b). Predicting corresponding colors with changes in adaptation. *Color Res. Appl.*, **4**, 143–55.
Berridge, M. J. and Irvine, R. F. (1984). Inositol trisphosphate, a novel second messenger in cellular signal transduction. *Nature*, **312**, 315–21.
Best, R. P. (1981). Computer match prediction in laboratory and factory. In: *The Golden Jubilee of Colour in the CIE*, The Society of Dyers and Colourists, Bradford.
Billmeyer, F. W. and Saltzman, M. (1981). *Principles of Color Technology* (2nd edn), J. Wiley & Sons, New York.
Birth, G. S. (1978). The light scattering properties of foods. *J. Food Sci.*, **43**, 916–25.
Bowmaker, J. K. and Dartnall, H. J. A. (1980). Visual pigments of rods and cones in a human retina. *J. Physiol.*, **298**, 501–11.
Boynton, R. M. (1979). *Human Color Vision*, Holt, Rinehart and Winston, New York.
Brimelow, C. J. B. (1987). Measurement of tomato paste colour: Investigation of some method variables. In: *Physical Properties of Foods—2: COST 90bis Final Seminar Proceedings*, Elsevier Applied Science Publishers, London, pp. 295–317.
Brown, J. E., Rubin, L. J., Ghalayini, A. J., Tarver, A. P., Irvine, R. F., Berridge, M. J. and Anderson, R. E. (1984). Myoinositol polyphosphate may be a messenger for visual excitation in *Limulus* photoreceptors. *Nature*, **311**, 160–3.
Brown, J. L. (1965). The structure of the visual system. In: *Vision and Visual Perception*, C. H. Graham (Ed.), J. Wiley & Sons, New York.
CIE (1971). Colorimetry: official recommendations of the International Commission on Illumination. *CIE Publication No. 15 (E-1.3.1)*, Commission Internationale de l'Eclairage, Paris.
CIE (1978). Recommendations on uniform color spaces—color difference equations, psychometric color terms. *Supplement No. 2 to CIE Publication No. 15 (E-1.3.1) 1971/(TC-1-3)*, Commission Internationale de l'Eclairage, Paris.
CIE (1986). Colorimetry (2nd edn). *CIE Publication No. 15.2*, Commission Internationale de l'Eclairage, Vienna.
De Valois, R. L. and De Valois, K. K. (1975). Neural coding of color. In: *Handbook of Perception, Vol. 5, Seeing*, E. C. Carterette and M. P. Friedman (Eds), Academic Press, New York, pp. 117–66.

Eagerman, B. A. (1978). Orange juice color measurement using general purpose tristimulus colorimeters. *J. Food Sci.*, **43**, 428–30.
Estévez, O. (1982). A better colorimetric standard observer for color-vision studies: The Stiles and Burch 2° color-matching functions. *Color Res. Appl.*, **7**, 131–4.
Fein, A., Payne, R., Corson, D. W., Berridge, M. J. and Irvine, R. F. (1984). Photoreceptor excitation and adaptation by inositol 1,4,5-triphosphate. *Nature*, **311**, 157–60.
Francis, F. J. and Clydesdale, F. M. (1975). *Food Colorimetry: Theory and Applications*, AVI Publishing Co., Westport, Connecticut.
Guild, J. (1931). The colorimetric properties of the spectrum. *Phil. Trans. R. Soc. (Lond.) Ser. A.*, **230**, 149.
Halstead, M. B. (1978). Colour rendering: past, present and future. In: *AIC Color 77*, Adam Hilger, London, pp. 97–127.
Hård, A. (1970). Qualitative attributes of colour perception. In: *AIC Color 69*, Musterschmidt-Verlag, Göttingen, Vol. 1, p. 351.
Hood, D. E. and Riordan, E. B. (1973). Discolouration in pre-packaged beef: measurement by reflectance spectrophotometry and shopper discrimination. *J. Food Technol.*, **8**, 333–43.
Huggart, R. L., Barron, R. W. and Wenzel, F. W. (1966). Evaluation of Hunter citrus colorimeter for measuring the color of orange juices. *Food Technol.*, **20**, 677–9.
Hunt, R. W. G. (1977). Specification of colour appearance. II. Effects of changes in viewing conditions. *Color Res. Appl.*, **2**, 109–20.
Hunt, R. W. G. (1978). Colour terminology. *Color Res. Appl.*, **3**, 79–87.
Hunt, R. W. G. (1982). A model of colour vision for predicting colour appearance. *Color Res. Appl.*, **7**, 95–112.
Hunt, R. W. G. (1987a). A model of colour vision for predicting colour appearance in various viewing conditions. *Color Res. Appl.*, **6**, 297–314.
Hunt, R. W. G. (1987b). *Measuring Colour*. Ellis Horwood, Chichester.
Hunt, R. W. G. and Pointer, M. R. (1985). A colour–appearance transform for the CIE 1931 standard colorimetric observer. *Color Res. Appl.*, **10**, 165–79.
Hunter, R. S. (1958). Photoelectric color difference meter. *J. Opt. Soc. Am.*, **48**, 985–95.
Hunter, R. S. (1975). *The Measurement of Appearance*, J. Wiley & Sons, New York.
Hurvich, L. M. (1981). *Color Vision*, Sinaver Associates, Publishers, Sunderland, Massachusetts.
Jacobs, G. H. (1981). *Comparative Color Vision*, Academic Press, New York.
Jameson, D. and Hurvich, L. M. (1968). Opponent-response functions related to measured cone photopigments. *J. Opt. Soc. Am.*, **58**, 429–30.
Judd, D. B. and Wyszecki, G. (1975). *Color in Business, Science and Industry* (3rd edn), J. Wiley & Sons, New York.
Kent, M. (1987). Collaborative measurements on the colour of light-scattering foodstuffs. In: *Physical Properties of Foods—2: COST 90bis Final Seminar Proceedings*, Elsevier Applied Science Publishers, London, pp. 277–94.
Kubelka, P. (1948). New contributions to the optics of intensely light scattering materials. *J. Opt. Soc. Am.*, **38**, 448–57.
MacAdam, D. L. (1942). Visual sensitivities to color differences in daylight. *J. Opt. Soc. Am.*, **32**, 247.
MacAdam, D. L. (1970). *Sources of Color Science*, MIT Press, Cambridge, Massachusetts.

McDonald, R. (1980). Industrial pass/fail colour matching. Part I—Preparation of visual colour matching data. *J. Soc. Dyers Col.*, **96**, 372–6.

McDonald, R. (1985). The CMC color difference formula and its performance in acceptability and perceptibility decisions. In: *AIC Mondial Couleur 85, Monte Carlo*, **1**, 46.

MacDougall, D. B. (1970). Characteristics of the appearance of meat. I—The luminous absorption, scatter and internal transmittance of the lean of bacon manufactured from normal and pale pork. *J. Sci. Food Agric.*, **21**, 568–71.

MacDougall, D. B. (1981). Visual estimate of colour changes in meat under different illuminants. In: *AIC Color 81*, Deutsche Farbwissenschaftliche Gesellschaft, Berlin.

MacDougall, D. B. (1982). Changes in the colour and opacity of meat. *Food Chem.*, **9**, 75–88.

MacDougall, D. B. (1983). Instrumental assessment of the appearance of foods. In: *Sensory Quality in Foods and Beverages: Its Definition, Measurement and Control*, Ellis Horwood, Chichester, pp. 121–39.

MacDougall, D. B. (1984). Meat Research Institute probe for stressed meat detection. *Anal. Proc.*, **21**, 494–5.

MacDougall, D. B. (1986). The chemistry of colour and appearance. *Food Chem.*, **21**, 283–299.

MacDougall, D. B. (1987). Optical measurements and visual assessment of translucent foods. In: *Physical Properties of Foods—2: COST 90bis Final Seminar Proceedings*, Elsevier Applied Science Publishers, London, pp. 319–30.

MacDougall, D. B., Francombe, M. and Whelehan, O. P. (1985*a*). Visual descriptive profiling of meat under different illuminants. In: *AIC Mondial Couleur 85, Monte Carlo*, **1**, 85.

MacDougall, D. B., Jenkins, K. M. and Pritchard, S. E. (1985*b*). The development of opacity in bacon as a function of temperature during curing. *J. Sci. Food Agric.*, **36**, 1339–40.

McLaren, K. (1981). The development of improved colour-difference equations by optimisation against acceptability data. In: *The Golden Jubilee of Colour in the CIE*, The Society of Dyers and Colourists, Bradford.

McLaren, K. (1986). *The Colour Science of Dyes and Pigments* (2nd edn), Adam Hilger, Bristol.

Mori, L. and Fuchida, T. (1982). Subjective evaluation of uniform color spaces used for color-rendering specification. *Color Res. Appl.*, **7**, 285–93.

Mudgett, P. S. and Richards, L. W. (1972). Multiple scattering calculations for technology, II. *J. Colloid Interface Sci.*, **39**, 551–67.

Nayatani, Y., Takahama, K., Sobagaki, H. and Hirano, J. (1982). On exponents of a nonlinear model of chromatic adaptation. *Color Res. Appl.*, **7**, 34–45.

Nayatani, Y., Takahama, K. and Sobagaki, H. (1986). Prediction of color appearance under various adapting conditions. *Color Res. Appl.*, **11**, 62–71.

Normann, R. A. and Werblin, F. S. (1974). Control of retinal sensitivity 1. Light and dark adaptation of vertebrate rods and cones. *J. Gen. Physiol.*, **63**, 37–61.

Parkhurst, C. and Feller, R. L. (1982). Who invented the color wheel? *Color Res. Appl.*, **7**, 217–30.

Pointer, M. R. (1980). The concept of colourfulness and its use for deriving grids for assessing colour appearance. *Color Res. Appl.*, **2**, 99–107.

Pointer, M. R. (1982). Analysis of colour-appearance grids and chromatic-adaptation transforms. *Color Res. Appl.*, **7**, 113–18.
Polyak, S. L. (1941). *The Retina*, University of Chicago Press, Chicago.
Robertson, A. R. (1977). The CIE 1976 color-difference formulae. *Color Res. Appl.*, **2**, 7–11.
Rodieck, R. W. (1979). Visual pathways. *Ann. Rev. Neurosci.*, **2**, 193–225.
Rummens, F. H. A. (1970). Color measurement of strongly scattering media, with particular reference to orange-juice beverages. *J. Agric. Food Chem.*, **18**, 371–6.
Rushton, W. A. H. (1965). A foveal pigment in the deuteranope. *J. Physiol.*, **176**, 24–37.
Saunderson, J. L. (1942). Calculation of the color of pigmented plastics. *J. Opt. Soc. Am.*, **32**, 727–36.
Smith, C. V. & Pokorny, J. (1975). Spectral sensitivity of the foveal cone photopigments between 400 and 500 nm. *Vision Res.*, **15**, 161–71.
Stell, W. K. (1972). The morphological organization of the vertebrate retina. In: *Handbook of Sensory Physiology*, Vol. 11/2, M. G. F. Fuortes (Ed.), Springer-Verlag, New York, pp. 111–213.
Stiles, W. S. (1946). A modified Helmholtz line element in brightness-colour space. *Proc. Phys. Soc. (Lond.)*, **58**, 41.
Stiles, W. S. (1978). *Mechanisms of Colour Vision*, Academic Press, London.
Stiles, W. S. and Burch, J. M. (1959). N. P. L. colour-matching investigation: final report. *Opt. Acta*, **6**, 1–26.
Stiles, W. S. and Crawford, B. H. (1933). The liminal brightness increment as a function of wave-length for different conditions of the foveal and parafoveal retina. *Proc. R. Soc. (Lond.)*, **113B**, 496–530.
Tarrant, A. W. S. (1981). The nature of colour—a physicist's viewpoint. In: *Natural Colours for Food and Other Uses*, J. N. Counsell (Ed.), Applied Science Publishers, London, pp. 1–25.
Trezona, P. W. (1973). Tetrachromatic color measurement. In: *AIC Color 73*, Adam Hilger, London.
Wald, G. (1968). The molecular basis of visual excitation. *Nature*, **219**, 800–7.
Wright, W. D. (1928/29). A re-determination of the trichromatic coefficients of the spectral colours. *Trans. Opt. Soc.*, **30**, 141.
Wright, W. D. (1967). *The Rays Are Not Coloured*, Adam Hilger, London.
Wright, W. D. (1980). *The Measurement of Colour* (5th edn), Adam Hilger, London.
Wright, W. D. (1981). 50 years of the 1931 CIE standard observer for colorimetry. *Die Farbe.*, **29**, 251–72.
Wright, W. D. (1982). The golden jubilee of colour in the CIE 1931–1981. *Color Res. Appl.*, **7**, 12–15.
Wyszecki, G. (1981). Uniform colour spaces. In: *The Golden Jubilee of Colour in the CIE*, The Society of Dyers and Colourists, Bradford.
Wyszecki, G. and Stiles, W. S. (1982). *Color Science; Concepts and Methods Quantitative Data and Formulae* (2nd edn), J. Wiley & Sons, New York.
Zeki, S. (1980). The representation of colours in the cerebral cortex. *Nature*, **284**, 412–18.

Chapter 5

SENSORY DIFFERENCE TESTING AND THE MEASUREMENT OF SENSORY DISCRIMINABILITY

J. E. R. FRIJTERS

Department of Marketing and Marketing Research and Department of Food Science, Agricultural University, Wageningen, The Netherlands

1. INTRODUCTION

Sensory difference tests (Peryam, 1958) form an indispensable group in the total number of methods available for sensory analysis and evaluation of foods and beverages. Some years ago, it was shown in two independent surveys, one in the USA (Brandt and Arnold, 1977) and one in the UK (Muller, 1977), that of all the procedures in use sensory difference tests were used most frequently. There are at least two reasons for the apparent popularity of these methods. The first is undoubtedly the simplicity of the experimental procedure involved. Sensory difference tests can be carried out quickly, and moreover most of these tests can be easily performed by subjects. The second, and probably more important, reason for the widespread and frequent use of sensory difference tests is that these tests have a specific purpose which is distinct from the purposes of other sensory methods. Sensory difference tests are comparative procedures for use in the study of sensory discriminability of similar types of stimuli. These methods have not been developed, and therefore are not suitable, for the scaling of sensory intensities. With the exception of the duo test, these procedures are also not suitable for the study of sensory preferences. However, some of the basic sensory difference tests have been modified or extended in order to make these procedures suitable for other purposes than the study of sensory differences (Amerine *et al.*, 1965). These modifications have mainly arisen from the practical need to obtain information as easily as possible, and only occasionally are the basic principles of perception and judgement taken into account when modifications are proposed. For this reason the various modified forms of basic sensory difference tests are not discussed in this chapter.

In the literature on sensory evaluation and analysis that has appeared since the Second World War, a number of comparative methods have been labelled as 'sensory difference tests'. Although this name suggests that these procedures are only of relevance to sensory analysis, this is certainly not the case. On the contrary, a survey of the literature in disciplines such as biology, psychology, economics and market research indicates that there is a general and traditional interest in issues related to and procedures developed for comparative judgement and choice. For example, Davidson and Farquhar (1976) illustrate the emphasis given to this basic method by scientists in many disciplines.

It therefore seems appropriate to consider sensory difference tests from a theoretical point of view, thereby crossing the ill-defined boundaries of the field of sensory evaluation as it was developed in the context of food research and the area of psychological research. The aim of this chapter is to show that in addition to their traditional use for testing statistical hypotheses, sensory difference tests can be used for the measurement of sensory differences. In the remainder of this chapter a distinction is made between the term 'sensory difference test' and the term 'sensory difference method'. When a procedure is used for statistical testing purposes it is called a test; when it is used for measurement it is called a method.

2. TWO BASIC PRINCIPLES OF SENSORY RESPONSE BEHAVIOUR

There are many situations in which the question arises as to whether two types of sensory stimuli (foods, beverages, solutions, odours, etc.) can be perceived as being sensorily different; that is whether they are sensorily discriminable. For example, a food scientist engaged in the development of a new product may want to know whether the new product is perceived as being sensorily different from an existing product, or whether it is perceived as sensorily different from a similar product of a competitor. Another example is that for the purpose of quality control, it may be necessary to know whether products of different batches have the same sensory qualities.

The first basic problem about sensory response behaviour to be considered is the question of sensory discriminability of two different stimuli. It can be answered only by a sensory examination. The stimuli must be tasted, smelt, felt or perceived by some other sense. The final results of such a sensory examination should lead to a clear decision regarding the equality or inequality of the stimuli. A paired comparison of stimuli would seem a natural and straightforward procedure for the assessment of similarity or dissimilarity, and many people do not foresee any complications when this investigation is carried out. However, in order to show that even a simple comparison can be quite complicated, a few basic principles of sensory response behaviour must be considered.

It is well known that the sensitivity of all human senses is limited. Firstly, there are limits to the physical magnitude of a stimulus that can be perceived. This is expressed in the concept of absolute sensitivity. Secondly, there are also limits to the magnitude of a physical difference between two stimuli which can be perceived. This is known as differential sensitivity. An absolute differential sensitivity implies that two stimuli (stimulus objects) which are not identical in physical composition may be identified as being sensorily equal when these stimuli are simultaneously or successively presented to a human subject. For example, two solutions of sugar in water need not be of equal concentration in order to elicit two identical sensory responses from a subject who has been asked to taste these solutions with the instruction to judge their sweetness. Thus, from this well-established behavioural observation a general principle may be derived: *Two physically different stimuli can evoke two identical sensory responses.* It is sometimes worthwhile to determine differential sensitivity since samples of a particular food product, which are not completely identical in physical composition, can be perceived to be identical when they are consumed.

The second basic problem about sensory response behaviour to be considered is another contradictory behavioural phenomenon. It is well known that each of these senses is, in layman's terms, unreliable. That is, sensory responses to a particular physical stimulus are not always identical on all occasions, but may show variation. Instead of the term *unreliable* the term *variable* is frequently used to denote this phenomenon. For example, when a particular sugar solution is repeatedly presented to the same subject who is unaware that he is being presented with the same stimulus, a variable sweetness response is likely to be obtained. Thus a second behavioural principle may be derived: *The same stimulus can evoke a variety of sensory responses on different occasions.*

In conclusion, it may be said that sensory response behaviour is governed by two contradictory behavioural principles. On the one hand, two physically different stimuli can each give rise to an identical sensory response, and on the other hand, the same stimulus can evoke different responses on different occasions. Thus it is obvious that, because of these principles, it is impossible to draw a valid conclusion about the sensory discriminability of two stimuli from a single paired comparison.

Any valid theory on the general relationship between a stimulus and its corresponding sensory effects should take into account the paradoxical situation resulting from these principles. This requirement was clearly recognised in the development of sensory difference tests in the period between 1940 and 1960. First, reliance was not placed on the subject's willingness to judge two or more stimuli as similar or dissimilar when the stimuli were mutually compared. Instead a subject was required to make a definite choice. Secondly, reliance was not placed on responses obtained in one single trial; repeated measurements had to be made. As a result of these two experimental measures it was possible to make use of the basic principles of statistical probability theory in order to evaluate the data obtained experimentally. These principles are mentioned briefly in the following section in which the main sensory difference tests are described.

3. SENSORY DIFFERENCE TESTS

Sensory difference tests are used in situations in which there are two types of stimuli, say A and B, and the investigator wants to know whether these stimuli are sensorily discriminable with respect to a particular sensory effect or class of sensory effects, that is a particular taste, odour, flavour, or any other type of sensory experience. When stimuli are randomly drawn from their respective sample A and B, pairs of A and B can be presented to one or more subjects with the instruction to select the stronger (sweeter, more pungent, harder, etc.) stimulus in each pair, and to make a guess when the subject is in any doubt or when no difference between the stimuli can be perceived. This procedure is the simplest sensory difference test and is known as the *duo test*. In actual fact it is a two-stimulus rank order method, since the sensory attribute or dimension with which the subject is concerned has been previously specified by the experimenter.

It is not always possible to specify such an attribute. In most studies in which the stimuli are foods, and not artificial stimuli entirely manipulated by the experimenter, it is impossible to specify *a priori* the sensory attribute of stimuli A and B which might differ. In such a situation the experimenter is forced to use another procedure. The most suitable alternative is the *duo–trio test*. In this procedure in each trial the subject is first presented with a standard stimulus. This reference stimulus may be either a stimulus A or a stimulus B. The subject is then presented with a pair of stimuli A and B with the instruction to select the stimulus from the pair which is more different from the standard stimulus than the other one of the pair.

Instead of using the duo–trio test, the experimenter could use another procedure known as the *triangle test*. With this procedure, sets of three stimuli are prepared: half are composed of two stimuli A and one stimulus B; the other half are composed of two stimuli B and one stimulus A. In each trial, the subject is presented with a tri-stimulus set ('triangle') with the instructions to select the 'odd' stimulus, that is to identify the stimulus that is different from the other two. The sensory nature of 'oddity' is not specified, so that it is left to the subject to identify the sensory attribute(s) which is relevant. As in the other tests, the subject must select a stimulus from each triangle presented even though he may not be able to differentiate between the three stimuli.

On the basis of frequency of use, the main types of sensory difference tests are the duo, duo–trio and triangle tests. These tests are also the simplest of the whole group of sensory difference tests which in principle is infinite. For example, it is possible to present a subject with a set of four stimuli with the instructions to select the odd stimulus. This means that sets of stimuli AAAB and BBBA are prepared and presented. Subjects could equally well be presented with a set of four stimuli with instructions to select two pairs of stimuli which are equal. In this case, sets of stimuli of the type AABB are prepared. Both types of the four stimulus procedure are variations of the *tetrad test* (Renner and Roemer, 1973).

Recently, the basic principle of the triangle test has been extended by increasing the number of stimuli from which the odd stimulus has to be selected (Basker, 1980). Instead of selecting the odd stimulus out of three stimuli, it has to be selected out of n stimuli. For example, in an hexagonal test, sets of six stimuli are presented in each trial. Each set contains five stimuli A and one stimulus B, or five stimuli B and one stimulus A. Again, there are several other variations of the hexagonal test. For example, sets of six stimuli can be prepared each containing two stimuli A and four stimuli B, or four stimuli B and two stimuli A. In this test, the subject is not instructed to select the odd stimulus, but to identify the two stimuli which are different from the other four. These and similar paradigms are all forms of what has been called polygonal testing (Basker, 1980).

More than two types of stimuli can be used in a set of stimuli. For example, a set of AABBCC may be prepared and presented to a subject with the instructions to select three pairs of identical stimuli. Many variations based on the same principle are possible with the result that complex paradigms and experimental tasks can be devised. In general, where there are more than two types of stimuli in one set, the test should be classified as a form of polyhedral testing (Basker, 1980). These types of tests are not considered in this chapter.

Even though Basker has recently discussed the use of oddity instructions in combination with tests other than the triangle test, this option has already been considered by Lockhart (1951). This author discussed the basic combination rules which lead to the generation of a series of sensory difference tests. In addition to this he tabulated the chance probabilities of obtaining a correct response as a result of guessing for a number of tests. Since his table contains some errors, a revised one is given in Table 1.

TABLE 1

PROBABILITIES OF A CORRECT RESPONSE AS A RESULT OF GUESSING FOR A NUMBER OF SENSORY DIFFERENCE TESTS. THE NUMBER OF STIMULI IN EACH OF THE TWO SUBSETS IS DENOTED BY a AND b RESPECTIVELY, SO THAT $a + b = n$ IS THE TOTAL NUMBER OF STIMULI IN THE SET PRESENTED IN EACH TRIAL

b	a	1	2	3	4	5	6
1		1/2	1/3	1/4	1/5	1/6	1/7
2		—	1/6	1/10	1/15	1/21	1/28
3		—	—	1/20	1/35	1/56	1/84
4		—	—	—	1/70	1/126	1/210
5		—	—	—	—	1/256	1/462
6		—	—	—	—	—	1/924

All sensory difference tests are forced-choice procedures, so that 'equal' responses (sometimes called a neutral vote) cannot be obtained. This type of instruction is given on the basis of the assumption that subjects will choose a stimulus completely at random from a number of alternatives even when it is impossible to discriminate between these stimuli. As a result of the forced-choice conditions, the responses made can be classified as either correct or incorrect, except in the duo test in which responses are given as A is stronger than B, or B is stronger than A. Given the number of alternatives in the set of stimuli and given the composition of the set of stimuli, the theoretical probability of a correct response can easily be determined. This probability is equal to the probability of obtaining a correct response as a result of a guess on the part of the subject (see Table 1). For example, in a three out of seven test, the probability that the three stimuli A are correctly selected from a set AAABBBB, or that the three stimuli B are correctly selected from a set BBBAAAA, on the basis of chance is: $3/7 \times 2/6 \times 1/5 = 1/35$. Thus, if the subject is not able to discriminate between the seven stimuli in each set, the probability is 1/35 that he selects the stimuli purely by guessing. If the subject is presented with, for example, 105 of these sets of stimuli, the expected number of correct responses is three. When, after the experiment, it is found that a subject has responded correctly in 15 out of the 105 trials, the decisive question arises as to whether the difference between the expected three correct responses and the obtained 15 correct responses is due to chance; that is, is this difference the result of sampling variability? This question should be answered using a statistical test. In this discussion it is sufficient to note that the most suitable test is the binomial test (on proportions), although the chi-square test (on raw frequencies) can also be used. In order to facilitate the evaluation of experimental results, several statistical tables have been published (Roessler et al., 1948, 1956, 1978; Basker, 1980).

One aspect of the use of statistical tests on sensory difference test data which requires further attention is that two types of error can occur. The first type of error called Type 1, or α error is the probability that the alternative hypothesis (H_a) is accepted when in actual fact the null hypothesis (H_0) is true. This probability is called the significance level of the test. It is convention to adopt a significance level of 0·05 or 0·01. The Type 2, or β error is the probability that the null hypothesis is accepted when in fact the alternative hypothesis is true.

With the use of sensory difference tests in sensory analysis it has become practice to consider only the Type 1 error and to leave the Type 2 error out of consideration. In fact, it is common when planning a study that the experimenter starts by determination of the number of trials, denoted as N. When the experiment has been completed, he determines a certain value for α. If the probability as determined by the statistical test is equal to or less than the significance level adopted it is concluded that the null hypothesis has to be rejected. On the other hand, if the obtained probability is larger than the significance level adopted, the null hypothesis should be accepted. In the first case the conclusion is drawn that the stimuli **A** and **B** are sensorily different; in the latter case it is concluded that these stimuli are not sensorily different. It is clear that the Type 2 error is extremely important, since accepting the null hypothesis does not automatically imply that it is true.

Radkins (1957) has pointed out the risks involved in this bad, but widely accepted, statistical practice. He has suggested that an experimenter should begin by specifying the magnitude of the Type 1 and 2 errors which he is prepared to accept, instead of starting by specifying N, the number of trials in the experiment. By specifying α and β, the number N is fixed. This follows from the fact that both α and β are dependent on N. It should also be recognised that Type 1 and Type 2 errors are mutually dependent. Given a particular value of N, specification of a small Type 1 error results in a large Type 2 error, and vice versa.

From the foregoing it is clear that sensory difference tests are based on statistical comparison of the distribution of correct and incorrect responses and the theoretical distribution that can be expected when the subject's responses are completely random. Therefore it can be concluded that these procedures are not based on principles of sensory perception, but on principles of guessing behaviour in combination with the basic rules of statistical probability theory. For this reason sensory difference tests have been called tests and not methods.

4. SENSORY PERCEPTION, PSYCHOPHYSICS AND PSYCHOMETRICS

Strictly speaking, the only conclusion that can be drawn from a statistically significant result obtained by a particular sensory difference test is that there must have been some factor that has caused the non-random response pattern. The statistical test does not allow for the conclusion that this factor must be a difference in a particular sensory attribute, and that no other factor can have been the causal agent. Non-random response distributions can be obtained, for example, in experiments where the necessary precautions have not been taken, as a result of temporal or spatial position preferences or stimulus code biases. The reverse may also be true. When a statistical test does not show a significant result, it cannot be logically inferred that the stimuli were not sensorily discriminable. Other factors may mask this effect. For example, sensory adaptation, fading of sensory traces in short-term memory, inconsistent decision-making, systematic differences between subjects when more than one subject is used, and lack of changes in motivation of the subjects may contribute to a distribution that cannot be distinguished statistically from a random distribution. Indications that factors such as these are operating cannot be obtained from a sensory difference test. Other experimental procedures must be used if these factors are to be investigated.

The statistical testing approach should be replaced by a psychological measurement approach if sensory and judgement processes occurring during the subject's performance of a sensory difference test are to be taken into account. In order to explain this, the chain of causation leading to sensory perception should be discussed. Naturally, every sensory process begins with the presence of a stimulus object. Energy released from this object strikes the receptive sites of a particular sensory system. For example, the stimulus 'object' may be a particular concentration of a volatile compound in the air inhaled by a subject. Some of these molecules will reach the olfactory receptor cells situated in the olfactory epithelium. As a result of the interaction between molecules and receptor sites, neural impulses are generated in the sensory fibres. In general, although not directly, fibres of all senses project to some part of the cortex, where the physiological activity is converted into a sensory perception. This is considered to be a complex element of the consciousness of the perceiving subject. On the basis of the nature of this perception, the subject may be motivated to make a motor or verbal response through the efferent part of the nervous system.

From the point of view of the study of sensory perception, there are three main elements in this input–processing–output chain. These are: the stimulus object; the sensory perception; and the sensory response evoked. In addition, there are two types of relationships which are of special interest. The first is the *psychophysical* relationship between the physical attributes of the stimulus object and the sensory characteristics of the perception which results from perceiving the stimulus object. The second is the *psychometrical* relationship between the sensory characteristics of the perception and the overt response behaviour of the subject. It is clear that the study of psychophysical and psychometrical relationships is complicated because one of the entities considered in such a relationship is not directly observable. The sensory perception resulting from a particular stimulus can only be observed by the subject himself. The other two elements, the stimulus object and the response behaviour, are directly observable. Physical attributes of the objects can be readily measured by instrumental methods, and parameters of response behaviour can be assessed by observation or by other means. The conclusion which can be drawn from the fact that sensory perceptions are private events is that all types of sensory measurement must be classified under the heading of derived measurement.

Although not often stated, there are two different types of measurement in sensory analysis and psychophysical research. Psychophysical measurement aims at relating one or more physical attributes of the stimulus to the attributes of the internal perception. Elements of a complex sensory perception are often called sensations and are thought to be basic structures of perception. It is clear that psychophysical models relate the input part of the input–processing–output sequence to the processing part of that sequence. In contrast to psychophysical models, psychometrical models relate parameters of sensory response behaviour to the attributes of the internal sensory perception resulting from sensory stimulation; that is the outer part is related to the processing part of the sequence mentioned before.

In the following section, the use of sensory difference tests for psychometrical measurement of sensory discriminability of stimuli and the use of these procedures for the psychophysical scaling of sensory discriminability will be discussed. However, the classical psychophysical approach to the assessment and scaling of sensory discriminability should be considered first.

5. THE PSYCHOPHYSICAL RELATIONSHIP BETWEEN STIMULUS DIFFERENCE AND SENSORY DISCRIMINABILITY

The basic problem of the measurement of sensory discriminability was first encountered in the 19th century when psychophysics was established. Research on perception was carried out with simple stimuli which could be completely manipulated by the experimenter. These stimuli were pure auditory signals, solutions of a particular compound, pure odorants, etc. The use of simple stimuli makes it possible to control the experimental conditions, a basic requirement when attempting to determine regularities in perceptual processes. However, it is often difficult to extrapolate the results of these investigations to problems with complex stimuli such as food.

The problem of the measurement of sensory discriminability can best be approached by means of a simple example based on a few assumptions. Assume that the internal sensory effect in a human observer resulting from stimulation by a particular stimulus S_s is always identical to a fixed value I_s. Further assume that there is only one particular attribute of the stimulus S_s that contributes to the arousal of I_s. By allowing another stimulus S_j to decrease in physical strength so that it approaches the strength of stimulus S_s, the corresponding sensation I_j will approach I_s. If the values I_s and I_j were directly measurable, a simple relationship between $S_{js}(=S_j-S_s)$ and $I_{js}(=I_j-I_s)$ could be established because S_{js} can be simply expressed in physical units. Such an expression is a simple general description of the functional relationship between a unidimensional stimulus difference and the corresponding sensory difference or sensory discriminability.

However, as already stated, the foregoing contains two unrealistic assumptions. The first is that the values of I_s and I_j are fixed, and the second is that these values are directly measurable. In actual fact, both I_s and I_j do vary on repeated presentations and these internal sensory responses cannot be measured directly. Weber (1846) tried to bypass both problems in one step by defining the concept of 'just-noticeable-difference' (JND). Two sensations I_j and I_s were defined as differing by JND when one of the sensations was judged to be stronger than the other in exactly 75% of the trials in which the stimuli S_s and S_j were presented in pairs to the subject. Thus, although the means of the internal sensory responses remained unknown, they were defined as being 1 JND different in such cases. For every stimulus S_j it is possible to find another stimulus S_n which is exactly 1 JND stronger. Therefore it should be possible to establish a simple psychophysical function between I_{js} expressed in JNDs and S_j expressed in physical units. However, there is a complication. In search of a general principle on which to base the construction of this function it was found that a sensory difference of the magnitude of 1 JND is not only dependent on the absolute stimulus difference corresponding to the sensory difference, but is also dependent on the absolute level of the stimulus difference. A particular stimulus difference at the lower part of the stimulus continuum results in a larger sensory difference than the same magnitude difference at the higher part of the stimulus continuum. Weber (1846), therefore, introduced a correction for the absolute level of the stimulus difference. He determined the fraction S_{js}/S_s which is equal to 1 JND. From the evidence available at the time, he concluded that this fraction was constant over a large part of the range of a particular stimulus. Thus, the principle which he extracted from this finding was that in order to obtain a sensory difference of the size of 1 JND a certain stimulus had to be increased in physical strength by a constant fraction of its own magnitude.

The next step in the development of a general stimulus-anchored psychophysical law was made by Fechner (1966). His basic assumption was that as all JNDs are psychologically of equal magnitude, a JND could be used as a psychological unit of measurement for the discriminability of stimuli. If the stimulus S_j is judged to be stronger than S_s in 75% of trials, and the stimulus S_n is judged to be stronger than S_j in 75% of trials, then S_j is 1 JND stronger than S_s, and S_n is 2 JNDs stronger than S_s. Thus with simple integration of the Weber fraction, a psychophysical function can be obtained as follows:

$$I = k \log S + c \qquad (1)$$

In this equation I denotes the sensory magnitude expressed in JNDs, S is the physical stimulus intensity, and k and c are unknown constants to be estimated. This equation, which represents the Fechner Law, relates the internal sensory continuum to the external stimulus continuum.

6. THE PSYCHOMETRICAL RELATIONSHIP BETWEEN OVERT RESPONSE BEHAVIOUR AND SENSATION

The approach followed by Weber (1846) and Fechner (1966) can be characterised as unidimensional because they both used stimuli that differed only in the degree in which they possessed a certain physical attribute. On the other hand, there was also only one type of sensation which corresponded to the physical attribute. Sensations resulting from different stimuli were assumed to be potentially different with respect to their mean sensory intensities and not with respect to their sensory qualities. The Fechner function represents a one-to-one correspondence between one physical and one sensory continuum.

The applicability of the Fechner Law in sensory analysis is extremely limited because of several factors. First, most foods and beverages are complex from the point of view of physical chemistry and must be described in terms of many physicochemical attributes. In addition, physical differences between products of a similar type are mostly not restricted to one particular physical attribute. Secondly, the complexity of a food as a sensory stimulus presents another problem. It is rarely possible to identify one particular physical attribute which corresponds over a large range to one particular sensation of a specified quality. For example, in almost all odour mixtures it is impossible to identify one single compound as being the causal agent of the sensory quality of the mixture. Thus it can be concluded that an approach such as that of Fechner to the study of sensory discriminability of complex stimuli is not possible.

Awareness of these limitations led to the development of a response-oriented approach to sensory measurement. Thurstone (1927) defined one of the main concepts, the *discriminal process*. Each stimulus presented to a subject can evoke a discriminal process on a psychological continuum (e.g. sensory continuum). Without specifying its precise nature and origin the discriminal process was defined as a means '*by which the organism identifies, distinguishes, discriminates or reacts to a stimulus*'. No postulates were made about the relationship between the physical stimulus attributes or sensory physiological processes and the nature of the discriminal process. It is essentially a psychological concept. Due to a number of influences, for example of motivational or neural origin, the discriminal process is subject to variation. Thus a particular stimulus S_s gives rise to a discriminal process x_s. On repeated observations of stimulus S_s, the corresponding discriminal process varies and this is the reason for the variation in the behavioural response, as discussed earlier. Thurstone

(1927) referred to the variation in a particular discriminal process as *discriminal dispersion*. It may be assumed that the variation in the discriminal process can be described by a normal distribution with mean μ_s and variance σ_s^2. A similar stimulus S_j can give rise to another distribution of discriminal processes when this stimulus is repeatedly presented to a subject. The mean of the distribution of x_j can be denoted as μ_j and its variance by σ_j^2. When a pair of stimuli is presented to the subject, two discriminal processes will occur, one for each stimulus. The momentary values of these processes, x_s and x_j respectively, can be considered as having been randomly drawn from the normal distributions $N(\mu_s, \sigma_s^2)$ and $N(\mu_j, \sigma_j^2)$ (see Fig. 1). If the subject has been instructed to select the stronger

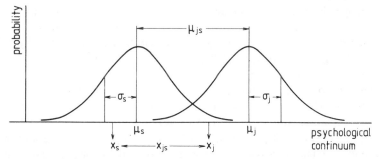

FIG. 1. Discriminal dispersions for stimuli S_s and S_j. The means of the distributions are μ_s and μ_j, and the standard deviations are σ_s and σ_j. The difference between the two means is μ_{js} which represents the psychological difference between S_s and S_j. The difference between two momentary discriminal processes x_j and x_s is denoted by x_{js}.

stimulus of the pair of stimuli with respect to a stated attribute (as in the duo method), the stimulus S_j will be selected as the stronger of the pair if $x_j > x_s$. As both variables x_j and x_s are normally distributed, the difference x_{js} between pairs of momentary values x_s and x_j is also normally distributed. The variable x_{js} has a mean μ_{js} and a variance that is equal to

$$\sigma_{js}^2 = \sigma_j^2 + \sigma_s^2 - 2r\sigma_j\sigma_s \tag{2}$$

Equation (2) contains three unknown parameters: the variances σ_j^2 and σ_s^2, and the coefficient of correlation r. In order to obtain a solution, some additional simplifying assumptions should be made. First, it is assumed that the discriminal processes generated by the stimuli S_j and S_s are mutually independent. The consequence of this assumption is that the covariance term in eqn (2) will disappear because r has become equal to zero. Secondly, it is assumed that the variances of all discriminal processes generated by a class of similar stimuli S_i are equal. This implies that σ_j^2 and σ_s^2 are equal, and that, therefore, these variances can be set equal to 1. The ultimate result is that the variance of the distribution of discriminal differences, as expressed by eqn (2), becomes equal to 2. Since the mean difference between the values x_j and x_s is identical to the difference between their means, it follows that $\mu_j - \mu_s = \mu_{js}$. Finally it can be shown that

$$z_{js} = \mu_{js}\sqrt{2} \qquad (\mu_{js} > 0) \tag{3}$$

In eqn (3) the mean discriminal difference is expressed simply in a standard normal deviate z_{js}. The unknown parameter μ_{js} can be assumed to be a measure of the mean sensory effects resulting from the stimuli S_s and S_j, respectively. A particular value of μ_{js} is obtained when the corresponding value z_{js} is known. As will be shown, this value has to be estimated from the distribution of responses obtained experimentally. It is essential to note that z_{js} is a behavioural parameter and that μ_{js} is a psychological parameter, i.e. a sensory parameter.

The importance of eqn (3) is that it relates the internal sensory continuum to the external behavioural continuum. A particular value of z_{js} can be obtained from a set of data experimentally (see Bock and Jones, 1968, pp. 19–21). If a pair of stimuli S_s and S_j is presented as a pair in a number of trials, the final experimental data will consist of two binary proportions (Fig. 2). These are the proportions of trials P_{sj} in which x_s generated by perception of the stimulus S_s was judged to be stronger than the corresponding value x_j evoked by S_j and the complementary proportion P_{js} ($P_{sj} + P_{js} = 1$). Using the cumulative distribution function

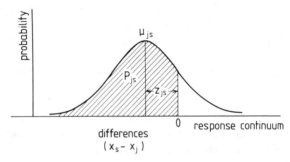

FIG. 2. Normal distribution of differences between discriminal processes. The proportion P_{js} denotes the proportion that $x_j > x_s$ as determined in a particular experiment with the stimuli S_s and S_j. The distance z_{js} has been derived from P_{js}.

of the standard normal distribution $N(0, 1)$, the values of z_{sj} and z_{js} corresponding to each of these proportions can be obtained by consultation of the appropriate table (e.g. Bock and Jones, 1968; Finney, 1971). These values are numerically identical but different in sign, except when they are equal to zero.

Measurement of discriminability of two stimuli is impossible if these stimuli are never confused by the subject. If there is never confusion it means that the sensory process x_j generated by S_j is greater than x_s caused by S_s in all trials. Then the proportion $P_{js} = 1$ and the proportion $P_{sj} = 0$. These proportions correspond with standard normal deviates of $+\infty$ and $-\infty$ respectively. In general, this prerequisite is not a problem since there is no reason to carry out the experiment if the two stimuli are *a priori* readily discriminable.

As Thurstone's approach to the measurement of psychological discriminability is response-oriented, the procedures developed can be applied to all stimuli for which a common psychological attribute can be specified. In contrast to the stimulus-oriented approach, with this approach a physical attribute which corresponds to the psychological continuum of interest is not required. However, sometimes the experimenter has available a series of stimuli S_s, S_j ($j = 1, 2, \ldots, n \neq s$) which differ with respect to the degree of a particular physical attribute they possess. This situation is identical to those created by the early psychophysicists, as already discussed. In such a situation the psychometrical approach can be extended to a psychophysical one in order to study the functional relationship between sensory discriminability of the stimuli and their physical difference (or ratio). If the standard stimulus S_s has been presented together with each of the stimuli S_j ($j = 1, 2, \ldots, n \neq s$), a series of estimates z_{js} and the corresponding values μ_{js} can be determined. By simple linear regression, a psychophysical function can be constructed of the following general form:

$$\mu_{js} = k \log (S_j/S_s) \qquad (4)$$

in which μ_{js} is the magnitude of the sensory difference between the stimuli S_j and S_s; S_j and S_s are physical stimulus values; and k is the regression coefficient to be estimated. Note that this function is fitted through the origin. This is done because if S_j is set equal to S_s, then $\log (S_j/S_s)$ is equal to zero, and the corresponding value of μ_{js} is also equal to zero. This latter value is to be expected because two physically identical stimuli should be sensorily indiscriminable, which means the sensory difference must be equal to zero.

7. THE MEASUREMENT OF SENSORY DIFFERENCES USING THE DUO, DUO–TRIO OR TRIANGULAR METHOD

From the brief discussion of the main concepts in Thurstone's theory of paired comparisons (1927) it can be seen that the description is fully applicable to what is called the duo method in sensory analysis. It is possible to analyse the duo–trio and the triangular method in a similar way. Sensory and judgement processes involved in these tasks can be described in terms of the same basic concepts. In this section a change in notation is made; the sensory difference $\mu_j - \mu_s = \mu_{js}$ is denoted as d' ($d' \geq 0$).

In the duo–trio method the subject is presented first with a standard stimulus, either S'_s or S'_j. As a consequence, a discriminal process with a given value of either x'_s or x'_j is generated. The subject is then presented with a pair of stimuli S_s and S_j. These stimuli generate two discriminal processes with the values x_s and x_j respectively. The subject will correctly select the stimulus different from the standard if

1. $|x_s - x'_s| < |x_s - x_j|$, when the stimulus S_s has been selected as the standard stimulus, or
2. $|x_j - x'_j| < |x_j - x_s|$, when the stimulus S_j has been used as the standard stimulus.

In the triangular method, the subject is presented with a triangle composed of two stimuli S_s, denoted as S_s and S'_s respectively, and one stimulus S_j; or alternatively, the triangle may be composed of the stimuli S_j, S'_j and S_s. As in the duo–trio method three discriminal processes are generated. The odd stimulus will be correctly selected if

1. $|x_s - x'_s| < |x_s - x_j|$ and $|x_s - x'_s| < |x'_s - x_j|$, when the triangle presented contains the stimuli S_s, S'_s and S_j, or
2. $|x_j - x'_j| < |x_j - x_s|$ and $|x_j - x'_j| < |x'_j - x_s|$, when the triangle is composed of the stimuli S_j, S'_j and S_s.

Both values x_j and x'_j (or x_s and x'_s) are drawn from the same normal distribution with mean μ_j and variance σ_j^2. The value of the third discriminal process x_s (or x_j) has been drawn from the other distribution with mean μ_s and variance σ_s^2. The means of the distributions are the distance d' apart.

Given these internal representations of discriminal processes and the decision rule the subject uses, which is specific to each method, it is possible to specify the relationship between the probability of a correct response, P_c, and d', the measure for sensory discriminability or sensory difference

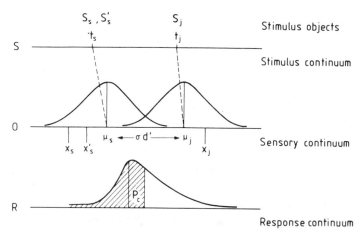

FIG. 3. Stimulus–Organism–Response (S–O–R) model for the triangular method and 3-AFC procedure. The stimuli S_s, S'_s and S_j taken here as an example (without loss of generality) have physicochemical values t_s, t'_s and t_j, respectively. On a particular trial these stimuli give rise to the sensations x_s, x'_s and x_j. In fact the sensations are random values drawn from the respective distributions $f(\mu_s)$ and $f(\mu_j)$. These distributions have equal variance σ^2, and their means μ_s and μ_j are located $\sigma d'$ from each other. Given the Triangular or the 3-AFC Method the value of d' can be obtained from a proportion of correct responses resulting from a number of trials. The parameter d' represents a measure for sensory discriminability or sensory difference. (Reprinted from *Journal of Food Science*, 1982, **47**, 139–43. Copyright © by Institute of Food Technologists.)

(Fig. 3). When using the duo method, the probability that $x_j > x_s$, and the probability that $x_s > x_j$ can be classified as P_c and P_{inc} respectively, if it is known prior to the experiment that $\mu_j > \mu_s$. This knowledge can be derived from the psychophysical relationship between the physical attribute of the stimulus and the sensory attribute as specified in the instructions; the stimulus with the higher physical value should have a higher value of μ.

If the proportion of correct responses obtained in the experiment is used as an unbiased estimate of P_c, the following equations can be specified on the basis of the same assumptions as made in the previous section:

Duo method:

$$P_c = \Phi(d'/\sqrt{2}) \qquad (5)$$

Duo–trio method:

$$P_c = 1 - \Phi(d'/\sqrt{2}) - \Phi(d'/\sqrt{6}) + 2\Phi(d'/\sqrt{2})\Phi(d'/\sqrt{6}) \qquad (6)$$

Triangular method:

$$P_c = 2 \int_0^\infty \{\Phi(-u\sqrt{3} + d'\sqrt{(2/3)}) + \Phi(-u\sqrt{3} - d'\sqrt{(2/3)})\}$$
$$\times \exp(-\tfrac{1}{2}u^2)/\sqrt{2\pi}\,du \quad (7)$$

In the psychometric functions given by eqns (5) to (7), Φ is the cumulative distribution function of the standard normal distribution $N(0, 1)$. With these equations the sensory difference, expressed as d', can be estimated from P_c, the proportion of correct responses in an experiment. In order to facilitate conversion of a particular value of P_c into the corresponding value of d', tables of d' have been produced by Ura (1960), David and Trivedi (1962) and Bradley (1963). More recently expanded tables of d' for the triangular method have been published by Frijters (1982).

The parameter d' can be estimated by each of the three sensory difference methods. Theoretically, the value obtained for d' should be independent of the method used; that is, it should not make any difference which method is used to determine d'. While present data are not available on the equality of the three methods for estimating a particular value of d', research has been carried out on the stability of d' when its value has been estimated independently by other procedures, and is discussed in a later section.

The sensitivity measure d' is of a parametric nature; it has been derived on the assumptions that the underlying sensory distributions are normal, and that these distributions have equal variances. Signal detection theory provides the theoretical possibility of determining the value of d' in the case of unequal variances, but when the distributions depart seriously from normality it is more appropriate to use a non-parametric measure such as $P(A)$ (Pollack et al., 1964). The use of a non-parametric measure related to $P(A)$, which has been called the 'R-index' (Brown, 1974), has been strongly advocated by O'Mahony (1979). This author proposed the use of non-parametric measures because in his opinion determination of values of d' is a complex and time-consuming affair. This view is shared by others, for example Brown et al. (1978), but has recently been challenged by Morrison (1982). Further discussion of non-parametric measures is not possible within this chapter, but the interested reader is referred to the sources given.

8. MULTIPLE ALTERNATIVE FORCED-CHOICE PROCEDURES AND THE THEORY OF SIGNAL DETECTION

The principles of Thurstone's theory of paired comparisons (1927) are rather similar to those of the statistical decision theory for sensory measurement, known as the theory of signal detection (Green and Swets, 1966). The latter has been developed by auditory and visual psychophysicists quite independently of Thurstone's ideas and concepts. Discussion of this theory, however, will be restricted to the measurement of sensory discriminability using multiple alternative forced-choice procedures.

In the duo method, a subject is presented with a pair of stimuli with the instruction to select the stronger stimulus with respect to a particular sensory attribute. The subject could also be presented with three stimuli with the directed instruction to select the strongest stimulus of the three. For example, if a subject is presented with a three-stimulus set composed of two stimuli of weak sugar solutions and one stimulus of a stronger solution, the appropriate instruction would be to select the sweetest stimulus. The subject could also be presented with a three-stimulus set composed of two stimuli of distilled water (blanks) and one stimulus of a solution of sugar in water. In this case the appropriate instruction is to select the sweet stimulus. Both types of instruction are directed. Instead of three stimuli, the number of stimuli in a set can be increased to four, or five, or m. Discrimination paradigms in which the subject has to identify one stimulus out of a set of m stimuli under directed instructions are thus logical extensions of the duo method. In an m-alternative forced-choice procedure (m-AFC), the subject has to select out of m stimuli the strongest stimulus S_j, or alternatively the stimulus that contains the 'signal'. The other $m - 1$ stimuli S_s are physically identical. In spite of the fact that all these stimuli S_s may be readily perceived they are referred to as 'noise' in the language of signal detection theory. Thus, these stimuli do not necessarily need to be blank stimuli.

When the sensory and judgement processes are expressed in the notation used in the previous section, a stimulus S_j in a set of m stimuli generates a normally distributed internal sensory response x_j. Each of the remaining stimuli S_s elicits a sensation of a particular value x_s, so that $m - 1$ responses occur. Each value x_s of the $m - 1$ values is likely to be different since the variable x_s is continuously distributed.

It is assumed that in a particular m-AFC procedure a correct response is produced by the subject if, in a given trial, the momentary value x_j is larger than each of the $m - 1$ momentary values of x_s. In other words, the subject will identify the stimulus S_j as the strongest, if the corresponding sensory value x_j is larger in that trial than the largest value of each of the values x_s generated by the other stimuli S_s ($x_j > x_{smax}$). It will be obvious that the value of x_{smax} will increase when m, the number of stimuli in the set, increases. The ultimate consequence is that the probability of a correct response will decrease, while the value of d' remains constant (see Fig. 4).

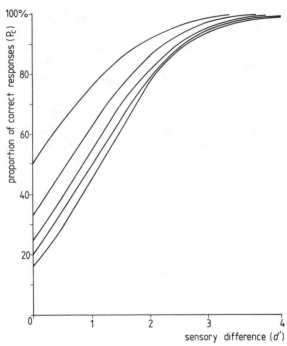

FIG. 4. Functional relationship between proportion of correct responses P_c and the sensory difference d' for a number of multiple alternative forced-choice procedures. From the top to the bottom the functions are given for $m = 2, 3, 4, 5$ and 6 respectively.

A general principle can be given for the relationship between the number of alternatives in an m-AFC procedure and the variable d', the measure for sensory difference between stimuli S_s and S_j. Again, let the means of the normally and independently distributed variables x_j and x_s be equal to μ_j and μ_s and let their variances σ_j^2 and σ_s^2 be equal so that these variances can be set equal to 1. The magnitude of sensory difference between stimuli S_j and S_s is equal to $d' = \mu_j - \mu_s$ ($\mu_j > \mu_s$). Under these assumptions the general equation that relates the probability of a correct response P_c to d' can be specified as follows:

$$P_c = \int_{-\infty}^{\infty} \Phi^{m-1}(u)\phi(u - d')\, du \qquad (8)$$

In eqn (8), Φ is the cumulative distribution function of the standard normal distribution and $\phi(u - d')$ is the normal density function with mean d'

above the mean of the other function. The number of alternatives is denoted by m.

Equations of this type must be solved numerically. This was done by Elliot (1964) who produced tables of d' for seven m-AFC procedures ($m = 2, 4, 16, 32, 256$ and 1000). On the basis of the same calculation procedure as used by Elliot, expanded tables of d' have been published by Hacker and Ratcliff (1979). They have produced tables for $m = 2$–$12, 16, 24, 32, 256$ and 1000. If the proportion of correct responses obtained in an experiment is taken as an unbiased estimate of P_c, the value of d' corresponding to a particular value of P_c can be obtained from these tables.

As already stated, a true measure of sensory difference or sensory discriminability should be independent of the particular method used to obtain this value. However, this basic requirement has not been extensively investigated. Recently (Frijters, 1980) it was shown that d' can be determined equally as well with the triangular method and the 3-AFC method. Since d' can also be determined with a single stimulus yes/no procedure, this procedure has been compared with a number of m-AFC procedures and the results obtained support the notion that the value of d' is independent of the method used for its determination (Swets, 1959; Swets et al., 1961; Green and Birdsall, 1964).

9. CONCLUDING REMARKS

In this chapter an attempt has been made to broaden the theoretical basis on which sensory difference tests can be evaluated, and to point out briefly how these procedures can be used for the psychometrical measurement of sensory discriminability in addition to their traditional use in testing statistical hypotheses. Both the psychometrical and psychophysical approach to sensory difference testing have been discussed. Especially, the differences in approach to the measurement of sensory discriminability between the classical psychophysical approach, which is stimulus-oriented, and the psychometric approach developed from Thurstone's theory, which is response-oriented, were discussed. The psychometric approach is also closely related to the measurement approach inherent in signal detection theory.

Recommendations on the use of sensory difference tests have not been made, with one exception. When a sensory difference test is used for testing a statistical hypothesis, the statistical evaluation should include consideration of the Type 2 error, in addition to the Type 1 error. Neglect of the

Type 2 error together with the use of a limited number of experimental trials could easily result in a substantial risk that a particular null hypothesis, stating 'no difference', is erroneously accepted.

Recommendations could have been made with regard to criteria for the use of a particular sensory difference test. However, discussion of such criteria has been omitted because these criteria are largely dependent on the aim of a particular study and the properties of the stimulus objects to be investigated.

In more complex sensory difference tests in which the number of stimuli in each set is large, for example in the three out of seven test, sensory adaptation that occurs within a trial can be quite large. This results in a decrease of sensitivity in the course of the trial. These effects can be large and can occur quickly especially for taste and smell, whereas recovery from adaptation is generally slow. The degree of adaptation which can occur in a particular trial is dependent on many factors, such as duration of stimulation of each stimulus, the order of perceiving the stimuli, the physical intensity of the stimuli, the period of time between the respective stimuli in the set, etc. In addition, the probability that residues of the stimulus are effective after the trial has been terminated increases with the number of stimuli. This is especially so when the stimuli have to be placed in the mouth, as in taste and texture studies.

Another effect which can easily occur in complex sensory difference tests is that the internal sensory responses fade when they have to be stored in the memory for a long period. In general, it can be stated that with an increase in the number of stimuli in a set, the average storage time increases. Further, the larger the number of coded internal sensory responses, the higher the probability that contrast and assimilation processes take place during storage in the memory.

As the sensory difference test becomes more complex, the decision making process becomes more complex too. A subject selects a certain overt response on the basis of the sensations stored in coded form in his short-term memory. With an increase in the number of stimuli in a set, the number of stored entities increases. This results in a higher probability that the subject will not correctly use the specific decision rule for the sensory procedure in use, which was generated by the instructions given before the experiment.

On theoretical grounds there is no need for the use of complex sensory difference tests. When the sensory attribute of relevance can be defined for the subject, a simple duo test (2-AFC) or a 3-AFC method can be used. If the attribute cannot be stated in the instructions, a duo–trio or triangle test can be applied. More complex sensory difference tests do not provide any other type of information.

The main argument for using complex sensory difference tests is based on higher statistical efficiency. As the probability of a correct response as a consequence of guessing decreases with an increase of the number of stimuli in a set, complex sensory difference tests must be preferred (cf. Amerine *et al.*, 1965). This argument is only valid when the procedure is to be used for testing of a statistical hypothesis and not when the procedure is to be used for the measurement of sensory discriminability. However, the question remains as to whether or not this is a real advantage. As pointed out, sensory and judgement bias can occur more easily in complex tests, and the final question as to whether the statistical advantage outweighs the negative effects of these biases remains.

REFERENCES

Amerine, M. A., Pangborn, R. M. and Roessler, E. B. (1965). *Principles of Sensory Evaluation of Food*, Academic Press, New York.
Basker, D. (1980). Polygonal and polyhedral taste testing. *J. Food Qual.*, **3**, 1–10.
Bock, R. D. and Jones, L. V. (1968). *The Measurement and Prediction of Judgement and Choice*, Holden–Day, San Francisco.
Bradley, R. A. (1963). Some relationships among sensory difference tests. *Biometrics*, **19**, 385–97.
Brandt, F. I. and Arnold, R. G. (1977). Sensory difference tests used in food product development. *Food Prod. Develop.*, **11**(8), 56.
Brown, D. G. W., Clapperton, J. F., Meilgaard, M. C. and Moll, M. (1978). Flavor thresholds of added substances. *J. Am. Soc. Brew. Chem.*, **36**, 73–80.
Brown, J. (1974). Recognition assessed by rating and ranking. *Brit. J. Psychol.*, **65**, 13–22.
David, H. A. and Trivedi, M. C. (1962). Pair, triangle, and duo-trio tests. Technical Report No. 55, Department of Statistics, Virginia Polytechnic Institute, Blacksburgh, Virginia.
Davidson, R. R. and Farquhar, P. H. (1976). A bibliography on the method of paired comparisons. *Biometrics*, **32**, 241–52.
Elliot, P. B. (1964). Tables of d'. In: *Signal Detection and Recognition by Human Observers*, J. A. Swets (Ed.), Wiley, New York.
Fechner, G. T. (1966). *Elements of Psychophysics*, Vol. 1 (Translated by H. E. Adler), Holt, Rinehart and Winston, New York.
Finney, D. J. (1971). *Probit Analysis*, 3rd edn, Cambridge University Press, Cambridge.
Frijters, J. E. R. (1980). Three stimulus procedures in olfactory psychophysics: an experimental comparison of Thurstone–Ura and 3-alternative forced choice models of signal detection theory. *Perception and Psychophysics*, **28**, 390–7.

Frijters, J. E. R. (1982). Expanded tables for conversion of a proportion of correct responses (P_c) to the measure of sensory difference (d') for the triangular method and the 3-alternative forced-choice procedure. *J. Food Sci.*, **47**, 139–43.
Green, D. M. and Birdsall, T. G. (1964). The effect of vocabulary size on articulation scores. In: *Signal Detection and Recognition by Human Observers*, J. A. Swets (Ed.), Wiley, New York.
Green, D. M. and Swets, J. A. (1966). *Signal Detection Theory and Psychophysics*, Wiley, New York.
Hacker, M. J. and Ratcliff, R. A. (1979). A revised table of d' for M-alternative forced-choice, *Perception and Psychophysics*, **26**, 168–70.
Lockhart, E. (1951). Binomial systems and organoleptic analysis. *Food Technol.*, **5**, 428–31.
Morrison, G. R. (1982). Measurement of flavour thresholds. *J. Inst. Brewing*, **88**, 170–4.
Muller, H. G. (1977). Sensory quality control: report on a survey. In: *Sensory Quality Control: Practical Approaches in Food and Drink Production*, H. W. Symons and J. J. Wren (Eds), Institute of Food Science and Technology and Society of Chemical Industry, London.
O'Mahony, M. A. P. D. (1979). Short-cut signal detection measures for sensory analysis. *J. Food Sci.*, **44**, 302–3.
Peryam, D. R. (1958). Sensory difference tests. *Food Technol.*, **12**, 231–6.
Pollack, I., Norman, D. A. and Galanter, E. (1964). An efficient non-parametric analysis of recognition memory. *Psychonomic Sci.*, **1**, 327–8.
Radkins, A. P. (1957). Some statistical considerations in organoleptic research: triangle, paired and duo–trio test. *Food Res.*, **22**, 225–34.
Renner, E. and Roemer, G. (1973). Der Tetraden-Test als aussagefähige sensorische Methode 1. Methodik und Anwendung. *Z. Lebensm. Untersuch. Forsch.*, **151**, 326–30.
Roessler, E. B., Baker, G. A. and Amerine, M. A. (1956). One-tailed and two-tailed tests in organoleptic comparisons. *Food Res.*, **21**, 117–21.
Roessler, E. B., Pangborn, R. M., Sidel, J. L. and Stone, H. (1978). Expanded statistical tables for estimating significance in paired-preference, paired-difference, duo–trio and triangle tests. *J. Food Sci.*, **43**, 940–3, 947.
Roessler, E. B., Warren, J. and Guymon, J. F. (1948). Significance in the triangle taste test. *Food Res.*, **13**, 503–5.
Swets, J. A. (1959). Indices of signal detectability obtained with various psychophysical procedures. *J. Acoust. Soc. Am.*, **31**, 511–13.
Swets, J. A., Tanner, W. P. and Birdsall, T. G. (1961). Decision processes in perception. *Psychol. Rev.*, **68**, 301–40.
Thurstone, L. L. (1927). A law of comparative judgement. *Psychol. Rev.*, **34**, 273–86.
Ura, S. (1960). Pair, triangle and duo–trio test. *Rep. Statist. Appl. Res.*, JUSE, **7**, 107–19.
Weber, E. H. (1846). Der Tastsinn und das Gemeinfühl. In: *Handwörterbuch der Physiologie*, Vol. 3, R. Wagner (Ed.), Vieweg, Braunschweig.

Chapter 6

SCALING AND RANKING METHODS

DEREK G. LAND

Taint Analysis and Sensory Quality Services, Norwich, UK

and

RICHARD SHEPHERD

Agricultural and Food Research Council, Institute of Food Research, Norwich, UK

1. INTRODUCTION

Scaling and ranking methods of measuring or comparing the sensory attributes of food and drinks, and of measuring attitudes or liking, are widely used in industry as quality control procedures, for product development, and in research (Muller, 1977). The methods, in their many varied forms, are almost certainly the most important element of the practical application of sensory measurement techniques. However, they cover a wide range, from the almost totally arbitrary to the sophisticated and complex. In common with all sensory measurements, they use the responses of people to controlled stimuli as measurements. These responses are a form of behaviour, use of which requires an input from the behavioural sciences. However, as most people frequently respond to food and drink, the operations of making judgements on appearance, texture, taste and smell are often regarded as not needing this behavioural science input.

Historically much confusion has arisen; methodology has been developed in one area, but largely used empirically in another, with a major communication and conceptual gap between them. These areas are the behavioural sciences, in particular perceptual psychology and psychophysics, from which most of the real conceptual and methodological development has come, and food science, which has little behavioural science background, where most of the demand for application lies. Over the last 15 to 20 years there has been a great improvement in communication and integration as reflected in the authorship of this book. In this chapter, the authors, who have food science with some behavioural science and behavioural science backgrounds respectively, are attempting to put into perspective, for food scientists and others involved in sensory measurement of food and drinks, the development and use of concepts and methods of measurement of perceived characteristics. It is by no means a comprehensive review; many of the references cited are illustrative and many others could also have been cited had more space been available.

One prerequisite for effective communication of ideas and concepts is to use words in an unambiguous and generally agreed manner. We therefore start by defining and discussing some of the more important terms used later in this chapter.

2. DEFINITIONS AND USAGE

Terms relevant to sensory measurement have been defined in the USA (ASTM, 1981), the UK (BS 5098, 1975) and internationally (ISO 5492, 1977–79, 1981).

Scaling (unqualified) has not been defined, but is used here to cover a range of methods to determine perceived magnitude, usually by means of scales. In psychology, scaling is often used to mean construction and calibration of a scale.

A scale has been defined as a graded arrangement, used in reporting assessments: it is divided into successive values, which may be graphic, descriptive or numerical. The successive values may be implicit, as in a graphic analogue scale, and the assessments are usually of magnitude. Scales may be unipolar (zero at one end) or bipolar (opposite attributes at each end), e.g. hedonic (like/dislike). The scale is the instrument used by the assessors to make explicit their perceptions. The scales have different levels (or 'strength' or power) of measurement, which may be classified into four divisions (nominal, ordinal, interval and ratio), the properties of which determine the methods of data analysis which may be validly used (Stevens, 1951, 1961).

A *nominal scale* specifies only class affiliation or identification and no more, i.e. the classes have no quantitative relationship (e.g. apples, pears, plums, etc., or sweet, sour, bitter). Strictly speaking, its use is classification or grading, not scaling.

An *ordinal scale* specifies more or less of an attribute or class without any defined quantitative implication about size. For example it may rank or order samples in increasing sweetness level. It may also specify amounts such as none, trace, small, medium, large, where there is no *a priori* information on the intervals between each category. It may also take the form of a line where only the ends (none: large or extreme) have verbal anchors. The appropriate measures for comparison are the median and percentiles, i.e. non-parametric statistics, unless the data are shown to have properties of a higher scale (see Chapter 9).

An *interval scale* specifies that successive categories or unit intervals on the scale are equal and that the origin is arbitrary, i.e. not a real zero. Many sensory scales used as interval scales, however, do have a zero, i.e. absence of a perceived attribute. A good, but non-sensory, example is the centrigrade scale for temperature. Many sensory scales are assumed to have interval properties, but it is rare for these properties to be demonstrated. The appropriate measures are the arithmetic mean and standard deviation and other parametric statistical techniques such as analysis of variance or t-tests.

A *ratio scale* specifies equal ratios between successive unit intervals and has a true zero. A non-sensory example is the Kelvin scale for temperature, where any values can be transformed in strict ratios of the numbers in degrees Kelvin ($x' = Ax$) without loss of information, whereas on the centigrade scale such a transformation takes the form $x' = Bx + C$. Note that the Kelvin ratio scale also has interval properties. Most ratio scale methods depend upon the ability of the assessor to follow instructions to construct and use a true ratio scale. This, for example, may take the form of: 'If a reference sample has a sweetness of 10, a sample one-tenth as sweet is called 1 and a sample ten-times as sweet is called 100, and so on'. It therefore follows that a zero score is inadmissible. The appropriate measure of comparison is the geometric mean. Parametric statistical techniques may be used to analyse the data.

Ranking is a method in which a series of three or more samples is presented at the same time and arranged in order of intensity or degree of some designated attribute. It is only an ordinal process, giving no direct information on the size of differences.

Rating is a method of classification into categories on an ordered scale, i.e. the positions on the scale are ordinal but the perceived differences between categories are not defined, although if they are spaced equally on a response sheet, they may be used by the assessor as if they are at equal intervals, and indeed the data may prove to have interval properties.

Scoring is a form of rating using a numerical scale where the numbers form an interval or ratio scale, i.e. the different scores have a defined and demonstrated mathematical relationship to each other. The term 'scoring' is frequently used erroneously where the scores have not been shown to have such a defined relationship.

The different scales or methods are always used by people (or subjects or judges) who have been defined as follows:

Assessors (unqualified) are any people taking part in sensory tests; *selected assessors* are those tested and chosen for their proven ability to carry out the particular test.

An *expert* is a person with considerable experience and proven ability in sensory assessment of a given product under specified conditions; experts rarely operate under normal conditions of sensory panels, but often work alone or in very small groups (e.g. tea or wine testers).

A *panel* is a group of assessors chosen to participate in a sensory test. In general, other than in hedonic tests where selected assessors should not be used, high discrimination, sensitivity and consistency in measurement depends upon the use of trained and selected assessors.

The above definitions have largely emerged from the practical application of sensory assessment in a food science context. However a rather different definition of scaling is used in the behavioural sciences, covering all means of sensory measurement including the use of sensitivity measures (differential thresholds or just noticeable differences—see below)—and difference tests (see Frijters, Chapter 5). This area of investigation in psychophysics has seen a strong development of theoretical models and hypothesis testing, an approach almost totally absent in the food science approach. The historical development of these concepts and methods is covered below and an attempt is made to show how and where this approach to methodology can and is being applied in the food science context.

The second important area of difference is in the development of scales. The vital importance of knowing the properties and limitations of a measuring instrument can hardly be denied by most natural scientists. There are a multitude of different scales for sensory measurement in use in food science, but very few of these have ever been validated; the few examples which can be cited (see below) have almost always been developed in extensive series of tests by behavioural scientists. It is a strange use of dual standards which cannot be allowed to continue.

The development of scaling methods in the behavioural sciences illustrates the way in which the use of concepts and models has been used to develop validated techniques. This involves some discussion of the historical background to scaling.

3. HISTORICAL BACKGROUND OF SCALING

The idea of scaling the sensation of a person against the physical intensity of stimulation goes back to Fechner (1860). He based such a scale on the idea that equal discriminability of stimuli meant equal difference in sensation (or equal psychological difference). Such a scale is based on the fact that when confronted with stimuli varying on some physical continuum (e.g. intensity of light) people would give more variable responses when the physical difference was small than when it was large. The size of difference necessary for a person to distinguish correctly the stimuli on a given percentage of trials can be measured. In 1834 Weber had shown that this difference was proportional to the overall intensity of the stimuli judged, and so at low intensities the difference is small, but as the overall intensity increases, the difference necessary for the stimuli to be distinguished increases. This is known as Weber's Law and holds reasonably well for middle range stimuli, but does not hold well for stimuli close to detection level or for very high intensity stimuli. The difference necessary for a person to distinguish correctly a pair of stimuli on a certain proportion of trials is called the just noticeable difference (JND). It is really a function of variability of the responses, being small when the assessor is both discriminating and consistent and large when he or she is not. It is the size of the physical stimulus difference necessary to give a specified error level in discrimination. Fechner, using mainly data obtained with himself as subject, assumed the size of this perceived difference was constant over the whole sensation range and constructed a scale of JNDs logarithmically related to the physical intensity of stimulation:

$$S = a \log(I/I_0)$$

where S is the magnitude of the sensation, I is the physical intensity of the stimulus, a is a constant and I_0 is the threshold value.

This method of developing a scale based on discriminability between stimuli was extended later by Thurstone (1927). In this type of procedure, stimuli are not presented singly for the subject to give a response based upon memory standards or experience, but are presented in pairs for comparison and the proportion of times each stimulus is judged greater on some dimension is measured. The dimensions need not relate to physical continua, e.g. attractiveness, seriousness of crimes, provided that the subject can order the stimuli along the dimension. Thurstone argued that each stimulus would give rise to a 'discriminal process' along some psychological continuum (the process by which the subject identifies or discriminates between stimuli), since there will be variation within subjects (due to motivation, physiological state, etc.) and between subjects; each individual stimulus if presented several times would give a 'discriminal dispersion' which would be a frequency distribution about the mean discriminal process (see Fig. 1).

Discrimination between the stimuli depends both on the separation of the means and the variance (or standard deviation) of the discriminal dispersions. If two stimuli are very close in comparison to the variances the subjects will frequently confuse them and judge the larger one greater on only slightly more than half the presentations. If, however, they are widely separated the subjects will judge them correctly on a much greater proportion of trials. Hence, knowledge of the number of occasions when subjects judge the stimulus order correctly gives a measure of the distance between the stimuli on the psychological continuum (see Chapter 5). In the general case this can only be done with knowledge of the two dispersions and the correlation between them using the mathematical relationship known as Thurstone's Law of Comparative Judgement. However, in general these values cannot be measured and so in practice the dispersions are assumed to have equal variance and to be uncorrelated; this is known as Thurstone's Case V model since Thurstone also put forward the relationship if other assumptions are made which then form the other cases.

In practice all the pairs of stimuli would be presented many times, and from the matrix of the probabilities that each stimulus of a pair would be

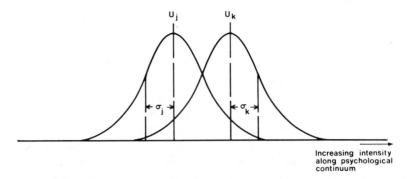

FIG. 1. Discriminal dispersions for two stimuli, j and k. Each stimulus can give rise to a range of different values along the psychological continuum centred on its mean value (u_j and u_k). On any individual trial there is the possibility that k will give rise to a lower value than j. The probability of this happening is a function of the overlap and so will depend on the difference between the means ($u_k - u_j$) and the standard deviations (σ_j and σ_k).

judged greater it is possible to calculate the distances between the stimuli in standardised (z-) scores (i.e. in terms of the number of standard deviations). Since the distributions are assumed to be normal, two stimuli would theoretically never be perfectly distinguishable, but in a real study with a finite number of comparisons this can happen and then the distance between such stimuli on the scale will be indeterminate. In scaling of this type the successive stimuli must be spaced closely in order to give sufficient variability in the responses for determination of the scale values.

Thurstonian scaling will give interval data since the relationships between the stimuli have been calculated and so some of the assumptions made in magnitude estimation and category scaling do not apply. However, assumptions do have to be made about the distributions which may not be valid. One major drawback in relation to indirect scaling methods is the large number of stimulus presentations required in order to construct the scale. This scaling procedure depends on the subjects being able to order the stimuli along some single dimension and in some cases, particularly with foods or drinks, this might not be easy to meet. An example of Thurstonian scaling will be presented later in the discussions of hedonic scaling. For more detailed discussion see Bock and Jones (1968).

Signal detection theory (SDT) (Green and Swets, 1966) is clearly related to Thurstonian scaling but includes cognitive factors in deciding judgements when stimuli are near threshold. As in Thurstone's model a stimulus gives rise to a normal distribution along some internal continuum, and the subject has to differentiate between overlapping distributions from stimuli of different magnitudes. The major difference between SDT analysis and those due to Fechner and Thurstone is that SDT assumes the subject's decision is influenced by a criterion fixed by expectancy and/or by the relative costs and benefits of missing real differences or deciding that there is a difference when there is none (i.e. false positives). The criterion can be calculated from the proportion of false positives and misses and a measure of sensory discrimination can be obtained which is free from response bias. As with Thurstonian scaling, certain assumptions need to be made about the distributions and the results are meaningful only around absolute or relative threshold since perfect discrimination does not give any information about the perceived distances between the stimuli. Thus the perceptual range, and therefore the practical utility of such scales, is extremely limited.

The construction of a scale from measures of discriminability has been termed an 'indirect' method of scaling because the scale is arrived at only after much manipulation of the originally observed data (the difference measures). There is another group of methods known as 'direct' scaling where subjects are asked to produce a scale more or less directly from the stimuli. These methods have been championed most strongly by Stevens (e.g. Stevens, 1956) but other related work took place before this. Plateau (1872) asked artists to paint a grey midway between a black and white and found an equal ratio of subjective judgement for an equal ratio of physical magnitude, thus giving a power law of the form:

$$S = aI^b$$

where S is the magnitude of the sensation, I is the physical intensity of the stimulus and a and b are constants. This contrasts with the logarithmic relationship found by Fechner. Plateau did not pursue this line of work, but in the 1930s there was a large resurgence of such work using this method which has come to be known as fractionation. From this body of work names were developed for subjective scales of various types, e.g. mel (pitch), brill (brightness). Although popular during the 1930s this method is rarely used now.

Fractionation has been largely superseded by a method known as magnitude estimation, where the subject is presented with a stimulus and asked to assign a number to it representing the subjective magnitude of the sensation. Richardson and Ross (1930) had subjects estimate the brightness of an after-image (rather than a true external stimulus) compared with '1·0' for the original image. The method came to be used more broadly with the work of Stevens (1956, 1974), who used it for a large number of different modalities and developed the concept of the power law (outlined above). He also promoted the use of log–log plots of rated subjective magnitude against physical intensity where the power law predicts a straight line relationship with a slope equal to the exponent in the power law equation and an intercept equal to the constant. Stevens also investigated cross-modality matching, where stimuli in different sensory modalities are matched for equivalent subjective intensity, and he found that in general the exponent for the cross-modality match was related to the exponents for the two modalities individually (see Stevens (1961) for a more detailed review).

The use of equal discriminability as the basis for a scale has been suggested as analogous in physics to error variability (converse of resolving power) which is clearly more related to the particular measuring instrument than to the physical continuum measured (Shepard, 1981). However, the approach of Stevens has also been criticised widely, mainly on the lack of any external validation of a subjective scale (see Shepard, 1981, for a detailed discussion), and although Stevens (1961) points to the use of cross-modality matching as a validation, this does not satisfy critics (e.g. Treisman, 1964). Stevens considers this method removes any need to infer mental processes and is related to the psychological tradition of behaviourism where only observable stimuli and responses are 'measured' with no recourse to attempting to 'observe' internal events. However, as Treisman (1964) argued, the idea of deriving a truly 'subjective' scale according to the power law formula is untenable since we cannot observe anything other than the response and this response will be related to the internal 'subjective' scale by some other function. Hence in this type of scaling the observed response is related to the stimulus by two functions, one relating to sensation and one relating to response. Stevens has tended to ignore the response side of this process, assuming that responding with cardinal numbers, strength of hand grip, cross-modality match, etc., are all equivalent. It is not possible in his formulation to disentangle the two components to get a true scale of sensation. In many practical applications these limitations may be of little importance.

It is worth noting that in the development of psychophysics most studies have used visual and auditory stimuli. This is probably because these two modalities are generally considered to be the most important, but also the stimuli in these modalities are often easy to generate reproducibly and to control, and the physical continua are easy to specify (although this need not always be the case). In the case of texture, taste and smell, the stimuli are often much more difficult to control adequately, and in complex media (e.g. food) possible interactions make physical continua difficult to specify.

All the methods considered up to now relate only to unidimensional responses, although not all are related to single physical continua. In some instances it is necessary to consider responses along several dimensions because the stimuli do not vary in only one single way. This can be taken into account in magnitude estimation and in category ratings by rating on several scales (as in profiling), although subjects may not be able to use the scales independently. Alternatively, a multidimensional scaling procedure may be used, as is described in detail in Chapter 10. In this sort of procedure some measure of psychological distance between stimuli is obtained by presenting all the stimulus pairs or triads, and these distances are then represented in a multidimensional space, where the number of dimensions is chosen so that it adequately represents the original distances. Although this method has drawbacks in terms of the number of trials needed, it does not require the subject to be presented with understood, named scales on which to rate stimuli, and the analysis procedure will handle ordinal as well as interval data.

Following this brief discussion of the historical development of scaling we now move on to consider types of scales in common use in sensory work.

4. SCALES IN PRACTICAL USE

Scales are the tools by which the size or extent of attributes of stimuli are made explicit by the assessors. They are many and varied in design. The importance of design lies in their efficiency in producing reliable and discriminating data about attributes under study. This can only be assessed by testing data produced, and in order to use the most powerful and valid statistical procedures, the properties of the scale should be known. In practice this is not easy to do and usually a pragmatic approach is taken by using a 'robust' analysis (see below). The most frequently used scales in practice are rating and scoring scales. They may take many forms, ranging from structured scales in which all the ordinal categories are labelled with words, through those in which numbers are also assigned, or which only have numbers, to graphic scales with only end anchor points which the assessor is instructed to use in what may be an interval or a ratio manner. The popularity of such scales is probably a reflection of their apparent simplicity of use by assessors, their flexibility over a wide range of applications and the apparently simple statistical analysis of results (Amerine et al., 1965). Guilford, a prominent psychophysicist, even admitted (Guilford, 1954) that rating scale methods have definite advantages (over pair comparisons and rank order methods) and results often compare very favourably with those from more accurate methods. He listed their advantages over difference tests and ranking methods as:

1. They require much less time than the other methods.
2. The procedure is much more interesting to the assessors.
3. They have a much wider range of application psychologically.
4. They can be used with psychologically naive assessors with a minimum of training.
5. They can be used with large numbers of stimuli.
6. They produce better aesthetic (hedonic) judgements of single samples than when comparative judgements are made.

However, as Harries (1960) has pointed out, these scales have no units of measurement and are necessarily arbitrary. Strictly speaking, the results should always be treated as only ordinal, unless the data produced can be shown to have interval or ratio properties.

4.1. Category Scales

Category rating scales are of two main types. These are bipolar scales such as the hedonic scale for measuring food likes and dislikes developed in the late 1940s–early 1950s (Peryam and Girardot, 1952; Jones et al., 1955) for which data on the properties are well established (see below) and unipolar scales for level or degree to which an attribute is present (Amerine et al., 1965). Hedonic scales are discussed below and details of the development of a bipolar scale of toughness–tenderness, based on that of Jones et al. (1955), are described by Raffensperger et al. (1956). In the latter case the bipolar scale was compared with two separate unipolar scales for toughness and tenderness, but there was little difference in their use. They did conclude that omission of a central, neutral category in the bipolar scale produced a more even distribution of responses than when it was present. The final scale used is shown below:

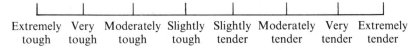

Extremely Very Moderately Slightly Slightly Moderately Very Extremely
 tough   tough   tough   tough   tender   tender   tender   tender

This sequence of adjectives is now very widely used in food category scales for many different attributes. It seems to be clearly understood by assessors and (except for hedonic scaling) apparently produces results with properties close to those of an interval scale although this has never been conclusively demonstrated. This series of adjectives is also widely used in unipolar scales of intensity of specific attributes. In this case it is always used with a zero or absent category and forms a five- or nine-category scale (e.g. for sweetness):

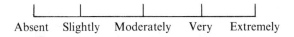

Absent Slightly Moderately Very Extremely

The linearity and equal interval properties of all but the extremes of these category scales can be demonstrated with perceptually simple systems. Sweetness is such a system, as elegantly demonstrated by McBride (1983b), using sucrose and fructose. He selected concentration ranges to avoid extremes of perceived intensity, and used ten concentrations of sucrose in three different incremental sets of four to six concentrations. Ratings were made on a 13-category scale, anchored with the descriptors above, and treated as scores, the panel means being plotted against the log sucrose concentration. The last data set (Fig. 2, circles) were on concentrations selected to minimise variation, and was very close to a straight line, with very small variance. It probably also reflects the effect of 'training' in the earlier sessions. Similar results were shown with fructose (Fig. 2, triangles) but not with glucose (squares). These results demonstrate what has long been felt by experienced users: with a trained, discriminating and consistent panel using familiar category (or other) scales, data of a quality suitable for parametric data analysis is usually produced. However, the extremes of the scale will be both non-linear and more variable (subject to bias—see below), as shown for glucose, where the lower concentrations were given mean ratings very low on the scale, i.e. many of the assessors would be using ratings near the scale end.

Such equal interval properties are much more difficult to demonstrate with more complex or less clearly defined stimuli, where several attributes may interact or where different assessors may be perceiving different attributes under the same label (e.g. off-flavour), and the psychophysical function is not known to be linear with log concentration. However,

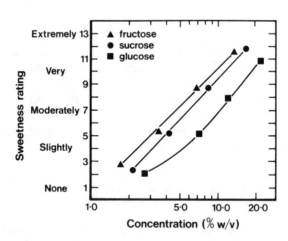

FIG. 2. Category scale data for fructose, sucrose and glucose with optimised stimulus spacing (reprinted from McBride, 1983*b*; with permission from The Psychonomic Society).

variants of the descriptive anchors used, although not strictly validated, are probably equally robust when used correctly. Two such sets are:

None, weak, moderate, strong, very strong.
None, very weak, moderate, strong, very strong.

The linearity of intensity adjectives for off-flavour was studied recently by Gacula and Washam (1986) in terms of a seven-category scale. This was found to be slightly non-linear, but some of the attitude data upon which it was based were extremely variable.

The general principle involved in these category scales is that the adjectives are those commonly used in everyday life to describe perceived attributes, and the use of them has therefore stabilised, i.e. people use them consistently. This does not mean that a period of training or familiarisation with the use of a particular scale and a particular application is unnecessary—this would be a very unwise assumption.

The scales shown so far are all used with the clear instruction to the assessor to select one category. Occasionally the five- or six-point scales are used with the comment that the boundaries between categories may be marked if the assessor feels the need for finer division. In effect this converts these scales into nine- and ten-category scales, with alternate categories unlabelled. An extension of these scales are graphic scales (see below) where the assessor has a very large number of positions from which to choose.

As the categories are always spaced with equal intervals the assessor is expected, and may be instructed, to use the scale as an equal interval scale with a true zero (i.e. like a ruler to measure the length of a line). However, he or she will be more strongly influenced by experience and training in the use of the word anchors than by this instruction and the scale may be used, albeit consistently, in a manner in which the intervals are not equal.

The number of categories for a scale has been extensively discussed (e.g. Guilford, 1954; Amerine *et al.*, 1965). The popularity of five points for unipolar and nine points for bipolar scales appears to originate from an extensive survey by Conklin (1923) of ratings by untrained assessors, but subsequent studies have clearly demonstrated that trained assessors can, with advantage, use finer divisions provided the attribute being rated is clearly defined. Too coarse a scale loses discrimination, but too fine a scale can introduce increasing error: there is an optimum for each case which could be determined, but 'fortunately there is a wide range around the optimal point in which reliability changes very little' (Guilford, 1954). Thus the popular use in food assessment of scales shown above is probably within the optimal range, with the proviso that although in clearly defined situations where highly trained assessors are measuring a single unambiguous attribute there may be advantage in training them with a larger, or graphic scale. Where multiple and less clearly defined attributes are concerned, use of such a scale could produce increased errors.

4.2. Multiple Comparison Procedure

In this technique a number of samples, including a coded standard, are compared with a designated standard for a specified attribute, the size of the difference being expressed on a nine-category scale with five verbal anchors—none, very slight, slight, moderate, large. In collaborative tests in 11 laboratories (Mahoney *et al.*, 1957) there was close agreement between all results (mean difference in least-significant difference for all laboratories less than 1%). Extensive experience has shown the technique to be generally easily used with good reproducibility and economy of effort. It provides

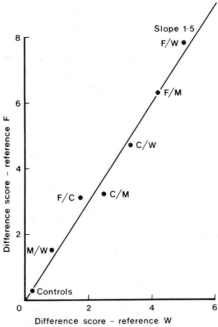

FIG. 3. Flavour difference scores for carrots grown at different centres (C, F, M, W) using F and W as separate reference standards. Category scales transformed to $x = 2^y$, where none = 1, very slight = 2, slight = 4, moderate = 8 and large = 16. Each point is the difference in flavour between the two samples measured by using each standard.

ready identification of any assessors not discriminating and because of the use of the standard, comparability from occasion to occasion. This is very difficult without the use of a standard.

The technique also offers a useful opportunity to test the properties of the scale and the effect of the standard. In studies on the effect of centre on carrot flavour, Land and Griffiths (unpublished observations) have used the multiple comparison procedure, first with the normal standard, and then by using the most different sample as a standard. When the differences between categories were treated as equal interval, the difference scores did not show more than ordinal agreement. However, after transformation to the power of base 2 ($x = 2^y$, where y is the original scalar difference and the assigned difference category scores became none $= 1$, very slight $= 2$, slight $= 4$, moderate $= 8$ and large $= 16$) the two sets of data were very highly correlated (Fig. 3). It is interesting to note that the standard influences the ratings; the slope of 1·5 indicates that the scale was used in a more extended form when the better flavour carrot was used as a standard than when the poor flavour carrot was used.

4.3. Hedonic Scaling

This is similar to a normal category rating scale except that it is not related to any particular physical continuum. Given its extensive use in sensory analysis it deserves special mention. Hedonics relates to pleasant and unpleasant states of an organism, and in hedonic scaling affective ratings of preference or liking and disliking are measured. Peryam and Pilgrim (1957) were responsible for much of the early development of this type of scale. Generally the scale is used with untrained assessors since trained ones are unlikely to give true affective responses. The general form of the scale is shown below:

Like extremely
Like very much
Like moderately
. Like slightly
Neither like nor dislike
Dislike slightly
Dislike moderately
Dislike very much
Dislike extremely

This is an example of a general category rating scale and certain questions regarding this or any other scale of this type are apparent. The first is whether the categories form equal intervals along a scale. Although the order of the categories is intuitively (and empirically) correct, is the difference between 'like very much' and 'like moderately' the same as that between 'like slightly' and 'neither like nor dislike'? If the intervals are equivalent (or approximately so) the data can be analysed using parametric statistics (e.g. t-test, analysis of variance) but if the intervals are not equal only non-parametric statistics are permissible.

In theory, category rating scales should always be tested as part of their development to assess their properties, although in practice this has rarely been done. However, the hedonic rating scale was tested by Jones and Thurstone (1955) and Jones et al. (1955). This was done by having a large number of subjects rate the category labels on a purely numerical scale from 'Greatest dislike' (-4) to 'Greatest like' ($+4$) and to look at the variability of each scale label. It is implicit in this work that such a scale will be interval if the categories are equally spaced and have equal variances. Jones et al. (1955) reported the distances between the category labels and the variances of the meaning of the labels. They also tested scales using different numbers of categories, either with a neutral or no neutral category, and with unbalanced scales which were not equivalent for the positive and negative sides. The category labels used were generally chosen to have small variances, i.e. to be low in ambiguity. Although the distances between the categories used were not exactly equal (Jones et al., 1955) the departure from linearity would be unlikely to be a major cause for concern. The discriminative ability of the scales was also tested and showed that in general this increased with increasing numbers of categories, and so although the five-category scale may be more linear than the nine-category scale, it is less discriminating.

Although in the study of the hedonic rating scale there does not appear to be great variability between the scales tested, this may not be true for other scales; they require testing if they are to be used with any confidence in their linearity.

The hedonic rating scale has been shown to be useful in terms of differentiating between foods and samples on a group basis (Peryam and Girardot, 1952). This is important since, no matter how well constructed a scale may be, it has to be sensitive enough to detect differences. The reliability of the scale has also been tested by administering it twice to the same people with the same samples (Peryam and Girardot, 1952; Jones et al., 1955). This type of assessment of scale performance is also of importance, since if there is large variability between the two presentations it is not possible to attribute exact meaning to responses on the scale.

The use of the hedonic scale has been extensive, including both consumer-type work and laboratory studies, using both the names of food categories as well as actual food samples.

Hedonic scales differ from other category scales in that the responses are not expected to be monotonic with increasing magnitude of some physical characteristic, but to show a peak (the maximally preferred magnitude) above and below which the rating will decline. Pangborn (1980) reported a study where lemonade with different concentrations of sucrose was rated on a hedonic scale and the results are typically that most subjects gave a peak in the hedonic/concentration plot, which might be taken as their 'ideal' concentration, although this varied greatly between subjects.

It is also possible to use magnitude estimation (see below) for hedonic scaling. Moskowitz and Sidel (1971) used both magnitude estimation and the usual category scale on the same samples and obtained very similar results for each.

Related to the hedonic scale is the food action scale developed by Schutz (1965), where the liking categories of the hedonic scale are replaced by how often the subject would like to eat the food. Schutz found group means on this scale to correlate highly with group means on the hedonic scale for different food types in a questionnaire. This scale (shown below) does not seem to have been widely used, and although it is highly correlated with the hedonic scale, it is not clear how it would relate to actual consumption, or how useful it would be at differentiating between samples of the same type of food.

I would eat this every opportunity I had.
I would eat this very often.
I would frequently eat this.
I like this and would eat it now and then.
I would eat this if available but would not go out of my way.
I don't like it but would eat it on occasion.
I would hardly ever eat this.
I would eat this only if there were no other food choices.
I would eat this if I were forced to.
Never tried.

4.4. Unstructured Scales

Whilst category scales have declined in use in recent years the popularity of unstructured scales has increased. An unstructured or graphic scale usually has anchors only at the ends, as shown:

None   Extremely

The assessor marks the line at the point he or she feels appropriate and the distance along the line is measured. There are no categories but the response is assumed to be a continuous function. The position is normally measured to the nearest mm, and so with a 100 mm scale this would give scores of 0 to 100 (equivalent to a 101 category scale). It has an advantage over a category scale in not having the coarseness of a fixed set of categories, and there is no problem about the number of categories included, or the labelling of such categories. An important similarity to the category scale is that the response is bounded and it would therefore be anticipated that it would be used as an interval rather than a ratio scale, as in magnitude estimation.

Baten (1946) compared palatability ratings using a seven-point scale with those on an unstructured 6 inch line. The latter responses were converted to numerical scores of 1 to 6 and were found to give better discrimination between samples than the ratings on the seven-point scale.

Stone et al. (1974) used unstructured scales in Quantitative Descriptive Analysis (QDA). The ratings are made on a 6 inch line with anchors at 0·5 inch from each end such as soft–hard, weak–strong. This has been widely used in industrial applications (Stone and Sidel, 1985).

Unstructured scales have received relatively little study of how truly interval the responses are, and with their increasing use this needs to be done.

4.5. Relative-to-Ideal Rating

One particular example of a graphic unstructured scale is the rating of samples relative to a person's ideal level of an attribute. This procedure has been used by Frijters and Rasmussen-Conrad (1982) and McBride (1982), who used a scale labelled as shown below:

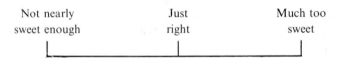

Plotting the responses against the logarithm of concentration gives a straight line and the point at which the line crosses the 'Just right' axis gives a measure of the individual's 'ideal' level of that attribute. The major difference between this type of scale and the conventional hedonic scale is that the responses are 'unfolded' rather then increasing up to a point and then decreasing. The concept of unfolding is discussed by Coombs (1979).

Responses on relative-to-ideal scales have been compared with conventional hedonic rating by McBride (1985b) and Shepherd et al. (1985). There is generally good agreement between the conclusions drawn using the two procedures, although there may be differences between the responses (Shepherd et al., 1985).

Generally this type of procedure has been used for rating simple attributes such as sweetness (Frijters and Rasmussen-Conrad, 1982; McBride, 1982, 1985b; Conner et al., 1987) or saltiness (Shepherd et al., 1984, 1985). It has also, however, been used successfully for rating more complex attributes such as yolk colour and flavour in eggs (Shepherd and Griffiths, 1987).

It is not clear whether the ratings relate directly to the intensity of the attribute. Shepherd et al. (1985) found that those subjects with a higher slope of the psychophysical function did not necessarily have a higher slope of the relative-to-ideal function. This might be because the labels used ('much too salty' and 'not nearly salty enough') have an evaluative element in addition to relating purely to intensity. Conner et al. (1987) used the alternative anchors of 'not at all sweet' and 'extremely sweet' and found good agreement between the intensity and relative-to-ideal ratings, but they did not report individual differences in responses.

4.6. Magnitude Estimation

Ratio scaling methods involve the use of scales in which successive points are in constant ratio to each other, i.e. the scale has ratio or geometric properties rather than, as intended with category scales, equal interval or arithmetic properties. In practice it means that if one stimulus is three times more intense than another the stimuli can be assigned any number in this ratio, i.e. 1 and 3; 4 and 12; 10 and 30; 100 and 300—the ratio is constant. The numbers assigned can be of any magnitude which is positive and greater than zero, although the scale has the property of having a true zero.

The most common form in practice is known as magnitude estimation. It was first introduced by Stevens (1953) and has been greatly promoted in sensory assessment of food by Moskowitz (1975). Detailed instructions for magnitude estimation are given by Moskowitz (1975). In outline there are two alternative approaches—the use or not of a designated standard. In the former, critical training of the panel consists of giving them a series of lines to assess for length, the first of which is arbitrarily given a value (e.g. 10) against which all the other lines are assessed as ratios. In the latter, less popular mode, each individual assessor is allowed to assign whatever length he wishes to the first line and to assess the remainder as ratios of this. There are no restrictions on the numbers used except that zero and negatives are not allowed. It is important that the range of line length should be at least 25:1. It is extremely important that the assessors understand the requirement to use numbers as ratios. The experimental samples should then be assessed using the same technique. The numbers assigned have no absolute meaning but are normalised as part of the analysis of results. The logarithm of the geometric mean (or mean of the logarithms of the numbers) is calculated for each sample and usually gives a straight line when plotted against the logarithm of the independent variable.

The method has been heavily criticised, mainly on the grounds that it does not produce true ratio results (see below), has not been externally validated and that high variance between assessors is readily masked by using log scales. However, it undoubtedly produces useful results.

4.7. Comparison of Different Scales

Arguments over the validity and relative merits of category scales and magnitude estimation have continued unabated from the extensive review of Stevens and Galanter (1957) to the present (Vickers, 1983; Lawless and Malone, 1986; Pearce *et al.*, 1986). Where carefully designed experiments have been conducted to test the two methods, each under their own optimum conditions, there appears to be little to choose between them. For example, Piggott and Harper (1975) used a range of known reproducible odour concentrations and found that with selected assessors the variance of estimation by the two methods was very similar.

Torgerson (1961) concluded that the proponents of each type of method would not convince each other of the value of the other technique because it was not possible to discover empirically what operation the assessors were using (i.e. internally—were they actually using a ratio or an interval scale?). He argued that the debate is empirically meaningless—a mere dispute over definitions. Birnbaum (1982) argued that neither type of scale provides 'direct' subjective values and argued for using difference judgements.

McBride (1983*a*, 1985*a*) has argued that category scales yield data which are comparable with those from JND scaling, whereas magnitude estimation yields responses which are not equivalent to either. One reason for this might be the response function for using numbers being logarithmic, and hence it yields the apparent power function in magnitude estimation.

Giovanni and Pangborn (1983) compared ratings on 100-mm lines with magnitude estimation in studies involving sucrose in lemonade and fat in milk. They found some advantages of the unstructured graphic scale for bipolar hedonic ratings but generally the results from the two types of scale were equivalent.

Shand et al. (1985) compared the use of category scales, unstructured scales and magnitude estimation when rating meat steaks. They found more differences between samples with the category scales and least with the unstructured scales. The panellists least preferred using magnitude estimation.

4.8. Use of Standards

One way in which variation and interpretation of scale values by the assessor can be reduced is by using standards (see Section 4.2). This approach was successfully used in developing the Texture Profile Technique (Szczesniak et al., 1963). For each attribute (hardness, brittleness, chewiness, gumminess, adhesiveness and viscosity) an experienced panel of nine assessors selected a series of food items corresponding to and defining each numerical scale point (category). The scales are claimed to have equal perceived intervals, although this has not been demonstrated. The standards were ranked and also shown to increase in definable ways with physical measures related to the perceived attributes. The major problem in using such a system is that the foods exhibit several other textural attributes, and training involves educating the assessor to recognise and isolate the particular attribute in the presence of others which may interact perceptually. This may require several months of training before a selected panel can consistently use such attribute scales in a discriminating manner. The use of some standards is essential in this process, although because of the difficulty of obtaining or preparing stable, unambiguous standards, they are most frequently used to illustrate the attribute rather than define the range of scale points, e.g. Williams (1975) and Clapperton (1975).

This problem is most acute in profiling techniques where an analytical approach is required to dissect out and measure the amount of the individual component attribute from a complex sensation which is perceived primarily as a whole, e.g. the flavour of whisky (Piggott and Jardine, 1979). In such cases it is frequently not possible to obtain standards for all attributes and the assessor has to learn to recognise the particular attribute from samples selected by experienced assessors showing variation in that attribute.

A more widely applicable use of a standard in a somewhat different manner is as a comparative standard for measuring the difference in attributes, as first suggested by Harrison and Elder (1950). It is known as the

Multiple Comparison Procedure (Mahoney et al., 1957) and was developed primarily for quality control in the US canning industry (see Section 4.2). It is also known as a Rating difference/Scalar difference from control test (Anon., 1981).

4.9. Data Analysis

The main consideration in analysing data from category or unstructured scales is whether the scales are truly interval and whether the data are normally distributed. If they are, then parametric statistics are appropriate, if not then non-parametric statistics should be used (see Chapter 9). Tests of whether scales are equal interval can be carried out as detailed by Guilford (1954) and Bock and Jones (1968). In a food context, Cloninger et al. (1976) tested category scales of varying lengths and transformed responses so as to normalise them. This showed some deviation from normality in the original data but conclusions from the statistical analyses of the data did not really differ, whether the data were normalised or not.

Only where data are very abnormally distributed will parametric procedures like analysis of variance or t-tests be greatly affected. The robustness of ANOVA to small deviations from normally distributed data is well known (e.g. Box, 1953, 1954; Winer, 1971) and hence doubt about its use with category or unstructured scale responses is probably not a major practical consideration.

In many applications more than one rating is performed on each sample (e.g. profiling). It is possible to analyse the resulting data by a series of tests on individual variables (e.g. ANOVA) but in general the responses on the different scales are related to each other and a better method is to employ some form of multivariate analysis on the whole set of data. For example, multivariate analysis of variance or discriminant analysis (Powers and Ware, 1986) may be used to test differences between conditions. In many instances it is desired to represent the data as samples in a space of say two or three dimensions rather than on the original large number of variables. This can be achieved using a procedure such as factor analysis or principal components analysis (Piggott and Sharman, 1986). Such procedures are very useful for data reduction and for visualising the relationships between the samples.

5. RANKING

Ranking is a process of ordering three or more samples for increasing or decreasing amounts of a specified attribute, overall quality or response (preference or acceptability) on the same occasion. Strictly speaking it is an extension of the paired comparison test to more than two samples. It is an old established method but is not now widely used in sensory assessment of food; it is considered to be most useful when samples can be assessed and reassessed with minimum time delay between samples, e.g. for visual assessment (ASTM, 1968) rather than for tasting where more confusion may arise. It has been suggested by Baker (1962) that ranking is best used when differences between samples are large, when a speedy assessment of many samples can be made, although Harrison and Elder (1950) consider the method to be less efficient than rating. However, the great advantage of ranking is that it makes no demands on the assessor to learn to use an external scale and then to produce ratings with it. Clearly the assessor must use an internal scale, based on experience, but as he or she is only required to place the samples in order, and not say how much of an attribute is present, the task is more readily carried out with confidence by untrained assessors. One of its main uses therefore is in preliminary screening of consumer products and in attitude measurement, where it offers a rapid and easy method for untrained assessors. It can also be used as part of the preliminary training and selection of assessors (Spencer, 1971; Anon., 1981).

The data can be analysed using Friedman's analysis (Chapter 9). Although there are published tables of rank sums (e.g. Kramer, 1963; Kahan et al., 1973), there are a number of problems with using these. The first relates to the mislabelling of some of the earlier versions, but more seriously the test is appropriate for testing whether one sample is different from the rest rather than whether there are differences between any of the samples. Since the normal purpose of ranking does not presuppose the existence of one outlier, the Friedman analysis is more appropriate and the results will differ from those of the Kramer test in some instances (ISO/DIS 8587.2, 1987).

This method of data analysis can be used on the rank scores of data obtained by rating, but the two different methods of data collection should not be confused.

The practical procedure is well described in ASTM (1968). The assessors must first understand and agree upon the attribute (e.g. uniformity of appearance, colour, size, freedom from defects, strength of flavour or odour) or reaction (preference or liking) to be judged. The samples are then presented to each assessor coded and in a random or, better, balanced order for assessment; usually the time to be allowed between samples is specified, together with other details such as whether the sample should or should not be swallowed. The assessor is normally instructed to assign a preliminary order and then to check this. The number of samples which can be given together depends upon the nature of the samples and the task—in tasting strongly flavoured samples the maximum may be three, whereas for visual assessment it may be as many as ten.

In conclusion, although ranking methods have been used in sensory analysis of food, their use is by no means as widespread as rating and scoring. However, they do have their place and they should not be totally dismissed as outmoded by other methods.

6. GENERAL ASPECTS OF EXPERIMENTAL DESIGN AND PROCEDURE

There are no hard and fast rules about the design of experiments since it depends upon the type of question to be answered. However, the general approach is always the same and this will be described briefly. If it is decided to test the relationship between events it is necessary to form an experimental hypothesis which describes the relationship. An experimental hypothesis may be that the assessors' ratings of some attribute will be higher in different conditions. The subjects are tested in these conditions of the independent variable and the dependent variable (rating) is measured. It is then necessary to compare the values of the dependent variable for the different conditions. Since there will be variability both between and within panelists, it is necessary to test a number of panelists doing the task several times, and then to compare the results with what might be expected if responses varied randomly and there were no real differences between conditions. This requires statistical analysis of the data (dealt with in more detail in Chapter 9) and involves comparing the results found against a null hypothesis, i.e. that there are no real differences and that any difference found is due to chance. Generally in sensory analysis of foods the questions relate to whether there are differences between samples and how great the differences are. Guidelines for work in this area are set out in British Standards (BS, 1980). Following is a brief discussion of some of the points which need to be taken into account when using rating and ranking procedures in sensory analysis.

In general assessors are trained in order to give less variable and more reproducible responses. There may also be an initial selection of assessors who are able to perform the task adequately. One exception to this is hedonic or preference assessment where the assessors should be representative of those people likely to buy or use the product and hence having trained assessors (or even assessors working in the industry) would be likely to produce unrepresentative results. Having untrained subjects and less control of the environment in hedonic assessment means that more assessors need to be included due to the variability in the responses.

There is a tendency for assessors not to use the end categories of a rating scale; this is called central tendency. It can be reduced with training and making the assessors familiar with the range of samples to be presented. Although it might appear that the end categories could be removed, e.g. reducing a nine-category scale to seven categories this would lead to the assessor not using the end categories of the new shorter scale.

Assessors should work in an environment with the minimum of external distractions. If samples are to be rated on one attribute it may be necessary to disguise other attributes, e.g. using coloured lights to disguise appearance differences when assessing for flavour.

Interactions between assessors should be kept to a minimum during testing. Procedures such as profiling may require preliminary discussion of the terms to be used but the actual assessments must be made independently by the assessors. Group discussions may reach a consensus but will be unduly influenced by individual panel members and most likely by the panel leader.

Samples must be coded and presented in random (or in balanced) order so as to remove expectancy effects.

6.1. Range–Frequency Effects

The ratings given to stimuli will depend upon the context in which they are presented and will be changed by both the range of stimuli presented and the frequency of occurrence of stimuli of different magnitudes. The effect of these may be explained using the model of range–frequency developed by Parducci (1974).

The effect of presenting stimuli in different stimulus ranges is that subjects tend to adjust the centre of the rating scale in the direction of the centre of the stimulus range. Hence, if subjects are rating the loudness of sounds they will rate the same stimulus lower if it is presented in a series of high intensities, than if it is presented in a series of low intensities (Poulton, 1987). The effect of range occurs also with both intensity and hedonic ratings for sweetness (McBride, 1985b; Conner et al., 1987) and for saltiness (Shepherd et al., 1984). It has important consequences if the rating for a particular stimulus presented in one series is assumed to be a true value related only to the stimulus characteristics, since in a different series the actual rating would be different; only the ratings relative to the stimuli in the same series will be equivalent. There may also be consequences where the stimulus range varies during the course of the experiment, e.g. storage experiments.

A similar phenomenon occurs if the stimulus is presented in the context of a greater number of low intensity stimuli, when it will be rated higher than if it is presented in the context of more high intensity stimuli. Hence, although the overall range of stimuli presented is the same, in the first case there is a skew towards presenting a greater number of low intensity stimuli and in the second towards high intensity stimuli. This was tested by Riskey (1982) using ratings of intensity of saltiness of soups where the distribution of concentrations of other presented stimuli affected the ratings of given concentrations. Again, pleasantness ratings were also affected by the context.

Both effects demonstrate the need to be aware of the influence of the context in which a stimulus is presented and to be careful in comparing results between conditions where these contexts differ greatly. Poulton (1977) has argued that bias can be avoided by having each stimulus judged by a different subject, but this would be impracticable in most situations. It is important to compare only data gathered in comparable contexts.

Booth has suggested that estimates of an individual's ideal level of an attribute can be obtained in an unbiased manner by centering the range of stimuli presented on each individual's own ideal (Booth et al., 1983). This method has been used in a number of studies (Conner et al., 1986, 1987) and has been found to give estimates which appear to be unbiased (Shepherd et al., 1984).

7. CONCLUSIONS

From the discussions of techniques in this chapter it will be apparent that although some of the specific applications to sensory analysis have been developed within the context of food science, much of the development (especially the original development) stems from the behavioural sciences. In order to use the techniques appropriately and to understand their limitations, it is necessary to understand something of their origins, and this may require looking to work in areas outside food science. In terms of using these techniques, however, it is not necessary to be involved in the development of scales, but to use scales which have been adequately assessed and to use them in appropriately defined circumstances.

The major limitation lies in data analysis. It is not easy to test the properties of data collected unless these are much more extensive than is usual in applications. It is very easy merely to assume that data collected have interval or ratio properties and use simple parametric statistical procedures to assess their significance. Although this frequently will not result in major errors in drawing conclusions, under certain circumstances it can. There is a need for better and easier procedures for testing the properties of small data collections. If there is any doubt it would be wise to also analyse data by non-parametric techniques, e.g. by analysing ranked ratings to test conclusions drawn from analysis of ratings by parametric techniques.

The versatility of the scaling procedures has been demonstrated and these are very popular in sensory analysis. Although ranking procedures are less versatile they do offer a useful alternative technique where untrained assessors are involved, providing that the number of samples is relatively small.

In all work on sensory analysis the design of experiments is of utmost importance since the data obtained will be strongly influenced by the exact procedures used, e.g. instructions to assessors, number and range of samples. The data are only comparable with data obtained in equivalent circumstances. In order to obtain meaningful results it is therefore necessary to conduct closely controlled experiments which will yield data capable of answering the questions posed.

Thus, if the best possible use is to be made of scaling and ranking methods of sensory analysis, it is necessary to have a clear understanding of the underlying basis of the methodology, of the factors which influence responses, and of the limitations of the techniques of data collection and analysis.

REFERENCES

ASTM (1968). *Manual on Sensory Testing Methods*, STP 434, American Society for Testing and Materials, Philadelphia.

ASTM (1981). Standard definitions of terms relating to sensory evaluation of materials and products. In: *Annual Book of Standards, Part 46*, American Society for Testing and Materials, Philadelphia, pp. 70–2.

Amerine, M. A., Pangborn, R. M. and Roessler, E. B. (1965). *Principles of Sensory Evaluation of Food*, Academic Press, New York.

Anon. (1981). Sensory evaluation guide for testing food and beverage products. *Food Technol.*, **36**(11), 50–9.

Baker, R. A. (1962). Subjective panel testing. *Ind. Qual. Control*, **19**(3), 22–8.

Baten, W. D. (1946). Organoleptic tests pertaining to apples and pears. *Food Res.*, **11**, 84–94.
Birnbaum, M. H. (1982). Problems with so-called 'direct' scaling. In: *Selected Sensory Methods: Problems and Approaches to Hedonics*, J. T. Kuznicki, A. F. Rutkiewic and R. A. Johnson (Eds), American Society for Testing and Materials, Philadelphia, pp. 34–48.
Bock, R. D. and Jones, L. V. (1968). *The Measurement and Prediction of Judgment and Choice*, Holden-Day, San Francisco.
Booth, D. A., Thompson, A. and Shahedian, B. (1983). A robust, brief measure of an individual's most preferred level of salt in an ordinary foodstuff. *Appetite*, **4**, 301–12.
Box, G. E. P. (1953). Non-normality and tests on variance. *Biometrika*, **40**, 318–35.
Box, G. E. P. (1954). Some theorems on quadratic forms applied in the study of analysis of variance problems. *Ann. Math. Stat.*, **25**, 290–302.
BS (1975). *Glossary of Terms Relating to Sensory Analysis of Food*, BS 5098, British Standards Institution, London.
BS (1980). *Methods for Sensory Analysis of Food. Part 1, Introduction and General Guide to Methodology*, BS 5929, British Standards Institution, London.
Clapperton, J. F. (1975). The development of a flavour library. *Proc. XV Europ. Brewing Conv., Nice*, Elsevier, Amsterdam, pp. 823–35.
Cloninger, M. R., Baldwin, R. E. and Krause, G. F. (1976). Analysis of sensory rating scales. *J. Food Sci.*, **41**, 1225–8.
Conklin, E. S. (1923). The scale of values method for studies in genetic psychology. *University of Oregon Publications*, **2**(1).
Conner, M. T., Haddon, A. V. and Booth, D. A. (1986). Very rapid, precise assessment of effects of constituent variation on product acceptability: Consumer sweetness preferences in a lime drink. *Lebensm.-Wiss. u. Technol.*, **19**, 486–90.
Conner, M. T., Land, D. G. and Booth, D. A. (1987). Effect of stimulus range on judgement of sweetness intensity in a lime drink. *Brit. J. Psychol.*, **78**, 357–64.
Coombs, C. H. (1979). Models and methods for the study of chemoreception-hedonics. In: *Preference Behaviour and Chemoreception*, J. H. A. Kroeze (Ed.), Information Retrieval Ltd, London, pp. 149–70.
Fechner, G. T. (1860). *Elemente der Psychophysik*, Breitkopf and Hartel, Leipzig. (Reissued 1964 by Bonset, Amsterdam).
Frijters, J. E. R. and Rasmussen-Conrad, E. L. (1982). Sensory discrimination, intensity perception, and affective judgment of sucrose-sweetness in the overweight. *J. Gen. Psychol.*, **107**, 233–47.
Gacula, M. C. and Washam, R. W. (1986). Scaling word anchors for measuring off flavour. *J. Food Quality*, **9**, 57–65.
Giovanni, M. E. and Pangborn, R. M. (1983). Measurement of taste intensity and degree of liking of beverages by graphic scales and magnitude estimation. *J. Food Sci.*, **48**, 1175–82.
Green, D. M. and Swets, J. A. (1966). *Signal Detection Theory and Psychophysics*, J. Wiley & Sons, New York.
Guilford, J. P. (1954). *Psychometric Methods* (2nd edn), McGraw-Hill, New York.
Harries, J. M. (1960). The quality control of food by sensory assessment. *Society of Chemical Industry Monograph No. 8*, Soc. Chem. Ind., London, pp. 128–37.

Harrison, S. and Elder, L. W. (1950). Some applications of statistics to laboratory taste testing. *Food Technol.*, **4**, 434–9.

ISO (1977, 1978, 1979, 1981). *Sensory Analysis—Vocabulary Parts 1–4*, International Standards Organisation.

ISO (1987). *Sensory Analysis—Methodology—Ranking, ISO/DIS 8587*, International Standards Organisation.

Jones, L. V. and Thurstone, L. L. (1955). The psychophysics of semantics: An experimental investigation. *J. Appl. Psychol.*, **39**, 31–6.

Jones, L. V., Peryam, D. R. and Thurstone, L. L. (1955). Development of a scale for measuring soldiers' food preferences. *Food Res.*, **20**, 512–20.

Kahan, G., Cooper, D., Papavasiliou, A. and Kramer, A. (1973). Expanded table for determining significance of differences for ranked data. *Food Technol.*, **28**(5), 61–9.

Kramer, A. (1963). Revised tables for determining significance of differences. *Food Technol.*, **17**, 1596–7.

Lawless, H. T. and Malone, G. J. (1986). The discriminative efficiency of common scaling methods. *J. Sens. Studies*, **1**, 85–98.

Mahoney, C. H., Stier, H. L. and Crosby, E. A. (1957). Evaluating flavour differences in canned foods, II. Fundamentals of the simplified procedure. *Food Technol.*, **11**(9), Suppl. Symp. Proc., 37–42.

McBride, R. L. (1982). Range bias in sensory evaluation. *J. Food Technol.*, **17**, 405–10.

McBride, R. L. (1983*a*). A JND-scale/category-scale convergence in taste. *Perc. Psychophys.*, **34**, 77–83.

McBride, R. L. (1983*b*). Category scales of sweetness are consistent with sweetness-matching data. *Perc. Psychophys.*, **34**, 175–9.

McBride, R. L. (1985*a*). Sensory measurement: An introductory overview. *CSIRO Food Res. Q.*, **45**, 59–63.

McBride, R. L. (1985*b*). Stimulus range influences intensity and hedonic ratings of flavour. *Appetite*, **6**, 125–31.

Moskowitz, H. R. (1975). Applications of sensory measurement to food evaluations II. Methods of ratio scaling. *Lebens.-Wiss. u. Technol.*, **8**, 249–54.

Moskowitz, H. R. and Sidel, J. L. (1971). Magnitude and hedonic scales of food acceptability. *J. Food Sci.*, **36**, 677–80.

Muller, H. G. (1977). Sensory quality control; report on a survey. In: *Sensory Quality Control: Practical Approaches in Food and Drink Production*, H. W. Symons and J. J. Wren (Eds), Soc. Chem. Ind., London, pp. 28–36.

Pangborn, R. M. (1980). A critical analysis of sensory responses to sweetness. In: *Carbohydrate Sweeteners in Foods and Nutrition*, P. Koivistoinen and L. Hyvonen (Eds), Academic Press, London, pp. 87–110.

Parducci, A. (1974). Contextual effects: A range–frequency analysis. In: *Handbook of Perception, Vol. II. Psychophysical Judgment and Measurement*, E. C. Carterette and M. P. Friedman (Eds), Academic Press, New York, pp. 127–41.

Pearce, J. H., Korth, B. and Warren, C. B. (1986). Evaluation of three scaling methods for hedonics. *J. Sens. Studies*, **1**, 27–46.

Peryam, D. R. and Girardot, N. F. (1952). Advanced taste-test method, *Food Eng.*, **24**, 58–61; 194.

Peryam, D. R. and Pilgrim, F. J. (1957). Hedonic scale method of measuring food preferences. *Food Technol.*, **11**(9), Suppl. 9–14.
Piggott, J. R. and Harper, R. (1975). Ratio scales and category scales of odour intensity, *Chem. Senses Flav.*, **1**, 307–16.
Piggott, J. R. and Jardine, S. P. (1979). Descriptive sensory analysis of whisky flavour. *J. Inst. Brew.*, **85**, 82–5.
Piggott, J. R. and Sharman, K. (1986). Methods to aid interpretation of multidimensional data. In: *Statistical Procedures in Food Research*, J. R. Piggott (Ed.), Elsevier Applied Science, London, pp. 181–232.
Plateau, J. A. F. (1872). Sur la mesure des sensations physique, et sur la loi qui lie l'intensite de ces sensations a l'intensite de la cause excitante. *Bull. Acad. Roy. Belg.*, **33**, 376–85.
Poulton, E. C. (1977). Quantitative subjective assessments are almost always biased, sometimes completely misleading. *Brit. J. Psychol.*, **68**, 409–25.
Poulton, C. (1987). Bias and range effects in sensory judgments. *Chem. Ind.*, 5th Jan, 18–22.
Powers, J. J. and Ware, G. O. (1986). Discriminant analysis. In: *Statistical Procedures in Food Research*, J. R. Piggott (Ed.), Elsevier Applied Science, London, pp. 125–80.
Raffensperger, E. L., Peryam, D. R. and Wood, K. R. (1956). Development of a scale for grading toughness–tenderness in beef. *Food Technol.*, **10**, 627–30.
Richardson, L. F. and Ross, J. S. (1930). Loudness and telephone current. *J. Gen. Psychol.*, **3**, 288–306.
Riskey, D. R. (1982). Effects of context and interstimulus procedures in judgments of saltiness and pleasantness. In: *Selected Sensory Methods: Problems and Approaches to Measuring Hedonics*, J. T. Kuznicki, A. F. Rutkiewic and R. A. Johnson (Eds), American Society for Testing and Materials, Philadelphia, pp. 71–83.
Schutz, H. G. (1965). A food action rating scale for measuring food acceptance. *J. Food Sci.*, **30**, 365–74.
Shand, P. J., Hawrysh, Z. J., Hardin, R. T. and Jeremiah, L. E. (1985). Descriptive sensory assessment of beef steaks by category scaling, line scaling and magnitude estimation. *J. Food Sci.*, **50**, 495–500.
Shepard, R. N. (1981). Psychological relations and psychophysical scales: On the status of 'direct' psychophysical measurement. *J. Math. Psychol.*, **24**, 21–57.
Shepherd, R., Farleigh, C. A. and Land, D. G. (1984). Effects of stimulus context on preference judgements for salt. *Perception*, **13**, 739–42.
Shepherd, R., Farleigh, C. A., Land, D. G. and Franklin, J. G. (1985). Validity of a relative-to-ideal rating procedure compared with hedonic rating. In: *Progress in Flavour Research 1984*, J. Adda (Ed.), Elsevier, Amsterdam, pp. 103–10.
Shepherd, R. and Griffiths, N. M. (1987). Preferences for eggs produced under different systems assessed by consumer and laboratory panels. *Lebensm.-Wiss. u. Technol.*, **20**, 128–32.
Spencer, H. W. (1971). Techniques in the sensory analysis of flavours. *Flavour Ind.*, **2**, 293–302.
Stevens, S. S. (1951). Mathematics, measurement and psychophysics. In: *Handbook of Experimental Psychology*, S. S. Stevens (Ed.), J. Wiley & Sons, New York.
Stevens, S. S. (1953). On the brightness of lights and the loudness of sounds. *Science*, **118**, 576.

Stevens, S. S. (1956). The direct estimation of sensory magnitudes—loudness. *Am. J. Psychol.*, **69**, 1–25.

Stevens, S. S. (1961). The psychophysics of sensory function. In: *Sensory Communication*, W. A. Rosenblith (Ed.), J. Wiley & Sons, New York, pp. 1–33.

Stevens, S. S. (1974). Perceptual magnitude and its measurement. In: *Handbook of Perception Vol. II. Psychophysical Judgment and Measurement*, E. C. Carterette and M. P. Friedman (Eds), Academic Press, New York, pp. 361–89.

Stevens, S. S. and Galanter, E. H. (1957). Ratio scales and category scales for a dozen perceptual continua. *J. Exp. Psychol.*, **54**, 377–411.

Stone, H. and Sidel, J. L. (1985). *Sensory Evaluation Practices*, Academic Press, New York.

Stone, H., Sidel, J., Oliver, S., Woolsey, A. and Singleton, R. C. (1974). Sensory evaluation by quantitative descriptive analysis. *Food Technol.*, **29**(11), 24–34.

Szczesniak, A. S., Brandt, M. A. and Friedman, H. H. (1963). Development of standard rating scales for mechanical parameters of texture and correlation between the objective and the sensory methods of texture evaluation. *J. Food Sci.*, **28**, 397–403.

Thurstone, L. L. (1927). Psychophysical analysis. *Am. J. Psychol.*, **38**, 369–89.

Torgerson, W. S. (1961). Distances and ratios in psychophysical scaling. *Acta Psychol.*, **19**, 201–5.

Treisman, M. (1964). Sensory scaling and the psychophysical law. *Quart. J. Exp. Psychol.*, **16**, 11–22.

Vickers, Z. M. (1983). Magnitude estimation vs category scaling of the hedonic quality of food sounds. *J. Food Sci.*, **48**, 1183–6.

Williams, A. A. (1975). The development of a vocabulary and profile assessment method for evaluating the flavour contribution of cider and perry aroma constituents. *J. Sci. Food Agric.*, **26**, 567–82.

Winer, B. J. (1971). *Statistical Principles in Experimental Design* (2nd edn), McGraw-Hill Kogakusha, Tokyo.

Chapter 7

CURRENT PRACTICES AND APPLICATION OF DESCRIPTIVE METHODS

JOHN J. POWERS

*Department of Food Science and Technology,
University of Georgia College of Agriculture, Athens, Georgia, USA*

1. INTRODUCTION

Since the publication of the first edition of this book, quantitative sensory profiling (QSP) has undergone a phase shift. Rather than being predominantly innovative today, QSP is more in a mode of maturation. During the early 1970s to the early 1980s several new pathways in sensory description analysis were instituted. The progenitor of most of the formal, systemised profiling methods, the flavour profile method (FPM) which Cairncross and Sjöström had first described in 1949, will long be a landmark in the field, but it is not without its limitations. Changes instituted in the early 1970s were designed to provide stricter quantitation than the FPM specified, to employ replication which had not been specified at all and to utilise statistical analysis, the need for which the originators had eschewed (Caul, 1957). In devising procedures which made the data amenable to univariate statistical analysis such as analysis of variance (ANOVA), the data were also often made suitable for the application of multivariate statistical analysis (MVA). Inasmuch as QSP yields multivariate sensory data, it was befitting that the application of MVA to QSP data be investigated. Study and development of the changes mentioned above—and others—are not as much in flux today as they were in the early 1980s. By no means has the quintessential method of QSP yet been developed, but there comes a time for consolidation of gains made. The accumulation of experience in application is often necessary to point out the next major development which should be sought. Today, QSP is at that stage. Until the next major advance is made, the publication of Laamanen and Jounela-Eriksson (1987) illustrates just how extensive experience now is. These investigators employed almost the complete array of 'right' procedures in their profiling study.

Regardless of how sensory results are acquired or analysed, the sensory technologist has to: (1) evoke, (2) measure, (3) analyse and (4) interpret the results. No method of sensory analysis calls for fuller or wiser use of each of these functions than does descriptive analysis. Moreover, no method calls for greater knowledge and experience on the part of the sensory leader than does the sound application of descriptive procedures. Apart from meriting an accolade for being the first of the QSP methods, the FPM called for a higher level of proficiency at meeting the four goals above than had other sensory methods up to its time.

Before moving on to discuss the various QSP methods prevailing, a glance backward is warranted. We in food science tend to forget that we owe a debt to early chemists who often employed their senses of taste and odour to characterise the sensory properties of chemicals. If we had to characterise today some of these same compounds (strychnine sulphate, mercuric chloride, arsenic trioxide), we might forego knowledge of their sensory properties. Prior to the development of clinical methods to diagnose disease, the nose was often used for that purpose. Moncrieff (1967), for example, describes typhus as causing those stricken with it to have a mawkish odour, and it is still recognised that the breath of those suffering from diabetes often has a sweetish acetone odour. Though physicians today do not rely chiefly on their sense of odour, knowledge of the relations between disease and odour is still extant. Recently Smith *et al.* (1982) published a fairly long list of forms of illness and odours which accompany them.

The characterising of chemicals by description has not gone entirely by the board. Wise gas chromatographers often use their noses to characterise gas chromatographic effluents, not merely to describe the odour but to aid in identification, for the odour of some chemicals is so characteristic that almost certain identification can be made. An illustration of the use of description *par excellence* to characterise sensory substances is that exhibited by expert perfumers and flavourists. Generally, not only can they identify a particular substance, but quite often specify its geographical origin even though the particular substance may be produced in five or more areas of the world.

Though use of sensory description to characterise substances has fallen into disuse by most chemists, other common uses are made of simple descriptive procedures. A complex procedure to decide whether a foodstuff has been justifiably returned for alleged defects or for having exceeded its normal market life may not be needed. The food may need to be merely sniffed or tasted. Furthermore, all of us as consumers use descriptive procedures. We may denominate a food as being delicious, nondescript or unacceptable, which alone is a matter of simple description, but to arrive at that ultimate decision we go through a complex process, subconsciously for the most part, assessing several individual characteristics to make our simply expressed decision.

As consumers, while we engage in such highly informal descriptive procedures every day, and, regrettably, some industrial firms—without proper sensory guidance—do the same, the real value of descriptive procedures resides in those of a systematic nature. From them we derive many practical benefits as well as fundamental knowledge of the relations between composition and sensory quality. The publications of Vuataz (1977) and Liardon et al. (1987) illustrate this. The descriptive procedures to follow will be considered from four points of view: (1) procedures available, (2) practices followed, (3) pitfalls to be avoided and (4) the potential for the development of uses or practices even more beneficial than those which exist at present.

2. MAJOR PROCEDURES

There are many variants of QSP procedures. In the first edition of this book, two of the major ones, the FPM of Cairncross and Sjöström (1950) and the quantitative descriptive analysis (QDA) method of Stone et al. (1974), were sometimes used in a generic sense. That is quite commonly done by practitioners in the field even though they recognise that the particular variant they are talking about does not conform exactly to the procedures specified by the originators of the respective methods. When speaking of quantitative-sensory-descriptive-analysis in a general sense, it seems desirable that FPM and QDA should not be used so loosely; thus QSP will be used here to denominate methods which possess the key attributes of quantitative-sensory-descriptive profiling, except descriptive was deleted from the acronym to shorten the abbreviation. Description is almost inherent in profiling. If the components of a profile are not designated or described, then the profile is not as information-bearing as it should be.

The various QSP methods existing today developed out of the FPM, for it was the first to set forth the philosophy that the examination of main modalities, such as flavour, is not enough. The components comprising flavour should themselves be evaluated as far as possible because together, logically, they must be the determinants of flavour. Methodology remained rather static for 25 years; then investigators began to append on to the profiling feature of the FPM alternative procedures which have made profiling more effective.

For several reasons, FPM will be described first though it no longer has a singular position in sensory analysis. It is, however, still an important method (Krasner *et al.*, 1985; Bartels *et al.*, 1986). FPM is losing ground to newer QSP methods; nonetheless it should be discussed in detail because many of the specifications the newer methods set forth are derived from considerations originally brought to the fore by Cairncross and Sjöström. A rather old procedure itself, the texture profile method (Brandt *et al.*, 1963), is based largely on concepts of the FPM. Even if many of the features of the FPM did not still apply, it should be described because of its venerability. There is an old Chinese saying that 'he who drinks from the well should bow in obeisance to the one who dug the well'. A debt is owed to those who first set sensory analysis upon the new path of structured, sensorially-sound profiling.

To speak of a method only 40 years old as being venerable may seem odd to some, but most of our major, sound methods of sensory evaluation are not much more than 45 years old. The triangular test, probably the procedure most widely used in the world, was described only 45 years ago (Bengtsson, 1943). Bengtsson actually set the pattern for one of the things present-day QSP methods almost invariably do, which the FPM fails to do, namely, make mandatory the coupling of statistical with sensory analysis.

2.1. Flavour Profile Method

With the exception of the long apprenticeship expert perfumers and flavourists undergo, FPM is the most stringent of sensory methods in the length of time and effort required to train members so that they can ultimately function truly as an analytical instrument. A major problem in applying all forms of sensory evaluation is that we humans are indeed individuals. In evaluating a product we tend to 'home-in' on different characteristics to describe a product or to discriminate among them. Similarly, we tend to score things in different ways. Some individuals are conservative and score samples accordingly. Others are more expansive in their performance. They use more of the scoring range or operate at a higher level along the scale. This leads to product–assessor interaction. The FPM attempts to dampen this tendency—in fact, tries to eliminate it entirely—by getting all panel members to agree upon the score values which should be assigned to weak, moderate and other intensity levels. This is the kind of performance one expects from analytical instruments. Even if cheap and expensive pH meters are compared with one another, we expect all of them to yield almost the same reading. Concordance in scoring is one of the things that the very long and arduous training involved in developing skills in flavour profiling is intended to achieve. Other methods do endeavour to reach this goal, but training is generally not as intensive and demanding as that for flavour profiling. The job of flavour profile assessors is not to express their likes or dislikes, though in judging 'amplitude' they are in effect making a judgement as to acceptability. Their task is to be able to (1) recognise fine nuances of odour, taste and feel, (2) assess these nuances quantitatively, and (3) apply the scaling system in such a way that there is not major disagreement among the judges regarding the score level which should be applied to a particular degree of intensity. Not only is the FPM designed to achieve concordance among panelists as they assess the intensity of each attribute, an attempt is made to reach agreement as to the order of perception of the different sensations. Before discussing the FPM further, let us first consider the steps involved in training panelists.

2.1.1. Training
One of the first steps is to assess whether the trainees are capable of recognising the four taste sensations. It is not enough that they are able to scale the intensity of attributes; they must also be able to recognise unequivocally that one sensation is of sourness or another is of bitterness. Usually, there are no problems in distinguishing sweetness and saltiness from the other sensations, but some individuals find it difficult to distinguish sourness from bitterness or bitterness from astringency. Astringency of course is not a taste or odour sensation; nonetheless it must often be included in flavour evaluation because it is a property common to many foods. Robinson (1970) has pointed out that an appreciable number of individuals are not able to distinguish between the sensations of sourness and bitterness.

To learn whether the would-be panelist can distinguish the four taste sensations from each other, Caul (1957) and ASTM Committee-E-18 (1981) recommend that solutions of the following substances, at the concentrations listed below, be presented to the trainees:

sucrose	2·0%	citric acid	0·07%
sodium chloride	0·2%	caffeine	0·07%

The intent is to have solutions which are not so strong in taste that they influence the taste of succeeding ones; thus all may be tested within a single session. The ASTM recommends that 30-ml amounts be presented in random order and that more than four specimens be presented so that some of the solutions are in duplicate; this minimises the opportunity for the trainees to guess the identity of a given solution. If a trainee does not recognise each specimen on the first test, the identities of the specimens should be pointed out and the trainees allowed to take the test again. The concentrations listed above are merely starting points. Concentrations should be changed by some geometric ratio so that most trainees are capable of distinguishing among the solutions; then the concentrations should be reduced as the trainees acquire experience to determine who among them are the most adept at identifying by taste the four kinds of solution. If a trainee cannot recognise the identity of the substances after having the identity pointed out to him or her and given opportunity to re-examine the solutions, that individual should be rejected.

Not only should the panelists be 100% successful in recognising the four basic tastes, they must also be adept at describing odour sensations. Generally, would-be panelists are presented with 20 or more odorants, some of which they would rarely encounter, to assess if they can identify the odour of common substances and if they can describe the odour of materials new to them in terms of some other substance whose odour is known to them. The purpose of testing them with odorants strange to them is to see if they are articulate at describing sensations previously foreign to their experience. Panelists must have a reasonable degree of facility with words, for often they will be called upon to describe a sensation new to them.

Both Caul (1957) and ASTM Committee E-8 (1981) suggest that the trainees may be graded according to performance by assigning 5 points to correct identification (cinnamic aldehyde for cinnamon), 4 points for apt association (red-hot candies for cinnamon), 3 points for characterising a substance by one of its secondary characteristics (sweet spice for cinnamon) and 1 point for attempted description (sweet for cinnamon). Assuming there are 20 odorants, a score of 70 should be required. Golovnya *et al.* (1981, 1986) have also suggested a uniform procedure for the selection of panelists based on their ability to perceive and describe sensory attributes.

Once the would-be panelists have demonstrated competency in recognising the basic tastes and in recognising or in describing odours, they then need to be trained to scale the intensity of stimuli so that ultimately each panelist brings his or her scaling system into line with the other members of the panel. This concordance is acquired through two different forms of training. ASTM Committee E-18 (1981) recommends that panelists be required to scale the intensity of bitterness, sweetness, saltiness,

TABLE 1
SOLUTIONS AND CONCENTRATIONS SUGGESTED TO PERMIT TRAINING OF ASSESSORS TO RANK, SCALE OR SCORE THE INTENSITIES OF THE FOUR BASIC TASTES
(From ASTM Committee E-18, 1981)

Taste	Concentrations (%) in odourless and tasteless water at room temperature
Bitter (caffeine)	0·035
	0·07
	0·14
Sweet (sucrose)	1·00
	2·00
	4·00
Salty (sodium chloride)	0·10
	0·20
	0·40
Sour (citric acid)	0·035
	0·07
	0·14

and sourness when presented with solutions such as those listed in Table 1.

Daget (1977) proposed a similar set of solutions (Table 2) except she included two extra substances, sodium carbonate and potassium alum, to yield an alkaline taste and astringency. While the weight of scientific opinion is that there are only four basic tastes, others have questioned this assumption. Schiffman and Erickson (1971) observed that there appears to be an alkaline taste as judged from multi-dimensional scaling of various

TABLE 2
SOLUTIONS RECOMMENDED BY DAGET TO TRAIN ASSESSORS TO RECOGNISE TASTE SENSATIONS
(From Daget, 1977)

Substance	Concentration (% weight per volume)						
Sucrose	0	0·15	0·35	0·55	0·75	1·00	1·50
Sodium chloride	0	0·02	0·04	0·06	0·08	0·12	0·16
Citric acid	0	0·001	0·001 5	0·003	0·005	0·01	0·02
Quinine sulphate	0	0·000 1	0·000 2	0·000 4	0·000 6	0·000 8	0·001
Sodium carbonate	0	0·005	0·01	0·02	0·04	0·06	0·08
Potassium alum	0	0·005	0·01	0·02	0·04	0·06	0·08

chemicals. For example, NaOH and Na_2CO_3 fell outside the tetrahedral structure envisioned by Henning (1916). Amerine et al. (1965) describe alkaline as 'a taste sensation usually attributed to a combination of sourness and bitterness (and possible tactile) stimuli'. Based on factor analysis, Anderson and Hartmann (1971) concluded that the primary taste of saltiness consists of two independent moduli. Yoshida (1963) also considered that two substances, potassium alum and monosodium glutamate, violated the tetrahedral structure. Schiffman et al. (1981) concluded that taste is not limited to the sweet, sour, salty and bitter domains. McBurney and Gent (1979) criticised the experiments of Schiffman and Erickson for not having eliminated olfactory stimuli entirely and for choosing to believe one form of multidimensional analysis rather than another.

Since this subject is not central to the training of assessors it should not be prolonged. It is appropriate however to point out that other groups also recommend that more than the four basic tastes be considered. The International Standards Organization (ISO) is in the process of establishing guidelines for choosing, training and monitoring the performance of assessors (ISO, 1987a). It specifies tannic acid, quercitin, or potassium aluminium sulphate as substances which illustrate astringency. The ISO is in the process of revising its Standard for the Determination of Detection Thresholds (ISO, 1987b). In the latter the possibility of using NaOH is under consideration. It is claimed that a solution of it illustrates the sensation of 'metallic' and of 'being the opposite of acid, that is alkaline'. O'Mahony and Tsang (1978) had monolingual Americans and bilingual Cantonese describe the taste sensations imparted by a solution of Na_2CO_3. Terms such as bitter, tasteless, ammoniacal and others were given, but the term, alkaline, was not. Saltiness was used only in English by the bilingual Cantonese panelists; it was not used by the monolingual American. In Cantonese, the term, *garn soyeu* (baking soda) was applied as was *fay dzo may* (soapy). The ISO report (1987b) further specifies that sodium glutamate should be included because the glutamate sensation 'is totally independent from the traditional tastes'. The basis for the last statement is not given, but it is in line with the report of Yoshida (1963), and somewhat so with that of Ishii and O'Mahony (1987).

Actually, for applied descriptive analysis, whether there are four basic tastes or there are more does not really matter. If assessors consider that saltiness does not really express all the dimensions of alkaline materials or salts, then a fifth term is useful, whether it represents taste or some other sensation. Caul (1957) pointed out that by adjusting concentrations of sweet, sour, salty and bitter chemicals in water, trained tasters could not distinguish components, only that they were low-body or high-body blends.

To return to the matter of training and intensity scores, if the scores assigned by the trainees do regress on concentration, the trainees, as already stated, may need further training to bring their scores into line with each other. The upper part of Fig. 1 illustrates the problem by showing scaling of the sensory attributes of tea by three assessors out of a panel of nine. All were 'good' panelists in the sense that their scale values differed according to the products, thus they were discriminating among the products on sensory grounds, but it would have been more desirable if all had used the same portion of the scale. Assessor 2, for example, was a good

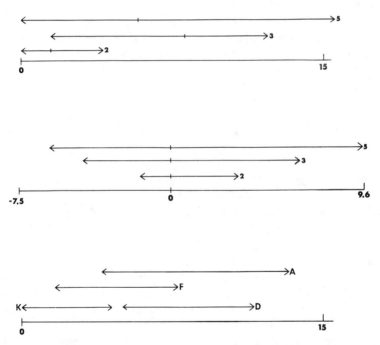

FIG. 1. Diagrams showing causes of interaction. The upper diagram shows the scale ranges used by three individuals being trained to evaluate tea. Each assessor was effective at discrimination, but by using different parts of the scale they caused product–assessor interaction. The middle diagram shows results for the same three assessors except their scale values have been standardised. The bottom diagram illustrates interaction caused by the assessors being both the subjects and the evaluators of the sensory attributes of sanitary pads.

assessor in that she could differentiate among the intensities of the attributes in the teas and was consistent in doing so. Similarly, assessors 3 and 5 were good assessors, but they used a larger portion of the scale. Discrepancies among the assessors in scaling is a problem common to all methods and it must be reduced to a minimum. Undoubtedly FPM is the method which places the greatest stress upon training assessors with the result that they do begin to act as analytical instruments rather than instruments somewhat variable in response.

2.1.2. Using the Product Itself
The second step in developing the facility to make intensity evaluations is to employ the food which itself is to be evaluated. If one were being trained to evaluate coffee, for example, coffee would need to be brewed by the various means commonly used, using water varying from soft to hard. All brands of coffee on the market should be purchased and examined to learn the range of intensities which exist among different kinds, blends and brands of coffee. Coffee should be made from freshly ground and 'flaked' coffee beans which have been slightly under-roasted, properly roasted, and over-roasted to demonstrate the influence of these variables on flavour and aroma. Possibly, samples should be spiked with chemicals known to be present in coffee to give the trainees the experience of evaluating the intensities of the different attributes against the background of all the other substances in coffee. The objective of this training is to show the panelists how intense or weak the products might be in each attribute being evaluated.

Sometimes reference compounds do not exist or are not known for particular odours (flavours); if this is the case, another substance may be needed to enhance a particular odour or flavour note even though this second material may impart other characteristics. Probably chicory would already be a component of some of the commercial coffee specimens purchased inasmuch as it is a common adjuvant. If not, it might be used.

Other substances which are sometimes mixed with coffee, such as roasted grains, might also be used. Just as for pure substances used as reference compounds, the sensory leader is attempting to illustrate to the trainees the kinds of flavours likely to be encountered, ranges of intensities, and when substances are added to food, possible interactive effects between the added substance and substances already present. Even pure compounds do this. There may be masking or synergistic effects between the added compound and those native to the food. Lawless (1986) is among the most recent to add to our storehouse of knowledge on this subject.

The foregoing presupposes that a set of descriptive terms for the different attributes of the sensory product being examined exists so that the assessors can be instructed about the specific characteristics to be scaled. Descriptors will be discussed later.

2.1.3. Subjective Decision

While exquisite attention is given to the training of flavour profile assessors and the conducting of the assessment itself, a major weakness is that judgement concerning differences in profiles is almost always purely subjective. In the hands of an experienced sensory leader, subjectivity is no more objectionable than is the evaluation of music or art by one qualified to judge composition, media and technique, but being so subjective, the decisions made are likely to be parochial to the particular group of judges and their leader. Proponents of FPM would of course dispute that statement. Sjöstrom (1956) reported that 21 panels trained by the Arthur D. Little Company (ADL) all obtained profiles that agreed, but Amerine *et al.* (1965) pointed out that no details were provided as to the commodity tested or to the degree of deviation in responses to individual attributes, and that it is difficult to define reproducibility without some statistical test. A manual prepared by ASTM Committee E-18 (1968) states that analysis and interpretation of flavour profile results are the responsibility of the panel leader, and that normally the results are not treated statistically although statistical assessment is possible. Søftoft (1974) employed replication and statistical analysis of the scores assigned. Naturally it is not complicated to apply statistical analysis to intensity scores; even the order of perception can be analysed as such or expressed as angles and then analysed. The originators of FPM disclaim the need—even the desirability—for statistical analysis (Caul (1957)). Caul states: 'Unlike difference tests, the profile is not concerned with precision to the extent of considering single qualities at a time in order to obtain results which can be analyzed statistically.' Nonetheless at a symposium honouring Sjöstrom, one of the originators of the method, a paper was presented on that very subject. Miller (1978) described a statistical method for the treatment of flavour profile data. Recently ADL itself (Hanson *el al.*, 1983) described a quantitative process called 'profile attribute analysis' which was stated to be an outgrowth of FPM but, being quantitative, permits application of robust statistical methods. To return to FPM, subjectivity in making a decision is not necessarily a fatal flaw. None of our sensory methods really gives us an absolute evaluation. Almost invariably, one thing is being compared with another or others. An experienced sensory leader may very well be able to interpret relative differences more sensibly than someone who has statistical output at hand but who lacks sensory experience. There is often a difference between importance and statistical significance. Furthermore, there are some things which defy statistical analysis. A quotation with respect to the savouring of wine is perhaps appropriate:

> Once beyond the point at which quality could be marred by faults, quality ceases to be measurable, ceases almost to be analyzable. When you look at a picture that is particularly fine or listen to a piece of music that's good or read something that impresses you, you don't measure the brush strokes, count the notes or parse the sentences. There is an effect, an impression. And with wine, too, we should allow ourselves to enjoy the effect, the impression.
>
> (Asher, 1976)

While FPM itself involves measuring the brush strokes and counting the notes, decisions about the things the brush strokes or the notes stand for depends upon experience and judgement. Since not only thorough training of the panelists is required but also wide and deep experience on the part of the sensory leader, judgement is likely to be sound, though subjective. Because judgement is so dependent upon the experience and common sense of the sensory leader, this does limit application of the method. The procedure is not for everyone. Other QSP methods which rely more upon statistical analysis to lend objectivity to assessment and to segregate variance by main and interaction effects do not require the same degree of excellence as is required if the FPM is to work at all. The above is not intended to denigrate the need for training when other QSP methods are utilised. The leader must be experienced and the assessors must be trained, but training for the assessors is not as exhaustive nor as broad as it has to be if FPM assessors are to do the job expected of them.

2.1.4. Amplitude

There is another side to flavour profiling which is subjective—assessment of the 'amplitude' of the product. Amplitude has been defined in several ways. ASTM Committee E-8 (1968) used a definition commonly employed:

> Amplitude is the initial overall intensity impressions including both the separately identifiable factors and the underlying unidentifiable part of the flavour complex. It is based on fullness, degree of blending, quality of separate factors appearing as either first or last impressions, and appropriateness of factors for the product.

Caul *et al.* (1958) state that panelists should assess the amplitude before they concentrate on the individually detectable character notes. Furthermore, she and her co-writers define amplitude as the 'breadth' of flavour. To return to the quotation from Asher (1976) above, amplitude basically is 'the effect, the impression'. The product is being judged whether it is too weak or strong in overall character, whether some note(s) is so strong as to lead to imbalance and whether the flavour is appropriate for the product. In essence, judgement as to fullness, balance or harmony and appropriateness of flavour are rendered as a composite decision.

2.1.5. Assessment Itself

Once the panel has arrived at a consensus as to a set of descriptors, has been sufficiently trained and reaches agreement about the order of perception of the different sensory notes, then it is ready to begin to evaluate the product. Flavour profiling differs from other methods of analysis in that the assessors usually do not act independently. They make their judgements separately, but then there is discussion among the panelists concerning the extent to which a particular note should be scored. While values are not averaged, there is an attempt to arrive at a consensus. Profiles are not usually constructed until this consensus is reached. At initial and intermediate stages, the scores for intensities of the various character notes and orders of perception may merely be tabulated. Once there is general agreement regarding assessment of the different products, then profiles are constructed. Caul (1957) cautions that the panel leader should translate a final profile in one or two sentences to the project leader. A stark profile with no explanation should not be presented since there is the risk that the project leader may attempt to refine interpretations beyond the limits of sensory evaluation. Table 3 illustrates the type of data acquired for two grape jellies and Fig. 2 shows their profiles. Note that the area of the hemisphere is larger for the jelly assigned a larger value in amplitude. This is one of the specifications of the method itself.

2.1.6. Further Considerations

Arriving at a consensus is possible in flavour profiling because the typical panel is made up of so few members. Since the training period is so long and costly, and since profiling itself, including the discussion time, takes the panelists away from other duties for rather lengthy periods, panels usually consist of approximately only five individuals. The cost of flavour profiling also means that the employer has to be fairly sure that the assessors will stay with the company for some time. Appel (1985) stated that McCormick &

TABLE 3
TYPICAL FPM TABULAR ASSIGNMENT OF INTENSITY VALUES

	Good grape jelly	Poor grape jelly
Aroma		
Amplitude	2·5	2·0
Grape, concentrated	2·0	1·0
Methyl anthranylate		2·0
Sweet, syrupy		1·5
Sour	1·5	
Sugar lag	1·0	
Grape mare	0·5	
Flavour by mouth		
Amplitude	2·5	1·0
Sweet	2·0	2·5
Grape, concentrated	1·5	
Methyl anthranylate		1·5
Sour	1·5	1·0
Astringent	1·0	
Drying		2·0
Gelatin		0·5
Aftertaste		
Sweet		Throat-catching
Grape		cloying sweet
Sour		

Company uses a formal 14-month employee training period and that 'judges... must be available for the future'. Replacing a panelist is difficult and expensive because flavour profiling depends upon all panel members acting as analytical instruments and being in good agreement with each other. However, the panel members do not have to agree with each other in their final decisions: each panel member needs to have a mind of his own. The assessors should neither be lions nor lambs. For a flavour profile panel to work, each member needs to be able to make his or her own decision, and not be led around by the nose by someone else who is more assertive. Having made his or her decision, the panelist at intermediate stages may still make some adjustments in the intensity values assigned if this seems reasonable to further a consensus. Even at the final stage some adjustment in score may still be appropriate to lead to greater unanimity of judgement provided a panelist does not feel he is compromising his own sense of judgement. An experienced and wise leader is necessary to bring about this accommodation and amalgamation of views.

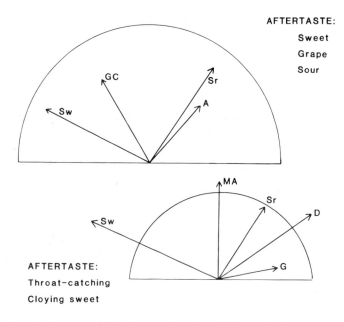

FIG. 2. Typical flavour profile. The symbols Sw, GC, Sr, A, MA, D and G stand for sweetness, grape concentrate note, sourness, astringency, methyl anthranylate note, dryness and a gelatin-like note, respectively. The area of the lower figure is smaller than the upper figure because the amplitude rating for the former was less than that for the latter.

2.2. Quantitative Descriptive Analysis

Next, QDA will be described to illustrate the requirements most present-day QSP methods specify. The various QSP methods differ from FPM in several important respects. They often involve the profiling of texture, appearance and noise attributes as well as odour and flavour. The texture profile method (Brandt *et al.*, 1963) will not be discussed in this chapter because texture evaluation is being considered elsewhere. In some respects, texture profiling is more difficult than flavour profiling. Instead of being able to use pure chemicals as reference substances, for the most part, commercial materials have to be used (Szczesniak, 1963, 1975; Szczesniak *et al.*, 1963). This generates problems. Particular commercial products are not always available, they may purposely not be manufactured identically in different countries, or identically in the same country at a different time, their use may be taboo for religious or cultural reasons. Muñoz (1986) proposed new reference materials to avoid some of the difficulties just mentioned. Another cause of difficulty, as Civille (1979) pointed out, is that a panel can be asked to give a rating to the intensity of overall flavour but not the texture.

2.2.1. Specifics of the QDA procedure
Unlike FPM, QDA makes replication and statistical analysis of the data mandatory, and QDA panels, unlike an FPM panel, are normally charged with the evaluation of only one type of product. Different panels are trained for different products. Many laboratories do not follow the QDA procedure exactly though they do adhere in general to its precepts. Before citing salient points of departure, perhaps a summary should be given of the differences that Stone *et al.* (1974) cite as distinguishing QDA from prior profiling techniques. They stated that it was their objective to originate a method which, rather than relying on the judgement of an expert (or the collective judgement of selected experts), would rely upon 'trained' assessors. This was intended to be accomplished on three grounds. The assessors were to be trained using the test products, not model systems; judgements were to be replicated; and, thirdly, assessors were to be elected based upon statistical examination of their performance. To achieve a good statistical analysis of the performance of the assessors, Stone *et al.* (1974) envisioned that as many as 12–16 replicate judgements per assessor might be needed. Since that time their experience (Stone *et al.*, 1980) and ours (Powers, 1981*a*) is that four to six replications usually suffice. Stone *et al.* (1974) stated that the panelists and the leader should participate in the development of the language for QDA but that the leader would not actively participate in product evaluation.

Another specification was that a line scale be used instead of category scoring. (They cite a few studies in support of the use of a line scale compared with other means of rating attributes for intensity.) An anchor point in the middle of the scale was originally included as an optional feature; it is considered not to be necessary today (Stone *et al.*, 1980). Unstructured line scaling is now quite common. The line scale is generally 15 cm in length with a vertical line at the left end and another vertical line about 1·3 cm from the right end of the line if the scale runs from 'absent' of intensity to the term used to designate the maximum; if the term to the left is 'slight' then the vertical line at the left is also 1·3 cm from the end. This provides a little leeway for an assessor who has used all the scale, then encounters a sample which he or she thinks should be rated stronger (or weaker) in intensity than any sample previously encountered. Arguments in favour of the line scale are that the data are continuous and the assessors are not constrained to make their perceptions match particular categories such as 1, 2, ..., 8, 9. Gacula and Washam (1986) are among the latest to reaffirm that assessors in fact do not hold to equal intervals in employing category scoring. Actually, while a category scoring system is described for FPM, there is nothing to stop an assessor from assigning a value intermediate to the designated score levels. Continuous data such as scale values have at least two advantages. First, as already mentioned, assessors do not have to fit their perceptions into specifically designated categories; second, continuous data are more likely to meet an assumption implicit for ANOVA, namely, the errors are normally distributed. A difference between QDA and FPM is that in QDA a new assessor should be able to be trained after no more than 20 h training (Stone *et al.*, 1974).

Apart from particular sensory features of the method, specific suggestions were made in some instances as to how the data were to be analysed statistically; in other instances comments were made regarding other kinds of statistical analysis which might be appropriate, and a means of graphically depicting differences among products was proposed.

In essence then, QDA boils down to screening would-be assessors for possible membership on a panel, developing a list of descriptive terms, training judges, using sufficient replication so that the performance of the assessors, the effectiveness of descriptive terms, product differences and possible interaction effects may be isolated and evaluated by statistical analysis, and expressing the results graphically as well as numerically.

3. PROCEDURAL STEPS THEMSELVES

In the following sections, the term QDA will be used where reference is specific to the QDA method, but most of its specifications apply to any of the QSP methods. If QDA is used for accuracy of citation, the reader should interpret the recommendation as applying to QSP methods in general. Where a statement is specific to QDA only, it will be so stated.

3.1. Screening Would-be Assessors

Zook and Wessman (1977) suggested that candidates for QDA panels be screened for their discriminating ability by applying the triangular test. They considered that a would-be trainee for descriptive analysis should be at least 70% successful in selecting the odd sample; otherwise the individual should be rejected.

Word and Gress (1981) similarly recommended the use of the triangular test except they coupled it with sequential analysis to lessen the amount of testing required. Gacula et al. (1974) also used sequential analysis. Candidates who demonstrate ability to discriminate are permitted to go to the next stage, which is training in descriptive analysis.

3.2. Training

Training for QSP involves examination of the test products themselves. Word and Gress (1981) trained their assessors to evaluate the properties of beer by having the trainees evaluate beers formulated to be different in their sensory attributes, by using competitors' products and by 'spiking' beer with chemicals to induce particular flavour characteristics. Although QDA specifies that training is on the test material itself, other systems sometimes include model test substances. Word and Gress (1981) stated, for example, that they used solutions of sucrose, NaCl, citric acid and caffeine to learn whether the would-be assessors could differentiate among common taste sensations and could rate their sensations for intensity. Wilkin et al. (1983) described in detail the training of quality-assurance assessors to evaluate 'on line' the organoleptic quality of rectified spirits. High-quality neutral spirits, 'strong-smelling' spirits and reference compounds were used. Many laboratories do not confine their training to the use of test materials. Noble and Shannon (1987), for example, used such things as one frozen boysenberry to illustrate a berry note in Zinfandel wine and a mixture of vegetable substances to illustrate a vegetative note. The latter consisted of $3\,mm^2$ (green?) Bell pepper, 3 ml brine from canned green beans, $\frac{1}{2}$ ml brine from green olives, 10 ml brine from black olives. Not only did they use the mixture above—and other reference materials—during the training phase; for aroma, they provided the reference materials at each of the formal sessions. Reference substances to illustrate flavour and textural attributes were not provided at the trial itself.

Whether only the material under test is used for practice or other substances are also used, practice has to be acquired in scaling of the sensory materials for the intensities of the different attributes agreed upon. For training, usually not less than four replicate sessions are held; more may be needed. As in training for FPM, the samples are brought closer and closer together in terms of their differences by blending samples or by selecting samples known not to differ greatly in their sensory qualities. ASTM Committee E-18 (1981) describes the procedures generally followed to train panelists for descriptive analysis.

3.3. Development of Descriptive Terms

One of the problems inherent in all methods of descriptive analysis is the development of a list of descriptors characteristic of the particular substance being evaluated. The vocabulary may have to be formulated from scratch as a part of the training, or there may have been prior development of a vocabulary which the assessors now have to be trained to use.

Regardless of how the set of descriptive terms is formulated, the objective is to have the set and the products interchangeable in identity. The characteristics of the products are the genesis for the set of terms. The set should be so pertinent and encompassing that once intensity values are attached to each descriptor, the descriptor set is now specific enough to identify a particular sample. While the intent is to develop a set of terms just as specific as possible, specificity can be carried too far. To a chemist or a flavourist, the odour note of some chemicals is so distinctive that the odour might be designated by the name of the chemical causing the odour. One industrial laboratory would not accept the term, phenylethanol, used by one of its panelists, a chemist. There is no question that the sensory leader or the panelist originating the term could have taught the rest of the panel to identify the odour as that of phenylethanol, but the sensory leader went the other way. Most of the panelists had used the term, 'lilac-like'. That was the term selected. 'Rose-like' probably would have been better, but the note reminded the panelists of lilacs; consequently it was used. The company wanted to couch its set of descriptors in terms consumers use. Sometimes it is important to foresee a later need to relate laboratory observations with terms consumers use. However, more often the opposite course of action is chosen. Bone (1987) stressed the desirability of devising 'nuance-rich' language to depict concepts for new products if management really hopes to have product developers arrive expeditiously at company goals. Szczesniak and Skinner (1973) examined the meaning laboratory panelists and consumers tried to convey in characterising textural properties. The same has been done for sensory characteristics of fish (Sawyer *et al.*, 1984, 1988).

Practices followed in settling upon a descriptor list naturally differ according to the purpose of testing and the QSP system used. In the example above, the firm had a good reason to make provision for possible correlation with consumer results still to be secured. Some QSP systems would proscribe the use of a term when a more specific one exists. Meilgaard *et al.* (1987) caution about this with an illustration consisting of vanilla and vanillin. Use of the term, vanilla, would be disallowed if the sensory note were really that of vanillin. The two substances differ in their sensory characteristics.

Liberty to use a term is also determined by the QSP system adopted. Spectrum™,* a procedure devised by Civille (Meilgaard et al., 1987), depends upon the assessors being taught to use the terms contained in a series of lexicons. They exist for the main sense modalities, i.e., colour, texture, etc., and for some particular products such as skin care items and fragrances. Reference substances are provided. Scale values are designated for the intensity of the different reference substances. The sensory leader is a teacher. The assessors are taught to use the system. They have little to say about the procedures. An objective of Spectrum is to lessen subjectivity in the procedure followed by providing a standard method rather than allowing the panelists to formulate part of the procedure (the set of descriptors) for themselves.

Other systems, such as QDA, were designed to have the sensory leader be more a facilitator in the selection of the descriptor list than to be a teacher of a fixed list. In an industrial laboratory where the QDA method is used, the practice is for the sensory leader to be responsible for the preparation and presentation of test samples known to differ, but the assessors develop the language by themselves. The leader facilitates the process of arriving at a consensus of terms which are appropriate and understood by the panelists, but rarely would she override decisions made by the assessors. That degree of liberty can be permitted only in a laboratory where the assessors are experienced and are developing a set of terms for a product new to them. If they were being indoctrinated to QSP for the first time, the sensory leader almost certainly would be involved in the development of the vocabulary. Normally a vocabulary has to be originated by sampling food and discussing the appropriateness and meaning of the terms among the assessors and with the sensory leader (Caul and Vaden, 1972). Whether a term describes a major or a minor character note, and whether it is a key to discrimination of the products, determine how appropriate and valuable it is.

Various attempts have been made by different groups to develop glossaries of terms intended to be adequate for all types of descriptive analysis and for various commodities. Most widely known is the list developed by Harper et al. (1968a,b). The list consists of 44 items.

---

* Trade Mark applied for, Sensory Spectrum, Inc., East Hanover, New Jersey 07936 USA.

Furthermore, Harper and his co-workers attempted to locate chemicals which would possess the odour characterised by each descriptor (Land *et al.*, 1970; Harper, 1975). While useful, the list has not entirely filled the objectives of the originators because the list of terms is not sufficient to embrace the 10 000+ kinds of odours which exist. Other groups have formulated lists for specific substances, e.g. tea (Werkhoven, 1974); beer (Stewart, 1961; Clapperton, 1973), or various schematic ways, e.g. flavour wheel for beer (Clapperton *et al.*, 1975; Meilgaard *et al.*, 1979*a*), concentric triangles for wine (Vedel *et al.*, 1972), a flavour wheel for wine (Noble *et al.*, 1987) and one for Scotch whisky (Shortreed *et al.*, 1979; Kluba, 1986). Piggott *et al.* (1981) examined 177 reference chemicals as odorants, then by principal component analysis (PCA) established relations between these chemicals and 44 descriptive terms.

Of the industries or groups above, brewers have done the best job of coordinating their efforts and in reaching agreement relative to a set of descriptive terms (Clapperton *et al.*, 1975; Meilgaard *et al.*, 1979*a,b*). Among the agencies involved were the European Brewing Conference, the Institute of Brewing (London), the American Society of Brewing Chemists and the Master Brewers Association of the Americas. Curin (1976) translated the terms into Czech, French and German. The terms have also been used in Russia (Konovalov, 1982). Spanish versions exist (Clapperton *et al.*, 1976; Meilgaard *et al.*, 1979*a*), as does a Finnish one. In fact the flavour diagram for beer is probably the set of terms translated into more languages and used most commonly the world over than any other set of terms. Whitear *et al.* (1979) selected 37 of the 122 terms for use in daily testing. Madsen (1980) added 22 terms to make a total of 144. Meilgaard (pers. comm.) states that he has taken part in the testing at the Carlsberg Laboratory and that the tasters are not unduly bewildered by the many terms although they admit that some are ill-defined and a handful have not yet been used. Olshausen (1957, 1971) has also proposed a list of terms. Macher (1957) suggested and defined several terms. Meilgaard *et al.* (1982) described purification methods and the sources of 27 chemicals studied as reference chemicals to illustrate the flavour note each term is intended to describe.

Just as the brewers initially did and the Scotch whisky and the California wine scientists did later, a sensory group in most instances has to formulate a list of descriptors for itself. Illustrations of this are: canned corn (Dravnieks *et al.*, 1973); apple juice (Dürr, 1979); apples (Williams and Carter, 1977); cider and perry aroma (Williams, 1975); axillary and vaginal odours (Dravnieks *et al.*, 1968, 1974; Dravnieks, 1975); water (Tuorila *et al.*,

1980). In examining a wide range of the product under investigation to ascertain the kinds of flavour or odour notes which exist, the panelists are asked to write down descriptions of the notes they detect or the things the product reminds them of. Round-table discussions are then held to decide whether a term suggested by one individual is equivalent in meaning to that suggested by someone else, and attempts are made to get the individuals who did not initially detect certain notes to develop the facility to perceive them once their presence is called to the panelists' attention. Reference substances may need to be used to do that. Unless the sensory leader and the panel is starting to work with an entirely new product, the list may not need to be developed *ab origine*. Preliminary trials and round-table discussions may merely be needed to augment a prior list or to refresh the memories of panel members with terms and characteristics already known to them.

An aid in the development and collating of a list of descriptor terms is to prepare extracts of the product and to have the would-be panelists sniff the various peaks at the effluent port of a gas chromatograph. In this way they familiarise themselves with the effects of concentration on odour character. If the panelists fail to detect a certain note in the food itself, it helps to let them familiarise themselves with the specific odour, free of others which exist in the product (Powers, 1981*a*). Once one has 'homed in' on the characteristics of a particular compound, one is then often able to detect the note in a food even though it might not have been initially detectable because of inability to single it out from amongst all the other odours of the product. Warren (pers. comm.) points out that the same procedure is being used in the training of expert perfumers and flavourists. The long apprenticeship can be shortened somewhat by having novice perfumers and flavourists sniff relatively pure substances at gas chromatographic effluent ports as well as on perfumers' sticks. Examination at the effluent port of a gas chromatograph should not be assumed to be free of error. Williams and Tucknott (1977) and Dravnieks *et al.* (1979) have commented upon some of the problems. Jellinek (1964) and Tilgner (1962) pointed out that dilution of the samples often permits particular notes to be made more demonstrable. Sjöström and Caul (1962) took a dim view of dilution relative to profiling itself. They pointed out that profiling is intended for the food as it normally exists.

Others have approached the problem of devising a glossary of terms in a different way. Instead of trying to synthesise a list, ASTM Committee E-18 worked from the opposite direction. It sought out from all kinds of industries producing materials with sensory properties (paints, glue, waxes, plastics, foods, beverages, toilet goods, perfumes, confections) terms which the particular industry uses. It then had teams of individuals take 50 of the 837 terms accumulated and scale these terms for degree of similarity to other terms suggested and for appropriateness. Ultimately the list of terms was boiled down to 146 (Dravnieks et al., 1978; Dravnieks, 1982). Over the course of three years, 160 individuals scaled the intensities of five chemicals monthly to establish which of the 146 descriptive terms applied to the characteristics of a given chemical and, if one or more did, how strongly. The profiles have been published by the ASTM (Dravnieks, 1985).

Few chemicals have one odour note only. Many of them may be reminiscent of several substances, but to varying degrees. There are often nuances of odour generated by concentration alone, as illustrated by dimethyl sulphide. At exceedingly low concentrations, many individuals find this odour reminiscent of cooked corn whereas at higher concentrations it is likely to be given a more chemical description such as 'sulphidic'. Forss (1981) states that trans-2-nonanal, just above its threshold of 0·1 ppb (parts per billion, where 1 billion $= 10^9$), possesses a woody character. Above 8 ppb it smells fatty, becomes unpleasant above 30 ppb, and in an aqueous solution at 1 ppm it has a strong flavour reminiscent of cucumber. Pino et al. (1986a) made similar observations with respect to concentration becoming too great.

Civille and Lawless (1986) have set forth several criteria descriptors should have. They pointed out that language plays a central role in determining the accuracy and potential benefits of evaluation (such as that of QSP). Lehrer (1983), who is a linguist, and has a special interest in the use of language to describe wines has pointed out (Lehrer, 1985) that the set of terms developed should be such that one would recognise the identity of a wine from the description of it.

There are procedures other than those summarized above to describe the sensory characteristics of foods. One of these is known as 'free-association' language (Moskowitz, 1982) or 'free-choice' profiling (Arnold and Williams, 1986). Williams has been a proponent of free choice instead of a set of fixed terms to describe sensory attributes (Langron et al., 1984; Williams et al., 1984; Williams and Arnold, 1985). Because of the widely varying configurations which result, Procrustean analysis or some other technique needs to be used to bring the word configurations into closer harmony with each other. Vuataz (1977) was an earlier user of the Procrustean technique and he in turn cited even earlier users such as Banfield and Harris (1975).

Free choice profiling is useful in basic studies such as Lehrer (1983) conducts where one of her objectives is to study lexical relations. It is also useful for exploratory applied research and possibly in the early stages of product development, but it is highly questionable as a fit alternative to a fixed set of terms for most industrial applications. There needs to be comparability between branch plants or even within the same plant at different times because of the volume of testing or as work shifts change. Variation in the profiling technique needs to be held to a minimum so that the minor differences which exist in most commercial products can be resolved.

Civille and Lawless (1986) have pointed out the importance of language in describing perceptions, and they gave criteria for the establishing of terminology. Lehrer (1985) has likened the network of decriptors generally required to a fishing net firmly fastened at its four corners to a wall, but at each intersection of the cord of the net there is play. Even when panelists agree fully upon the meaning of a term, there is usually a certain amount of play in the way they apply that meaning. In spite of lexical and evaluation difficulties, description has to be made comprehensive and particular for the products to be examined if profiling is to be as beneficial as it can be. Lyon (1987) utilised free choice in the early stage of her project; she followed that process later with factor analysis (FA) to cull the word list to a lesser number of words and factors. Her study will be described more fully when we come to MVA, for the procedures she followed are more common than trying to bring harmony to differing sets of terms through the use of Procrustean analysis. She did take advantage of the standardised lexicon of warmed-over-flavour descriptors for meat developed by Johnson and Civille (1986). Many of the words her panelists used were descriptors previously originated by others, but the panelists were free to use any associative or cognitive word which seemed appropriate to them.

3.4. Variations in Methods

Once descriptive terms have been settled on and the panelists selected and trained, then the next task is to evaluate the test products themselves. Once that is done, the burden of moving forward reverts to the sensory leader. He or she has to analyse the data using appropriate statistical means. Before undertaking a discussion of data analysis some of the variations in application of profiling and quantitation of responses not already referred to should be mentioned.

Søftoft (1974) used a profiling technique based on FPM with a fixed set of terms, a three-point category scale and replicated the assessors' judgements three times so that statistical analysis could be applied to the data. Pokorńy and Davídek (1986) and Peppard (1985) employed modified versions of FPM. Baines (1976) also used flavour profiling. He had his assessors train for examination of toothpaste by brushing their teeth with an unflavoured base and with various types of flavouring component added to the base to build up a vocabulary of flavour notes and to enable the assessors to gain practice in describing events in their mouths in terms of these notes. Instead of the usual round-table discussion with the sensory leader, the Delphi technique (Jolson and Rossow, 1971) was used.

Independently, the assessors scaled the attributes. The means and ranges were tabulated. The panelists then carried out their second replication except now they had before them information from the first replication regarding the order of perception, the means and the ranges. At this stage they could utilise this information in making any adjustments they wished in their decisions. The same procedure was followed for the third replication. Baines pointed out that three or four repetitions were usually enough to produce a satisfactory tight range of evaluations. From the data set, traditional flavour profiles were constructed. Cross et al. (1978) used an 8-point scale with some features drawn from flavour profiling and texture profiling (Cairncross and Sjöström, 1950; Brandt et al., 1963; Szczesniak et al., 1963) but with the statistical analysis pretty much in line with that of QDA. Piggott and Jardine (1979) used a QSP method, then applied MVA to their results. Moskowitz (1982) employed both 'free association' language, as already noted, and a fixed set of terms to profile lotions and creams. Moskowitz has been among those who regularly use magnitude estimation for profiling (Moskowitz, 1977, 1979). Many others have utilised it, among them being Jounela-Eriksson (1982) and McDaniel and Sawyer (1981a,b). Various category scales have been employed (Mecredy et al., 1974; Brown and Clapperton 1978; Fjeldsenden et al., 1981). Scaling against a reference has been utilised by Vuataz et al. (1975), Powers et al. (1977), Wu et al. (1977), Godwin et al. (1978), Noble (1978). Among those who have used untrained panelists are Palmer (1974), Wu et al. (1977), Clapperton and Piggott (1979a), Lundgren et al. (1979). Moskowitz (1979), Piggott and Jardine (1979) and Williams et al. (1986). The panelists employed by Wu et al. (1977) commendably handled quantitative descriptive profiling. Part of this success may have been because directions were carefully written (Powers, 1982a). Andersson and Lundgren (1981) commented upon this same point. Sidel and Stone (1976) stated that QDA is intended for trained assessors, but our experience (and apparently that of others) is that the innate skill of untrained panelists to respond consistently and quantitatively when carrying on QSP is often underestimated.

The papers cited immediately above are all by accomplished and knowledgeable sensory analysts. No one QSP method is clearly superior to all others. Quite naturally the originators of a particular method often feel that their method should remain inviolate, that any change detracts from its validity or effectiveness. Stone *et al.* (1974) state that QDA is a total system covering sample selection, judge screening, language development, testing and statistical analysis. Sjöström and Caul (1962) bridled when Tilgner (1962) chose to tamper with the FPM by suggesting that dilution of samples had its benefits. There have been no full-scale comparisons of QSP methods, only study of individual components such as scaling or scoring. Without question Pearce *et al.* (1986) conducted one of the most thorough studies. They compared two methods of scaling and one of scoring. Exquisite care was devoted to planning, instructing the assessors and in adhering throughout the study to sound sensory principles. They concluded that all the scaling procedures yielded results very much alike. Lawless and Malone (1986) compared four types of rating scale: 9-point category scales, line marking, magnitude estimation and a hybrid of category and line scale. All methods generated highly significant differences among products; however, category scaling generally had a modest advantage. Riskey (1986) pointed out that category ratings have the advantage that the manner in which they are affected by situational or contextural factors is well understood. The very last point is questionable, but he is probably right in claiming that such effects on category ratings are better understood than for most other methods.

4. EVALUATION OF DATA SETS

For any of the QSP methods, statistical analysis needs to be carried out to do the following: evaluate the performance of the assessors; determine whether the descriptors are merely that or are discriminators; ascertain whether differences in replicate treatment are a significant cause of experimental error; examine some or all of the first-order or higher-order interactions possible; and then, of course, determine whether there are significant differences among treatments (products). Replication is mandatory. Without it, the tests either cannot be made or they are of doubtful value.

4.1. Evaluating Assessor's Performance
Following training and also actual trials themselves, the performance of each assessor is tested statistically. One- or two-way ANOVA is used. For

each assessor and descriptive term, the F-value between treatments is calculated. The F-value itself is generally not used but rather the probability of the F-value having occurred by chance. Table 4 shows probability values for the performance of seven individuals being trained to evaluate tea as part of a class exercise to illustrate the application of QDA. Most computer programs print out the probabilities as well as the F-values themselves. If a microcomputer or a program is used where that is not done, then naturally recourse will have to be made to tables. Stone *et al.* (1974) stated that any assessor whose F-value for performance has a probability ≤ 0.50 is contributing to discrimination. That is so, but there should then be the question whether discrimination is sufficient at the 0·50 level to aid in sample differentiation or performance is so minimal that it is obscuring product differences rather than illuminating them.

In Table 4 only here and there does an assessor have an F-value, the probability of which is >0.50. There is no problem as to the trainees evaluating tea. So many of their probability values are below 0·50 that almost any laboratory would retain them.

If performance were considerably poorer, candidates for membership on a panel would be given additional training, probably confining it to those attributes which gave them difficulty. Their probability values would be compared with the members of present panels. Attempts would be made to bring their performance up to the range of the panel, or, if performance did not improve with additional training, they would be rejected.

The whole subject of decision-making with respect to assessors is not as

TABLE 4
PROBABILITY LEVELS ASSOCIATED WITH F-VALUES, AS A BASIS FOR DECISIONS AS TO WHETHER ASSESSORS ARE PERFORMING SATISFACTORILY

Terms	Assessors						
	1	2	3	4	5	6	7
Fish	<0·001	<0·001	<0·001	<0·001	<0·001	<0·001	<0·001
Flowery	0·218	0·019	<0·001	0·205	<0·001	<0·001	<0·001
Spicy	0·023	0·097	<0·001	0·002	0·002	0·001	<0·001
Aroma	0·157	0·735	0·009	0·453	<0·001	0·014	0·044
Pungency	<0·001	<0·001	0·005	0·244	<0·001	<0·001	<0·001
Briskness	0·882	0·086	<0·001	0·004	<0·001	<0·001	<0·386
Bitterness	0·007	<0·001	0·008	0·155	<0·001	0·183	<0·001
Astringency	<0·001	0·068	0·002	0·205	<0·001	0·437	0·053
Flavour	0·186	0·012	0·001	0·265	<0·001	0·989	0·055

definitive as it should be. There are good reasons for this. Some products are far more difficult to evaluate than others. Some attributes are more important than others (Barylko-Pikielna and Metelski, 1964; Powers, 1975, 1982a; Powers and Quinlan, 1974; Quinlan et al., 1974; Jackson et al., 1978) in determining the overall quality of the food.

Ideally the same weight should not be given to each attribute since some are more closely correlated than others or are more a major determinant of the overall sensation of flavour, texture or desirability. The contribution a particular attribute makes to the main modality can be calculated by multiple regression (Hwang-Choi, 1982), but weighing values for the different attributes are rarely calculated, much less applied. With respect to the selection or retention of panelists, in theory one panelist might be acceptable because he or she is a good judge for attributes meriting heavy weighting, though poor for attributes of minor importance, whereas another judge whose performance is the opposite might be rejected. That is the basic idea behind weighting attributes before making a decision relative to the performance of panelists. In practice, all the attributes are assumed to have the same weight. A saving grace for those in industry is that there is usually a fairly long record of experience with different assessors and products to guide the sensory leader in making a decision. Those starting a descriptive analysis program for the first time cannot fall back on that kind of experience. Calculating the F-values between treatments for each attribute and perusing the number of probability levels below 0.50 is only providing information ancillary to making a decision. The real question is: For how many attributes should performance be significant and at what level in order to be judged to be an acceptable assessor?

Cross et al (1978) proposed that the F-values of each assessor for each attribute be summed and that the assessors then be ranked according to the magnitudes of their sums of F-values. A cut-off level was not suggested. In essence, most methods proposed leave the ultimate decision to the subjective judgement of the sensory leader. There is a need, however, for a standard, even if its only purpose initially were to provide a reference with which actual decisions could be compared.

We have looked at the problem from two different points of view. One, which is quite simple, is to use the cumulative binomial table (Beyer, 1968) to decide whether the number of attributes for which the assessor is significant exceeds chance. Let us first illustrate the procedure, then point out its weaknesses. Assume that an assessor has 16 attributes to evaluate and we are willing to utilise a probability level of $p \leq 0.50$. For 16 trials, one observes that only at 13 or more events is the remaining cumulative probability below 0·05, the usual probability level considered to be significant. Those whose performance resulted in 13 or more F-values having probability levels ≤ 0.50 we would accept. Those who failed to attain that level, we would reject. This incidentally comes fairly close to the cut-off level Stone and his co-workers (1974) apparently used.

Their assessors evaluated 16 attributes. In discussing the matter of using the 0·50 probability level, Stone and co-workers tabulated results for 10 assessors. Thereafter, in explaining other features of the QDA procedure, only six assessors were used. It thus appears four assessors were eliminated. There were four who were not successful for 14 or more attributes. Stone and co-workers do not suggest how many attributes, or the percentage of attributes, an assessor must be significant for; a set level naturally cannot be specified for the reasons mentioned above. Laboratories which have past records of performance can, as already explained, use them as a guide, but benchmarks are needed for those without such records.

The flaw in our procedure is that binomial probability is based on independent events. Some of the attributes may be correlated with each other plus there is the likelihood that the judgement of the assessor for one attribute may be coloured to a certain extent by his judgement about another even though the two are not correlated. Furthermore, there is the problem of setting probabilities for Type I and II errors. A difference needs to be assumed, but even if one decides upon a difference for one attribute, it is unlikely that the same difference will prevail for the other attributes. Although the procedure is subject to valid criticism, it has nevertheless worked well in our hands (Powers, 1982a, 1984). Larmond (1981) has commented that she has looked upon the selection of assessors in somewhat the same way as we have. Figure 3 shows the number of attributes required to be significant at 0·50 if chance performance is to have a probability ≤ 0.05. Also shown is the percentage success required based on the total number of attributes evaluated.

4.2. Trained Assessors

The performance of assessors who are already members of a panel is analysed in the same fashion as candidates being considered for membership. For those newly trained, monitoring of performance is necessary to learn if they are living up to the standard expected of them. Once the routine of regular sensory evaluations sets in, some who did well in training may not perform as well because their interest lags. Sometimes it is necessary to have direct evidence of effectiveness for a particular task even though the assessors already have sustained records of efficient

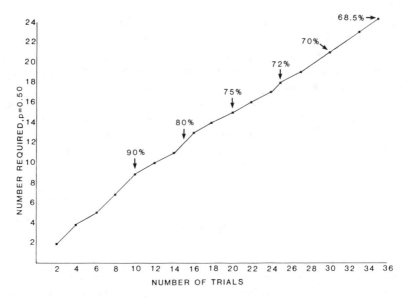

FIG. 3. Relationship between number of events and number required to be 'successful' at $p \leq 0.05$ based on the cumulative binomial table, assuming the probability of the event happening is $p = 0.50$.

service. Occasionally those in higher authority, not really understanding the nature of sensory examination, may be inclined to blame the panel when the results for products do not turn out the way they had hoped they would (Powers, 1984). Having a data base as to the mean and range of effectiveness of each assessor and of the panel itself permits each assessor's performance to be compared *vis-a-vis* his (or her) own prior record and that of the panel. For retraining at intervals or for routine evaluation of performance such a data base is invaluable. Although performance is monitored, the presumption should be that trained assessors are doing the task assigned to them. If the probability levels for F-values exceed 0·50, the inference should be that the products differ only in trivial ways; otherwise the assessors would have detected significant differences.

4.3. Effectiveness of Descriptive Terms

The next stage of analysis consists of pooling the scale values of the assessors followed by ANOVA to learn whether the descriptors are discriminatory of product differences and to learn whether there is product–assessor interaction. Palmer (1974) pointed out that not all descriptors are discriminatory. His choice of words will be used here because he considered 'discriminator' to apply to assessors. If the F-value for a term is non-significant, then the descriptor can only be characterised as being a descriptor. If the F-value is significant, then the term is discriminatory. Actually, the decision is not quite that simple. If the purpose of the trial is to evaluate descriptors for possible use and an F-value turns out to be non-significant, then it is logical to purge the list of terms demonstrated to be ineffective. If a term has already been selected and is in regular use, then failure to discern significance may merely mean that the products being examined do not differ in the respect the discriminatory item is supposed to make evident. In essence, at the main trial stage, the assumption is that the terms effect discrimination and that failure to detect differences there stems from the fact that the products do not differ significantly in that particular respect. Hereafter, 'discriminators' will be used either for terms or assessors unless 'discriminatory' is needed to avoid confusion.

A term may fail to be a discriminator because the attribute it represents exists in all the products at approximately the same intensity. In that case, it is a good descriptor, but it fails as a discriminator. A term can also fail to be useful because the attribute it is supposed to describe is absent from the food. Then of course it is just an inappropriate term. Interaction may also cause a term to fail. There is the possibility that the term does not mean the same thing to different assessors. This difficulty usually can be cleared up by training. Another cause of interaction is that only a few of the assessors may be sensitive to a particular note and thus be able to scale it consistently whereas the rest are not sensitive to it and thus are inconsistent in their scaling. This is a cause of failure of a discriminatory term where a decision is particularly difficult. The fact that some can use the term effectively shows that the term is apt. Unless there are already so many terms that the number is bothersome to the assessors, the term should probably be retained in the hope that, with additional experience, some of the assessors initially unable to use the term effectively will improve enough in their performance so that they can assess that characteristic effectively in succeeding trials and, secondly, there is always the possibility that the intensity of that attribute in some products evaluated later will be higher and thus within the sensitivity range of a greater percentage of the assessors.

Experience and other considerations need to be used in conjunction with statistical analysis. Wu *et al.* (1977) had 33 assessors examine red wine for different attributes. There were also 33 descriptors originally. The statistical output for a small part of the descriptors is listed in Table 5. The terms

TABLE 5
ILLUSTRATION OF THE POOLING OF ASSESSORS' RESPONSES TO DETERMINE WHETHER A TERM IS SIGNIFICANT AND THUS A DISCRIMINATOR; 14 RED WINES EVALUATED BY 33 ASSESSORS, TRIALS REPLICATED TWICE; 12 WHITE WINES EVALUATED BY 18 ASSESSORS WITH FOUR REPLICATIONS
(From Wu et al., 1977)

Term	F-values	
	Red wine	White wine
Tart	16·1	33·1
Dry	37·6	47·8
Astringent	19·8	28·1
Yeasty	0·9	1·3
Vinegary	0·5	10·1
Sulphurous	1·3	1·1

Reprinted from *Journal of Food Science* (1977) **42**, 944.
Copyright © by Institute of Food Technologists.

'vinegary' and 'sulphurous' were non-significant, but they were retained simply because these are terms which sometimes do apply to wine. In a second trial with white wine, the term 'vinegary' did become a discriminator. Wu *et al.* (1977) discarded, on the other hand, some discriminators even though they were statistically significant. Because of the large number of terms they wished to reduce the list to a more manageable number as far as the assessors were concerned. The significant terms dropped were those used sparingly by the assessors. Lyon (1987) did the same. Unless a term was used 40% of the time, she eliminated it. Some non-significant terms were retained by Wu *et al.* (1977) because experience indicated the term might be appropriate, though initially non-significant, while some significant terms were dropped so as to lessen the burden on the assessors.

4.4. Interaction
The results need to be examined for possible product–assessor interaction either concurrently or once decisions have been made about the effectiveness of the assessors and the descriptors. Table 6 shows the ANOVA for examination of the scale values assigned to the floral nuance of teas. If the interaction term is significant, as it is, then the mean square for interaction should be used in place of the mean square for error to

TABLE 6
ANOVA FOR FLORAL NOTE, FIVE TYPES OF TEA, SIX REPLICATIONS, SEVEN ASSESSORS

Source	df	Mean square for error	F (error)	p > F	F (interaction)	p > F
Total	209					
Products	4	34 100·81	34·43	<0·001	15·05	<0·001
Replications	5	1 898·04	1·92	0·093		
Assessors	6	11 389·50	11·50	<0·001		
Products × assessors	24	2 266·38	2·29	<0·001		
Error	170	990·31				

determine whether the F-value for treatment is significant. If the treatment F-value is still significant when tested by the interaction term, this means that the product differences are sufficiently great to override the 'noise' generated by the interaction. If the treatment F-value is not significant upon using the interaction term, all one knows is that a difference has not been detected.

If the interaction term is non-significant, then the error term itself should be used to test for product differences. All statisticians do not agree as to the next step. Most would say, probably, that it is permissible to recalculate the ANOVA, combining the sums of squares for interaction and for error (they having been shown not to be significantly different), for then the degrees of freedom (df) for error will be greater, the mean square for error smaller and thus the F-value between treatment will be correspondingly larger. A small minority would say that once a decision has been made to split out interaction and error, there should be no recombining of them, for this smacks of attempting to demonstrate significance, it perchance not having been detected at the first try.

A common cause of interaction between products and assessors is failure of the assessors to scale intensities in the same fashion. The upper part of Fig. 1 illustrates the situation regarding scaling of the floral note in teas. None of assessor 2's mean-scale values for any of the five teas exceeded 2·67 cm though the scale was 15 cm long. Assessors 3 and 5 used the entire scale range in making their distinctions. Their mean values ranged up to 12·5 cm. All three assessors were good judges. They could differentiate among the teas as shown by the probability values in Table 4, but they had not yet reached agreement concerning the scale value which should be assigned to different degrees of intensity. The conservatism of Assessor 2 in scaling was pointed out to her (this still being a training period), and she was requested to attempt to utilise more of the scale without disrupting the discrimination she had shown in distinguishing among the teas. By training, interaction arising from differences in scaling on the part of the assessors can generally be reduced, frequently, but not always, to a non-significant level.

4.5. Standardisation

Standardisation of the data is not a substitute for adequate and proper training, but a means of overcoming part of the variance caused by disagreement among the assessors about the scale values assigned.

Standardisation is normally used when different measurement values need to be put on a common basis. An example would be standardising the weights and heights of people so that one measurement does not carry undue influence in comparisons because of its scale being greater. If height were expressed in metres and weight in kilograms, quite obviously for most individuals weight would be considerably greater in magnitude. If however,

TABLE 7

F-VALUES RESULTING FROM APPLICATION OF ONE-WAY ANOVA,[a] BEFORE AND AFTER STANDARDIZATION OF SCALE VALUES ASSIGNED BY 10 PANELISTS EVALUATING FIVE GRADES OF TEA OF THE SAME TYPE, FIVE REPLICATIONS

Variable	F-values	
	Raw scale	Scale normalised
Flavour	1·72	1·64
Aroma	6·03	7·53
Colour	14·16	26·82
Briskness	9·05	11·53
Body	4·36	12·66
Grass-like note	0·34	0·24
Resinous	1·17	2·36
Malt-like note	3·99	11·91
Harshness	10·25	14·49
Bitterness	8·33	11·70
Pungency	4·20	5·05
Metallic note	2·58	3·13
Sweetness	0·50	0·58
Floral note	8·14	10·55

[a] Taken from Godwin (1984).

the mean value for each form of measurement is set at zero and the respective standard deviations are used as the scaling device, the two forms of measurement have been set to the same scale. When assessors scale the sensory attributes of a product, they of course are not using different scales, but as the upper portion of Fig. 1 shows they may use different sections of the scale. The idea behind standardisation is to bring the mean value for each assessor to the grand mean for the panel. In effect what one does is to slide the respective means for each assessor up or down the scale to the mean for the panel, designate the mean for the panel as zero, then re-scale the original values by dividing each assessor's values by his or her standard deviation, the effect of which is to make the scaling unit for each panelist be 1·0 standard deviation. The middle set of scales in Fig. 1 shows the uppermost set re-scaled (standardised). Because the error component caused by different assessors using different parts of the scale has been reduced, in most instances the F-value between the treatment is increased (see Table 7). The data are for a panel of trained assessors evaluating five grades of tea of the same type. Two-way ANOVA, or some higher form, detects product differences as readily. Standardisation helps to separate numerically error arising from differences in scaling from that due to scoring the products in a different order.

4.6. Replication

As should already be evident, replication is necessary if the effectiveness of individual panelists is to be assessed. As a part of the two-way (or higher) ANOVA to learn whether interactive effects exist, replication is also used to learn if differences in the samples or methods of presentation have induced appreciable experimental error. Assuming care has been taken in preparation and presentation of the samples, there should be little cause for differences among replicate samples or sessions for some kinds of products. Products which are stable over a reasonable period of time and which do not depend too much upon the 'art' of making the product usually do not present problems whereas evaluation of fresh fish or meats, apples or an air pollution sample may present difficulties. Even at the same session, often there is not enough material from the same specimen to furnish each assessor with a portion; consequently different specimens have to be used. Even when there is enough material from the same specimen, there may be differences. Meat scientists generally balance out portions served from the right and the left side of the carcass because not only are there differences between carcasses but also there may be differences between the right and the left side of the same carcass. Replication enables interactive effects to be examined to learn whether these effects need to be partitioned out as a source of error.

Sometimes product–assessor interaction or replication effects are almost impossible to avoid. The lower part of Fig. 1 shows differences in scaling on the part of women evaluating the sensory properties of sanitary pads. Quite obviously training for such a task cannot be handled on a short-term basis as can training for the evaluation of tea where the assessors can be served the same products at one or two sessions per day, several days in a row, to reach agreement as to the way the properties should be scaled. For some properties, simulated tests can be used upon a daily basis; for other properties, e.g. flow patterns, only actual use suffices. By discussion some agreement can be reached as to how different properties should be scaled, but a moderate amount of product–assessor interaction in scaling is probably less objectionable than the extraordinary length of time required for training if interaction were to be reduced to a negligible level based upon actual use. In fact because of all the variables involved, reducing interaction to a negligible level would be nigh impossible for any appreciable number of assessors. Fortunately, just as for the tea, the women involved in assessing the properties of the pads were quite effective as judges in spite of interaction. Their judgements were sufficiently consistent and discriminating to permit differences to override interaction effects.

The body-care industry is particularly fecund in presenting some of the most difficult of sensory tasks. In the illustration above, the women serve both as subjects (a form of replication in this instance) and as judges. In other parts of the industry, the two functions of being a subject and an assessor can be kept separate. An illustration of this is the assessment of the efficacy of such products as soaps, deodorants or antiperspirants to control axillary odours. Individuals are recruited who either are prolific in the production of strong malodours or who are willing to undergo a regimen intended to lead to the development of strong odours. The function of the assessor is to evaluate the intensity of different types of malodour. Quite naturally a large number of subjects (from 60 to 90) have to be used because of the great variability among humans. Like many other forms of sensory analysis, there are positional effects; consequently trials have to be balanced by applying the test substance and the placebo alternately to the left and to the right armpit. Not only must positional effects be accounted for, but experiments have to be designed so that more than one assessor will have evaluated the underarm odour of each one of the subjects. This is akin to having judges evaluate the flavour or texture of apple slices from more than one apple since all apples are not the same.

Because of replication, interaction—objectionable as it is—can be segregated and its influence estimated. Of course, replication alone is not enough; there has to be sound design and execution of the sensory trial. Once interaction is segregated and in spite of it being statistically significant, product differences often can be demonstrated because the variance between treatments is still larger than that of the interaction itself as shown in Table 6.

4.7. Product Differences

The primary purpose of the application of ANOVA following the main trial is to learn whether the mean values for the attributes differ significantly among the products. If the F-values indicate that differences do exist, then usually one of the range tests such as Duncan's or Tukey's is applied to learn which treatments or products differ significantly from the others. Profiles, such as those shown in Fig. 4, may be constructed to permit visualisation of the differences between products. The distance from the centre of the profile to the boundary line connecting each one of the attributes is in proportion to the means. The angles in Fig. 4 are all equal. Stone et al. (1974) showed a profile where the angles between attributes were not equal and they stated that correlations can be used to set the angles between attributes. The latter part is correct. Correlation co-efficients can be transformed into angles by taking the inverse cosine of the coefficient. The correlation table they showed, however, was that of correlations between each attribute and each of all the other attributes. One cannot assign directly from a simple correlation table the angles between each attribute and still keep true all the relations among all the attributes. In fact, even by more complicated methods, when several attributes (dimensions) are reduced to two or three dimensions, there is some distortion of the relations among the attributes. Since all the angles among the attributes cannot be properly set by simple means, the angles are generally made equal. Profiles are thus purely empirical. Profiling does have the advantage of permitting visualisation of differences, or in some instances of showing that the products are almost identical in sensory properties (see Hall (1984)). There are dozens, perhaps hundreds, of profiles diagrammed in the literature.

Salovaara et al. (1982) showed graphically, for example, the effect of lack of salt on the sensory characteristics of rye and wheat breads. When salt was not included in the formula, a pasteboardy note was drastically increased while on the contrary the bread was less palatable, sour, saliva-stimulating and somewhat less yeasty. Pokorný et al. (1981) diagrammed

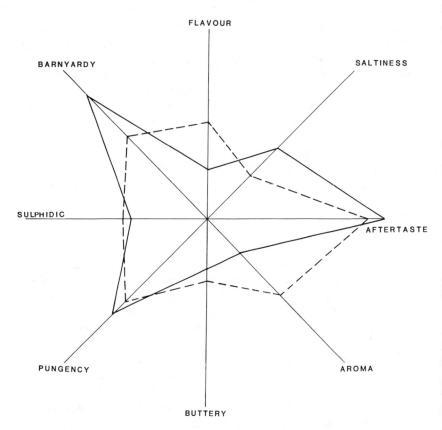

FIG. 4. A typical spider-web diagram of differences between products. The solid line is for Roquefort cheese; the broken line, for a Danish blue cheese. The distance from the centre of the diagram reflects the magnitude of the mean for that particular attribute and product.

differences in the perception of aroma and flavour notes of a Fernet appetiser as judged by men and women. Men scaled acidity and an aroma, characterised as new-mown hay, significantly higher than did women. Malek *et al.* (1982) incorporated a desirable feature in their profiles to illustrate differences in specimens of beer. Instead of using a simple line to connect the different attributes of the beer, they broadened the line into a band, the widths of the bands being in proportion to the standard deviations for each attribute. They pointed out that if there is a space between the bands this means that the products differ significantly in that attribute. They used a computerised system to 'read' and to log in the scale values assigned, make the calculations, generate a report and construct the spider-web diagram.

ANOVA is not the only method applied to QSP data to derive information. Some laboratories rank each judge's means for each attribute from most to least intense. The procedure quickly reveals whether judges differ in their response, if so, where, and very often explains significant product–assessor interaction. The frequency with which a particular product has been scored higher or lower than another product may also be examined to learn whether many of the assessors scaled a sample higher, though not high enough or often enough to cause statistical difference by ANOVA. Böhler et al. (1986) used a 7-point category scale in their QSP study of the cooking quality of potatoes, but they pointed out that category scoring sometimes needs to be supplemented by the 'somewhat simpler but more powerful ranking procedure'. They treated their data by non-parametric procedures. Notwithstanding the fact that QSP methods have their limits and there is not unanimity as to the particulars and the merits of the various methods, QSP has had ready acceptance throughout the sensory world. There is no doubt that QSP has assumed a major position in the armamentarium of sensory methods.

5. APPLYING MULTIVARIATE ANALYSIS METHODS

The procedures and practices discussed so far have been those commonly used by laboratories engaged in the profiling of sensory properties of materials. In addition to the univariate methods of analysis such as ANOVA, some laboratories subject their data sets to MVA. Applications involve canonical, cluster (CA), correspondence, discriminant (DA), factor (FA), multidimensional scaling, and principal component analysis and various regression procedures. Each one has some limitations, but often they provide insight to the relations among variables not obtainable from univariate analysis alone.

5.1. Using MVA to Examine Performance

A problem involved in examining the performance of each assessor for each attribute separately is that some assessors will be moderately effective for all the attributes whereas others will be quite effective for a few, barely acceptable for others and utterly worthless for still others. Decision-making would be simplified if judgement could be based on only one value

TABLE 8

NUMBER OF ATTRIBUTES OUT OF NINE FOR WHICH EACH ASSESSOR WAS SIGNIFICANT AT ≤ 0.30 AND ≤ 0.50 PROBABILITY LEVELS (FROM ANOVA) AND MULTIVARIATE F-VALUE (FROM MANOVA), FIVE KINDS OF TEA, EVALUATION REPLICATED SIX TIMES

Assessor	Number of attributes for which the assessor was significant at probability levels		F-Value MANOVA
	≤ 0.30	≤ 0.50	
1	8	8	0·025
2	8	8	0·002
3	9	9	0·003
4	8	9	0·063
5	9	9	0·008
6	7	9	0·001
7	8	9	0·012

For nine terms, six or more would have to be guessed correctly (be significant) if the probability level was 0·30 and eight or more terms would have to be guessed correctly if the probability level was 0·50 for the likelihood of success to be below 0·05.

instead of having to mentally weigh a whole group of probability levels or make crude judgements based on the frequency of statistical success. Multivariate analysis of variance (MANOVA) provides such a composite value, but MANOVA too has its flaws and difficulties.

Table 8 summarises the number of attributes for which the trainees who evaluated tea were significant at $p \leq 0.30$ and $p \leq 0.50$. We have sometimes used the 0·15 or 0·30 level instead of the 0·50 level as the first two demand a somewhat higher degree of performance. In the table, the second and third columns list the number of attributes for which the assessors were significant. The last column lists the probability level associated with the multivariate F-value. This table is presented chiefly to show the kind of probability levels MANOVA yields when the assessors are performing ably.

Piggott and Jardine (1979) also examined the performance of their assessors by MVA. They applied PCA to the QSP values their assessors had assigned prior to being trained and after training. Three out of four replicates showed greater dispersion between samples than within samples.

This showed that the assessors were detecting differences between the samples. Their MANOVA tests revealed that replications three and four were the closest pair indicating that performance improved with training and experience. As was pointed out in discussing training for flavour profiling, they had also had their assessors sniff reference odorants to develop the facility of recognising different odour notes. In comparisons involving 17 reference standards, the mean success in recognising the odorants was 78% with a range of 68 to 97%.

5.2. MANOVA and Trained Assessors

Table 9 is presented for two reasons. At least two of the difficulties in applying MANOVA may be pointed out from it and, secondly, the table illustrates the typical situation where a panel has been trained but the data for its first assignment as a regular panel cause one to wonder whether the panel has been sufficiently trained or the somewhat patchy performance is a consequence of the samples being so much alike that the assessor should not be expected to be successful for all or nearly all the attributes. The data

TABLE 9
FREQUENCIES WITH WHICH 12 ASSESSORS WERE SIGNIFICANT AT THE $\leq 0\cdot30$ AND $\leq 0\cdot50$ PROBABILITY LEVELS (BASED ON ANOVA), AND MULTIVARIATE F-VALUES FOR FLAVOUR TERMS SUBDIVIDED INTO THREE GROUPS BY FACTOR ANALYSIS AND FOR TEXTURE, THREE KINDS OF CHEESE, EVALUATION REPLICATED FIVE TIMES
(Adapted from Hwang-Choi, 1982)

Assessors	Frequencies significant at probability levels[a]		Multivariate F-values for three subgroups of flavour terms			Multivariate F-value for texture evaluation
	$\leq 0\cdot30$	$\leq 0\cdot50$	Factor 1	Factor 2	Factor 3	
1	17	18	0·63	<0·01	0·11	0·06
2	16	18	0·27	0·42	0·36	0·07
3	17	20	0·06	<0·01	0·02	<0·01
4	12	14	0·13	0·50	0·60	0·14
5	15	20	0·04	0·23	0·01	0·04
6	14	12	0·02	0·89	0·21	0·01
7	14	16	0·35	0·08	0·21	0·04
8	12	16	0·26	0·09	0·13	0·03
9	11	14	0·36	0·29	0·40	0·26
10	11	15	0·32	0·04	0·55	0·60
11	16	17	0·19	S	0·67	0·01
12	11	15	0·11	0·45	0·62	0·07

[a] From the cumulative probability table, correct guessing of 12 or more attributes would be necessary at 0·30 to be below the 0·05 probability level; at 0·50, 16 successful guesses would be required. S in column 5 shows that the matrix was singular.

are drawn from Hwang-Choi (1982). She carried out QSP to learn whether three cheese products differed in sensory quality. One cheese product consisted of natural processed cheese; the other two were synthetic cheese. One of the formulated products contained some dairy components whereas the other was derived from soy products. Both of course contained various flavourants, emulsifiers and other adjuvants to impart functional properties. The question really was: In what respect do the formulated products differ from the natural product so that steps can be taken to strengthen any sensory deficiencies or abnormalities? The cheeses were examined for 15 odour and flavour attributes and for seven textural characteristics. The textural evaluation was carried out at different sessions from those for odour–flavour. There was not a problem in applying MANOVA to the seven textural attributes because the number of degrees of freedom (df) needed for each term and the products (less one) did not exceed the total df—there being 15 df arising from the three products and the five replications. Except for Assessor 10, there is no question that all the assessors had performed effectively (last column, Table 9) in evaluating the textural properties of the cheeses.

A problem immediately arose in trying to apply MANOVA to the 15 flavour attributes. There were not enough df to permit all 15 terms to be evaluated at once. Hwang-Choi had carried out FA. The 15 terms could be reduced to three factors; so accordingly she subdivided the sensory terms into three groups to bring the number in each group within the limitation the df imposed. MANOVA shows that Assessor 9 is contributing to discrimination though ANOVA would suggest otherwise. Assessors 4, 10 and 12 are borderline, but helpful overall, in establishing product differences. The output for Assessor 11 presents a particular problem, which is one of the limitations of MANOVA. The matrix for that assessor was 'singular' because for one of the terms the assessor assigned the same value to all the products. One of the advantages of having carried out ANOVA first is that the sensory leader can look at the ANOVA output, see for which term the assessor was totally ineffective, delete that term and then carry out MANOVA with one less term. The resulting MANOVA probability value should, of course, be accorded less weight than if all terms had been used, but it still provides information. While in this particular study, the MANOVA output could not be presented as one value only, it does provide greater information on assessor performance than could be obtained by accepting or rejecting assessors solely on the basis of inspection of ANOVA output. Perusal of the 22 ANOVA probability values provides detailed information; MANOVA encapsulates it. Even though all of the 22 terms could not be analysed together, nor even all the 15 flavour terms, reducing 22 probability values to four does aid the sensory leader in deciding whether failure to detect differences is a result of poor performance or a result of the samples not being particularly different. The denouement as to whether Hwang-Choi finally decided some of her assessors were ineffective or the cheeses were so close in sensory characteristics they caused her assessors to appear to be poor judges will be withheld until cluster analysis (CA) has been discussed.

Before taking up CA, it is instructive to note the procedures Lyon (1987) followed in studying warmed-over-flavour in chicken patties. She permitted free choice description on the part of her panelists (see Section 3.3., this chapter). The free-choice process resulted in 45 terms being generated. She applied MANOVA to the samples, the sessions, and samples × sessions to ascertain whether differences existed at all. Sample and session differences were significant, but not the interaction of the two. Having demonstrated globally that differences did exist, she then turned to ANOVA to ascertain where the differences were. Initially, she had set an elimination criterion of 40% if a term were used less frequently than that by the panelists. She deleted terms which had an F-value, the probability of which was >0.05. Actually, she ran FA for all the terms. She pointed out that those not used 40% of the time had low loadings on the factors and all had F-value probabilities >0.20; Initially, she had set $p<0.30$ as a criterion for panelist performance. She now lowered that criterion to 0.20. The various elimination criteria reduced the number of useful terms to 22. She next applied PCA to ascertain panelist agreement and to detect outliers. Four of the panelists were outliers, chiefly because they used only certain portions (see Sections 2.1.1. and 4.4., this chapter) of the 10-cm, semi-structured line scale employed.

5.3. Cluster Analysis

The purpose of bringing in Lyon's study is to point out that there is more than one way of getting at the matter of panelists being outliers. She used PCA, and as pointed out earlier, Piggott and Jardine (1979) did the same. Cluster analysis is also used to classify panelists according to their agreement with each other. Laamanen and Jounela-Eriksson (1987) used a still different procedure. They applied Spearmann's rank correlation procedure to the results of each assessor for each descriptor. Three of the assessors were eliminated for having negative coefficients compared with the others. The problem Hwang-Choi (1982) encountered is a reflection of the fact that assessors are individuals. If product differences are minor, the performance of some assessors will be significant for some attributes while others will be significant for different attributes; consequently there is not a great enough cumulative effect to establish significance for the panel as a whole. If, however, the panel is partitioned into homogeneous subsets, significant differences hidden in product–assessor effects may often be discerned.

Table 10 shows that Hwang-Choi (1982) was 76·1% successful in classifying 180 of her cheese samples into the class to which the cheese really belonged. From her two-way ANOVA she knew that significant product–assessor interaction existed. She therefore carried out cluster analysis to learn if the assessors could be grouped into subsets which would be more homogeneous than the panel itself. Table 11 shows the way the assessors separated out. To classify the 180 samples (Table 10), she had used stepwise discriminant analysis (SDA). She re-applied SDA to the scale values of Assessors 1, 2, 4 and 8; to those of Assessors 5 and 7 and to those of Assessors 3, 6 and 11. The average percentage correct classification for each group was then, respectively, 85·3%, 100% and 90%, or a weighted average of 90%. In using CA to effect greater discrimination among the cheese samples, Hwang-Choi not only showed that the cheeses were different, but in doing so she established that her assessors individually had performed well. Their only flaw was that they were not in complete unison in their responses. Golovnja *et al.* (1981) and Palmer (1974) have likewise used CA to segregate assessors into homogeneous subgroups.

TABLE 10
CLASSIFICATION OF CHEESE SAMPLES BY STEPWISE DISCRIMINANT ANALYSIS (SDA), FLAVOUR AND TEXTURE JUDGEMENTS POOLED
(Adapted from Hwang-Choi, 1982)

Actual product	Number of cases	Predicted products		
		1	2	3
1	60	48 (80%)	0 (0%)	12 (20%)
2	60	2 (3·3%)	52 (86·7%)	0 (10%)
3	60	13 (21·7%)	10 (16·7%)	37 (61·7%)

The average percentage of cases correctly classified was 76·1%.

TABLE 11
SEPARATION OF ASSESSORS INTO CLUSTERS BASED ON THE INTENSITY VALUES EACH ASSIGNED TO THE 22 FLAVOUR AND TEXTURE ATTRIBUTES EVALUATED
(Adapted from Hwang-Choi, 1982)

Number of clusters	1	2	4	8	10	Panelists 12	5	7	3	6	11	9
12	*	*	*	*	*	*	*	*	*	*	*	*
11	*	*	*	*	*	*	*	*	*******	*	*	
10	*	*	*******	*	*	*	*	*******	*	*		
9	*******	*******	*	*	*	*	*******	*	*			
8	*******	*******	*	*	*	*	*************	*				
7	*******************	*	*	*	*	*************	*					
6	*******************	*	*	*******	*************	*						
5	**************************	*	*******	*******************								
4	**************************	*	*******	*******************								
3	*********************************	*******	*******************									
2	**	*******************										
1	***											

Godwin *et al.* (1978) applied CA to the scores assigned by 27 assessors to canned and frozen green beans. Their purpose was the same as that of Hwang-Choi: to learn whether sample differences could be demonstrated more conclusively if the panel were subdivided into groups more homogeneous than the panel itself. The assessors had scored the green beans for intensities of 20 attributes and at separate sessions they had evaluated the beans for overall acceptability, appearance, colour, flavour and mouthfeel. Acceptability had been evaluated both before and after conducting the intensity sessions. Normally, an acceptability decision is a non-cognitive, evaluative judgement. The investigators wondered whether being trained to make analytical, cognitive judgements for intensity would influence one's acceptability decision if it were made after the analytical phase. Hence there were 10 sets of acceptability values. Cluster analysis applied to the acceptability values showed that two major subgroups existed among the 27 assessors (Table 12). Thirteen of the assessors apparently liked the canned products more than the frozen products because their scores for the canned products were almost invariably higher, whereas eight other assessors scored the beans in the opposite way. Their scores tended to be higher for the frozen beans. Clustering provided information otherwise hidden because of offsetting responses. Table 13 shows the terms which were the most discriminatory when SDA was applied to the results of Hwang-Choi's panel and when it was applied to two of the subgroups. Here subgroups yielded essentially the same answer,

TABLE 12
MEAN SCORES ASSIGNED BY ASSESSORS WHO APPARENTLY PREFERRED CANNED OR FROZEN GREEN BEANS
(From Godwin et al., 1978)

Product assessors (13) who generally scored canned beans higher than frozen beans, subgroup C

	Acceptance	Appearance	Colour	Flavour	Mouthfeel
Canned	5·6	5·5	5·0	5·6	5·5
Frozen	4·5	5·4	5·4	4·4	4·7

Assessors (13) who generally scored frozen beans higher than canned beans, subgroup F

	Acceptance	Appearance	Colour	Flavour	Mouthfeel
Canned	4·9	4·7	4·4	4·8	5·0
Frozen	6·2	6·5	6·7	6·0	6·0

Reprinted from *Journal of Food Science* (1978), **43**, 1231. Copyright © by Institute of Food Technologists.

but they arrived at it via different pathways. Clustering was done on the basis of the intensity scores.

There is less likelihood for variation in response when intensity rather than acceptability is the basis for clustering; nonetheless, our experience is that the detection of differences can often be enhanced, as Hwang-Choi's analyses reveal. The reader should note however that improved resolution of sample differences may not occur. See the material to follow pertaining to Finnish sour rye bread.

Figure 5 depicts the divergence Powers and Quinlan (1974) observed in assessing a blueberry-whey beverage for acceptability. The beverage had been made with 19%, 26%, 33% and 40% blueberry pulp. The means for the panel showed significant differences between the lower three levels with acceptability increasing as the amount of pulp increased. Examination of the correlation coefficients, ANOVA and Duncan's range test revealed that there was a minority of panelists who considered the beverage with the two lower amounts of pulp to be most acceptable and their acceptability scores decreased beyond the incorporation of 26% blueberry pulp in the beverage. For commercial application one could be seriously misled if one considered only the panel means themselves.

Malek et al. (1986) similarly used correlation analysis to improve product

TABLE 13
ORDER OF INCLUSION (AND DELETION) OF TERMS USED BY 12 ASSESSORS, BY SUBSETS OF FOUR AND THREE ASSESSORS GROUPED BY CLUSTER ANALYSIS
(From Hwang-Choi, 1982)

Step	12 Assessors		4 Assessors		3 Assessors	
	Discriminator	F-Value	Discriminator	F-Value	Discriminator	F-Value
1	Dry texture	83.9	Firm texture	37.0	Dry texture	32.1
2	Smooth texture	41.1	Cheesy flavour	19.1	Firm texture	19.8
3	Cheesy flavour	30.5	Springy texture	15.8	Saltiness	17.2
4	Firm texture	24.9	Caramel flavour	15.3	Chewy texture	15.5
5	Nutty flavour	20.9	Cheddar flavour	15.6	Sweet	13.9
6	Chewy texture	18.3	Delete cheesy	16.5	Nutty flavour	12.7
7	Saltiness	16.4	Pleasant aftertaste	14.7	Acid taste	12.5
8	Overall preference	14.9			Delete saltiness	13.6
% correct classification		76.1%		85.3%		90%

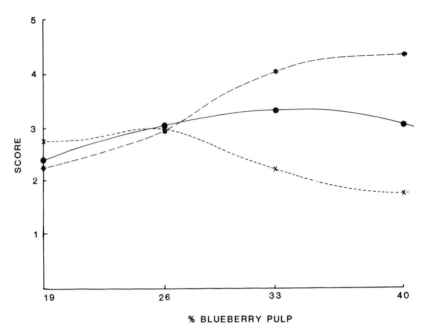

FIG. 5. Diagram showing panel means and two subsets of panelists whose preference for a blueberry-whey beverage differed according to the amount of blueberry pulp incorporated in the beverage. The solid line shows the panel response; the two broken lines show that one subgroup preferred the beverage more, the greater the amount of pulp incorporated, whereas the other subgroup found the beverage less acceptable as the amount of pulp was increased.

discrimination. Correlation analysis was more effective than selection of judges based on F-ratios. McDaniel et al. (1987) also employed correlation. They categorised panelists as being non-perceivers or non-correlators.

Cluster analysis is particularly appropriate as a means of classifying assessors. In forming clusters there is no implication that the things grouped together are related in any way. They just happen to be together because they have certain characteristics in common. In clustering assessors, the common element is the pattern of scores assigned. Once one has reduced heterogeneity, then there is greater likelihood of being able to detect differences. This may not make decision-making any easier. Subsets of responses have to be considered instead of responses *en masse*, but clustering of assessors can lead to the discernment of differences otherwise hidden in conflicting responses. The unwary might be content to accept the panel means at face value. Those more prudent or inquiring will look

DESCRIPTIVE METHODS OF ANALYSIS

WARD'S MINIMUM VARIANCE CLUSTER ANALYSIS APPLIED TO THE SCALE VALUES OF 20 ASSESSORS EVALUATING 12 ATTRIBUTES OF FINNISH WHOLEMEAL SOUR RYE BREAD (From Hellemann et al., 1986) behind the means for things otherwise hidden or, where significant differences between means have not been attained, to learn the cause of failure.

Several clustering processes exist. Table 14 shows clustering results when Ward's minimum variance clustering process (SAS, 1985) was applied to the scale values of 20 panelists who replicated their judgements four times in evaluating five types of Finnish wholemeal sour rye bread for 12 attributes. Note that the form of the cluster pattern differs from that shown in Table 11 which was based on the 1982 version of SAS. The chief thing to note, however, is how cohesive the panel was. It broke into two main clusters readily, but a considerable reduction in the semi-partial R^2-value had to occur before there was any further separation into still smaller clusters. The Finnish panelists were highly trained and experienced at evaluating rye bread. Table 15 sets forth the mathematical parameters. Table 16 illustrates the output from the VARCLUS program of SAS (1985). Its advantage is that clustering can be commenced from raw scores or a covariance or correlation matrix. Note that by either program, Assessors 7 and 12 were least like the others in their responses.

Unlike Hwang-Choi's experience, forming two groups of panelists more homogeneous than the panel itself did not lead to improved resolution of product differences. When panel results were pooled, differences between two of the five types of breads could not be resolved. The same was true for the two cohesive clusters. The panelists in the clusters were equally efficient. Discriminant analysis (DA) did show however that their results were obtained by somewhat different pathways.

5.4. Procrustes Analysis

Rather than merely segregating assessors into different clusters according to their agreement and disagreement with each other, others have attempted to minimise dispersion among the assessors by rotating their responses (Banfield and Harries, 1975). See also the references to Williams and his co-workers, Section 3.3., this chapter. Vuataz (1977, 1981) described and illustrated the technique. Basic publications are those by Gower (1966, 1971, 1975). Laslett and Bremner (1979) also applied Procrustes analysis. Vuataz et al. (1975) discussed the merits and limitations of different MVA methods as they applied to the examination of sensory data.

5.5. Deleting an Assessor's Results

Hwang-Choi attained greater precision by eliminating a few assessors who in essence were outliers (Table 11). The question then arises: Should any

TABLE 15
TABULAR OUTPUT OF WARD'S MINIMUM VARIANCE CLUSTER ANALYSIS

Variables	Eigenvalues of the covariance matrix			
	Eigenvalue	Difference	Proportion	Cumulative
1	796·987	313·406	0·429 565	0·429 56
2	483·580	254·313	0·260 643	0·690 21
3	229·267	132·981	0·123 572	0·813 78
4	96·286	22·873	0·051 897	0·865 68
5	73·413	25·975	0·039 569	0·905 25
6	47·438	9·109	0·025 568	0·930 81
7	38·329	7·307	0·020 659	0·951 47
8	31·023	6·098	0·016 721	0·981 63
9	24·925	8·704	0·013 434	0·081 63
10	16·221	5·445	0·008 743	0·990 37
11	10·776	3·687	0·005 808	0·996 18
12	7·089	·	0·003 821	1·000 00

root-mean-square total-sample standard deviation = 12·434 3
root-mean-square distance between observations = 60·915 3

Number of clusters	Clusters joined			Frequency of new cluster	Semipartial R-squared	R-Squared
19		3	12	2	0·010 001	0·989 999
18	CL19	11	14	2	0·011 672	0·978 327
17			7	3	0·014 082	0·964 245
16		4	18	2	0·014 111	0·950 134
15		8	15	2	0·014 587	0·935 547
14		1	13	2	0·016 947	0·918 600
13		2	6	2	0·019 158	0·899 442
12	CL15		9	3	0·024 987	0·874 454
11		10	CL18	3	0·025 403	0·849 052
10	CL17		5	4	0·027 946	0·821 106
9	CL14	CL10		6	0·029 373	0·792 732
8	CL13		17	3	0·030 646	0·761 086
7	CL16	CL11		5	0·034 871	0·726 215
6	CL8		20	4	0·041 325	0·685 890
5	CL9		19	7	0·054 941	0·684 890
4	CL7		16	6	0·060 708	0·569 241
3	CL5	CL12		10	0·062 688	0·506 553
2	CL6	CL4		10	0·189 603	0·316 950
1	CL3	CL2		20	0·316 950	0·000 000

TABLE 16
VARCLUS (OBLIQUE PRINCIPAL COMPONENT) CLUSTERING OF 20 PANELISTS BASED ON THEIR CORRELATION MATRICES (SAS, 1985)[a]

	Variable (panelists)	R^2 With		
		Own cluster	Next closest	$1-R^2$ ratio
Cluster 1	2	0·695 5	0·448 7	0·552 4
	4	0·672 7	0·521 7	0·684 2
	10	0·466 7	0·259 8	0·720 5
	11	0·561 8	0·338 5	0·662 5
	14	0·513 7	0·231 1	0·632 5
	16	0·665 2	0·390 6	0·549 5
	17	0·605 1	0·361 0	0·618 0
	18	0·699 0	0·398 1	0·500 0
	20	0·762 1	0·354 8	0·368 7
Cluster 2	1	0·487 3	0·206 0	0·645 8
	3	0·434 0	0·219 7	0·725 3
	5	0·532 6	0·325 7	0·693 1
	6	0·640 5	0·508 5	0·731 4
	8	0·447 6	0·281 8	0·769 1
	9	0·521 1	0·268 1	0·654 4
	13	0·526 7	0·345 5	0·723 1
	15	0·546 2	0·331 6	0·678 8
	19	0·507 6	0·240 1	0·648 0
Cluster 3	7	0·664 5	0·154 1	0·396 7
	12	0·664 5	0·189 8	0·414 2

Cluster	Members	Cluster summary for three clusters			
		Cluster variation	Variation explained	Proportion explained	Second eigenvalue
1	9	9·000 00	5·641 74	0·626 9	0·717 73
2	9	9·000 00	4·643 71	0·516 0	0·914 97
3	2	2·000 00	1·328 92	0·664 5	0·671 08

total variation explain = 11·614 4   proportion = 0·580 718

[a] Taken from Shinholser *et al.* (1987), with permission. Copyright © by Food and Nutrition Press Inc., Westport, CT, USA.

part of the results be eliminated in carrying out statistical analysis? In theory, this is taken care of during the training period. Assessors who are outliers are eliminated there, if they are detected, on the ground that one is endeavouring to secure a set of instruments, all of which respond alike. The assessors need not be fully representative of the population; they are just a very narrow stratum of it, but as pointed out earlier in discussing the FPM and subjective decisions, all one is really doing in using sensory analysis is securing judgements on one product relative to others. A narrow band of agreeing assessors permits this to be done. One of the basic premises of statistics is that one decides in advance on the number of tests to be carried out, the design of the experiments and other considerations of that sort. Having made those decisions, one should be willing to accept the results no matter how they turn out rather than attempting to find something 'significant' by deleting some part of the results. This is a sound rule, but where human beings are used as the analytical instrument there is some justification for deleting the results of an assessor if that assessor is clearly not performing as an analytical instrument or is an 'outlier'. If one were using a physical instrument and midway through an experiment a circuit board became defective, one would have no hesitancy in eliminating that instrument's results fron consideration. If an assessor is thoroughly an outlier or ineffective, that is justification for eliminating the results of that assessor, but calculations should be carried out both with and without the results of an assessor whose performance is in question. Quite naturally, a company might want the sensory results sharpened up as much as possible to know where potential differences lie. It can do so by deleting the scale values of assessors, too much of whose performance is non-significant or whose values are widely divergent from the majority of the panel, but on the other hand it should recognise that these same results yield valuable information. If a trained assessor cannot detect differences in certain attributes, then it is unlikely that those not trained, such as consumers,will detect a difference. The company should want to know where differences are not detectable. Once accepted as a panelist and especially when experienced, an assessor should not be removed from the panel unless he is consistently less sensitive than other panel members or consistently an outlier, but account may need to be taken of an individual being an outlier in particular cases.

5.6. Principal Component Analysis
Cluster analysis was discussed early among the correlative methods because its application logically goes with the first thing to be evaluated, the

performance of assessors, though its use need not be confined to that (Lisle et al., 1978; Clapperton and Piggott, 1979b; Pino et al. 1986b). Unquestionably, PCA is used more widely than CA in the evaluation of data sets. The function of PCA is to reduce a set of individual items into components such that the first component has maximum correlation with all the variables and accounts for the greatest amount of variance, the second for the second-largest amount of variance, etc., until as much of the variance has been accounted for as is reasonable.

There are many illustrations of the use of PCA in the literature: Piggott and Jardine (1979), Lyon (1987), Clapperton (1979), Clapperton and Piggott (1979b), beer; Noble (1981), Williams (1982), wine; Horsfield and Taylor (1976), Frijters (1976), meat and poultry products, respectively; Toda et al. (1971), gels produced by boiled plant proteins; Novais et al. (1982), texture of mashed potatoes; Martens (1985), white cabbage; Gaydou et al. (1987), differentiation of citrus species according to flavone content. Hashimoto and Eshima (1980) reduced 20 terms applied to beer to seven components. They commented that the first component had a large loading on terms such as overall, mild, fresh, bitter, body and seemed to represent balance or coordination of the individual sensory characteristic. The second loading component contained such terms as sulphidic, diacetyl, yeasty, young, grassy (from hops), hoppy and estery and contained information about off-flavours or characteristics abnormally perceptible. They commented that although the International System of Flavour Terminology (Meilgaard et al., 1979b) makes no provision for hedonic terms such as mellow, mild, smooth, acceptable and pleasing, these terms do have meaning to a given panel and are as necessary for a panel as terms applicable to identifiable notes.

A use of PCA which Powers and his co-workers (Powers, 1984; Godwin, 1984; Powers et al., 1984, 1985; Powers and Ware, 1986) have made is to construct graphs in three dimensions to show each assessor his spatial position *vis-a-vis* the others. Originally three-dimensional models, using moveable pedestals, were employed; today, three-dimensional graphs are displayed on the monitor or printed using a personal computer. While a visual display does not tell the assessor why he is an outlier, seeing one's location in relation to the others often stimulated assessors to try harder to get in line with the rest.

5.7. Factor Analysis

Factor analysis is also a data reduction process, but it differs from PCA in the way data reduction is accomplished. In PCA, variance is the element leading to inclusion of an entity within a component. In FA, covariance is the critical element. A consequence is that an entity need not be included within one factor only. According to its covariance, an entity may be partially in one factor and partially within one or more others. It is looked upon as a means of detecting underlying structure or order among variables, which might otherwise seem to have no elements in common.

There certainly is dichotomy of thought relative to FA. Piggott and Sharman (1986) briefly mention FA, then dismiss it: 'but the method [FA] is little used and will not be discussed further'. Chatfield and Collins (1980) were more blunt. They stated: 'in view of [its] disadvantages ... we recommend that FA should not be used in most practical situations'. Fischman et al. (1987) and Shinholser et al. (1987) on the other hand pointed out that the time has come when investigators should look at possible product–assessor interaction when making judgements about factor patterns. They claimed this is necessary because the application of FA is on the increase. Whether or not one thinks FA has value, there are several papers in the literature describing FA results. Among them are those dealing with attributes of toothpaste (Baines, 1976); texture of chicken meat (Lyon, 1980, 1987); chicken meat canned by different agencies (Lyon and Klose, 1980); odour qualities of meat products (Galt and MacLeod, 1982); meat texture (Harries et al., 1972); wine terms (Wu et al., 1977; Kwan and Kowalski, 1980); characterisation of carrots, swedes and cauliflower (Fjeldsenden et al., 1981); delineation of the degree of communality among terms used to describe different varieties of apples (McLellan et al., 1984) and comparison of FA and multidimensional scaling of odourants (Jeltema and Southwick, 1986).

Table 17 shows a flavour factor discerned in the study of Wu et al. (1977) and a texture factor detected in the study of Powers et al. (1977). Correlation analysis merely shows that correlation exists; FA suggests that members of the same factor have certain elements in common. These elements may be the antithesis of each other such as dryness and sweetness in wine. Note that the signs are different. Sweetness also is in the same factor as tartness, sharpness and bitterness, again because they are antipodean in nature. For the texture factor, coarseness, fibrousness and crispness would all seem to most individuals to be related. While we might not normally think of objectionable characteristics such as sliminess and sogginess being correlated with juiciness and tenderness, second thought does bring to mind some elements common to all these sensations. Factor analysis often aids the investigator in understanding interplay among sensory sensations.

Factor analysis also functions by facilitating the avoidance of

TABLE 17
ILLUSTRATION OF TERMS INCLUDED IN A FLAVOUR FACTOR AND IN A TEXTURE FACTOR

Flavour[a]		Texture[b]	
Tart	0·55	Coarse	0·57
Biting	0·52	Fibrous	0·57
Astringent	0·46	Crisp	0·53
Sharp	0·46	Slimy	−0·53
Bitter	0·42	Juicy	−0·55
Dry	0·35	Soggy	−0·63
Vinegary	0·35	Tender	−0·66
Coarse	−0·26		
Sweet	−0·33		

[a] From Wu et al. (1977). Reprinted from Journal of Food Science (1977) **42**, 944, with permission. Copyright © by Institute of Food Technologists.
[b] From Powers et al. (1977). Reprinted from ACS Symposium Series No. 51, 1977, with permission. Copyright © by American Chemical Society.

redundancy. Logically, Godwin et al. (1978) should have reduced the number of terms they used because the earlier study of Powers et al. (1977) had shown that some of the terms were redundant, as perusal of the texture factor in Table 17 reveals. Godwin et al. had reasons for keeping the two studies on green beans, conducted one year apart as parallel as possible, but where the same kind of a product may be evaluated time after time, redundant terms should be eliminated to lessen the burden on the assessors.

5.8. Orthogonal vs Oblique Rotation
Normally, FA is carried out to establish as much of the orthogonal relations among the variables as possible. In some instances, oblique FA is used. The intent there is to discern the associations among the variables. With reference to profiling of QSP data as spider-web diagrams, it was pointed out that profiling is empirical because of the impossibility of relating each attribute to all other attributes based upon calculation of simple correlation coefficients alone. Factor analysis is one way to discern some of the probable relations among the attributes, although some consider there is distortion of the relations in reducing them to factors; Vuataz (1977) considered that the relations are only incomplete, not distorted. If all possible combinations of the variables (dimensions) were to

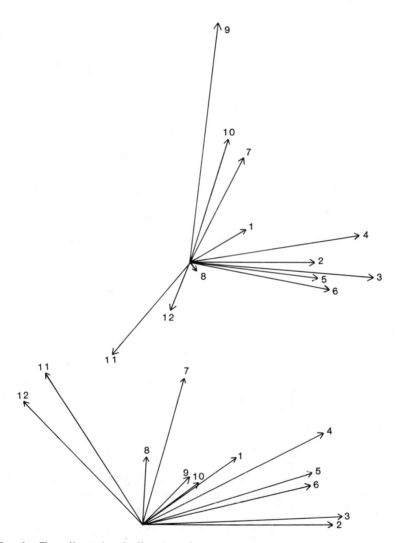

Fig. 6. Two-dimensional diagrams for tea sweetened with sucrose (upper diagram) and with saccharin (lower diagram). From 1 to 12, the numerals, respectively, stand for: intensity of aroma, flavour, pungency, briskness, bitterness, astringency, colour, brightness, fishiness, smokiness, sweetness and acceptability.

be examined by projecting them onto appropriate planes, full and true relations among the variables could be ascertained. If the investigator has access to a computer and one of the statistical packages available, calculation of a factor analysis is not particularly expensive or time consuming. Rogers et al. (1986) conducted FA to learn whether the sweetening of tea with sucrose and with saccharin resulted in differences in the tea. Using the triangular test, they had assessors examine tea sweetened with the two sweeteners until they could no longer differentiate between the sweetness of the teas. Having established equivalent sweetness, they then had the assessors scale several other attributes of tea. The investigators were really interested in learning whether the different sweeteners affect characteristics of tea other than sweetness (and bitterness in the case of saccharin (Hyvönen et al., 1978)). Figure 6 shows that the attributes were grouped in different ways for sucrose and saccharin as sweetening agents. Application of Hotelling's T^2 test had shown that the means of the attributes of the teas treated in the two ways were not significantly different. Bartlett's test showed that the correlation matrices were significantly different. The sucrose seemed to have a melding effect. Note that the angle (20·5°) from attribute 6 to 4 for the tea sweetened with sucrose is less than that (26·5°) from attribute 2 to 4 (the same attributes) for the tea sweetened with saccharin. For some of the other attributes, however, association was less close. For the saccharin-sweetened tea, the relation between acceptability and sweetness was very close, whereas acceptability was influenced more by the other attributes when sucrose was the sweetening agent, as shown by the smaller vector for acceptability. (The means for acceptability and sweetness of the two teas were not significantly different from each other.)

Quite often assessors perceive of products as being significantly different when discrimination tests are applied, but scaling results do not show significant differences between means. It is interesting that it is difficult to find literature citations supporting this statement although almost invariably sensory technologists accept the conclusion as gospel. Gormley et al. (1986) did report that paired comparisons were more effective in the detection of differences between tomato samples than hedonic scoring, but hedonic scoring was carried on only for single samples presented a day apart. Lin (1983) was able to differentiate between soy sauce samples more readily when paired comparisons were used, although the judges were permitted to declare no difference (Powers and Ware, 1976), than when line scaling was utilised. Bohler et al. (1986) considered that under some conditions ranking was more sensitive to differences than scaling. Not all of our perceptions of difference depend upon magnitude. The study of Petró-Turza et al. (1986) confirms that point. Apparently as we sense attributes we subconsciously make judgments about the ratios among sensations. Whether we do or do not, a major advantage to conducting FA examination of QSP data is that the relations between attributes are more realistically depicted than construction of profiles with the angles equidistant. The word 'realistic' is used with full realisation that any reduction of several dimensions to two or three may not be exactly in accord with reality.

Jounela-Eriksson (1981) likewise depicted relations in two different ways. She first showed a typical spider web diagram, then PCA diagrams of the three types of whisky she evaluated and commented upon the differences among the inter-relationships of the various attributes. Stone et al. (1980) state that they now recommend separate display of the means and the correlations among attributes. They showed a spider-web profile with unequal angles, but they did not specify the basis upon which the angles were set.

Three benefits result from the application of FA. First, the profiles are likely to be more true to actual relations among attributes than arbitrary construction of a spider-web profile. Secondly, insight is provided into interaction effects among components of a food or other sensory material. Similar effects of sucrose on aroma and taste qualities of blueberry and cranberry juices was observed by von Sydow et al. (1974) as Rogers et al. (1986) noted. Sucrose diminished harsh tastes, increased overall acceptability, augmented aroma notes associated with pleasantness and decreased those associated with unpleasantness. Though the authors stated that the action of sucrose was probably on a psychological rather than a chemical level, Moskowitz (1981), one of the authors of the von Sydow et al. (1974) article, reported that increasing the viscosity of a solution reduces sweetness. Ingate and Christensen (1981) showed by FA that there were two major mouthfeel dimensions, density/thickness and a chemical irritant effect, for the fruit-based beverages they examined and that sweetness ratings were negatively correlated with mouthfeel terms describing oral mucosal irritation. Kokini (1987) has likewise studied texture–taste interaction. The subject of interaction effects is beyond the scope of this chapter. Powers (1981b,c; 1982a,b) cited a few publications dealing with interactions; Forss (1981) has done likewise. Meilgaard (1982) has studied singly and in combination the effects of added chemicals on the flavour of beer. The publications of Lawless (1986), Frank and Archambo (1986) and De Graaf and Frijters (1987) also relate to interaction effects. The study of

Rogers et al. (1986) showed that only a fishy note and the intensity of aroma were significantly different when sucrose and saccharin were the respective sweetening agents, but the perceived associations among attributes were quite different. In some respects, the change is as if it were in the 'non-identifiable' fraction of the flavour profile (Caul, 1957) rather than among the identifiable attributes though they were the ones which were assessed.

The third benefit to the application of FA is a practical one. Enough of the means may not be sufficiently different to permit the investigator to conclude that this product is different from that product relative to the particular attributes examined, but if the FA patterns vary, this does suggest that the products are different though, figuratively speaking, the assessors cannot quite put their fingers on the difference. When a fixed list of terms is used, there may be attributes not being assessed, as Caul (1957) commented when she stated that trained tasters could not distinguish the individual components of a sweet-sour-salty-bitter solution, only that they were high-body or low-body blends. Perhaps it is necessary to examine attributes other than those initially examined to learn if they influence response.

A word of caution is needed about FA and all the internal analysis methods (correlation, PCA, CA, FA and multidimensional scaling). Internal analysis methods are frowned upon by some statisticians on the basis that the output is provincial to the particular assessors and products. However, there is an error which has largely been overlooked, but which now can be assessed because of the studies of Powers et al. (1985, 1987), Fischman et al. (1987) and Shinholser et al. (1987). Other than Harries et al. (1972) who sought to learn whether their panelists were responding sufficiently similarly to yield comparable individual factor patterns, many factor patterns in the literature, including those of Wu et al. (1977), Powers et al. (1977) and Godwin et al. (1978), are somewhat suspect because the investigators did not demonstrate that a substantial majority of the panel yielded similar factor patterns. If the factor patterns of the individual assessors are not almost alike, then the factor pattern for panel results pooled can be quite misleading, for it could be but the mean of a group of divergent patterns. Fischman et al. (1987), Powers et al. (1987) and Shinholser et al. (1987) have set forth a procedure to test either the correlation matrices or the factor patterns themselves for agreement. Their objective was to determine the extent to which the patterns of individual assessors or small subgroups of the panel are correlated with the pattern based on panel results pooled. Laslett and Bremner (1979) observed that the correlation matrices of each of their tasters varied. Hence, they concluded no single matrix could be constructed which would represent the whole panel.

5.9. Discriminant Analysis

Powers and Ware (1986) have described the basic principles behind DA, the various forms of DA and given illustrations of the application of DA to determine which terms are the chief cause of product–assessor interaction and to secure a composite judgement as to whether products differ. Quite often ANOVA indicates that one product differs from the rest in one component but not in others whereas a second product differs from the rest of the samples but for a different reason. MANOVA can be used to tell the investigator whether there is any statistical difference among all the samples, but it does not provide information as to where the differences are. Discriminant analysis is helpful in that the investigator can learn whether the differences cumulatively are sufficient to permit the samples to be partitioned into discrete classes. If so, some or all product differences have been resolved.

Stepwise discriminant analysis (SDA) is a procedure for (1) winnowing from amongst many variables those which have significant discriminating power and (2) ferreting out from amongst the discriminators the combination of variables with the greatest discriminating power. In effect, the most effective single discriminator is selected first, then a second discriminator is selected in such a way that the pair of discriminators adds to the efficiency of classification, then a third is added, etc., until a set of terms has been selected which permits full resolution of sample differences or else full classification fails because there are not enough effective discriminators (used in combination) to permit complete classification.

Multiple discriminant analysis (MDA) generally follows SDA. Once the most useful group of discriminators has been discerned, then a discriminant function may be calculated to permit samples to be classified according to their measurement values and the weighting value which goes with each term of the discriminant function. Goulden (1952) has described the process. The investigator need not go through the SDA stage to get to the MDA process, but often there are advantages to reducing the data set so that too much computer time will not be involved in calculating the discriminant function.

Sometimes one discriminant function suffices to resolve sample differences; at other times, more than one function may be needed to effect full resolution (Milutinović et al., 1970; Powers, 1970, 1976; Powers and Quinlan, 1973, 1974). There is one 'best' function for a given set of data.

There usually are several alternative ones, some of which come very close to the best function in terms of efficiency. The reverse unfortunately also applies. A given equation is tied very closely to a given data set. Note from Table 13 that the variables used differ according to who the assessors were. Persson and von Sydow (1972) cautioned that sensory trials should be carried out with separate sets of samples, preferably different panels and at times well apart to minimise the likelihood that effects observed are parochial to a given data set and are not universal effects. Powers et al. (1977) and Godwin et al. (1978) carried on their trials one year apart, with different products and with a different set of assessors. Meilgaard (1981) has called attention to a flaw which is nearly universal. He stated that almost all MVA publications are provincial in that they often exhibit good classification or regression, but the data are not of much value to others because equations leading to separation are not given. His objection is sound, though, as Powers (1982c) pointed out, it is the task of industry to overcome some of these limitations such as avoiding the basing of discriminant functions on too narrow a sample set. Industry is the agency which, as a part of its quality-assurance programs (Wren, 1972), is in the best position to acquire substantial amounts of data on the same products but at different times, in different plants and with different panels.

With respect to the last point, different panels are almost certain to require different discriminant functions (Powers and Rao, 1985). Because of a heavy work load in a particular plant, panels operating during different shifts or evaluation of a product which leads rapidly to fatigue, more than one panel may be required for a particular commodity. Panels at different branch plants quite obviously are going to be different. Even if the same person does the training and requires the same degree of efficiency for selection (or retention) as a panelist, there will be some differences in the way the panelists perceive the product. This subject was covered earlier. In some instances the differences are a consequence of culture. Lin (1983) and Paule (1986) used panels of Orientals and non-Orientals in their investigations. Lin was evaluating soy sauce produced fermentatively and chemically. Table 18 lists the 11 descriptors she used to evaluate the intensity of different components of odour of soy sauce and scale values assigned for overall desirability of the aroma. Her data were selected for three reasons. The first was to illustrate that the magnitude of scale values assigned differed significantly in most cases according to the cultural conditioning of the panelists. Second, the reader should notice that the two groups assigned nearly the same scale value to overall desirability of aroma, but notice third that their concepts of flavour relations among

TABLE 18
LINE SCALE VALUES ASSIGNED BY ORIENTAL AND NON-ORIENTAL PANELISTS TO AROMA ATTRIBUTES OF SOY SAUCE

Descriptor	Mean values	
	Oriental panel	Non-Oriental panel
Alcoholic	3·69	5·60
Aromatic	5·74	5·32
Acidic	3·53	3·48
Tingling	4·02	4·06
Beany	3·82	3·80
Burnt	2·31	3·80
Caramel-like	2·89	4·10
Molasses-like	2·42	3·56
Yeasty	2·99	2·88
Medicinal	1·83	2·87
Musty	3·08	3·79
Overall desirability	5·63	5·67

Any two values underscored by the same line are not significantly different.

different variables were not the same as shown in Fig. 7. That there should be differences in their perceptions is logical, for the magnitude of the scale values assigned to the different attributes were significantly different, unlike the study of Rogers et al. (1986) where only two of the individual attributes differed significantly. Hence one should not be surprised that the association diagrams (Fig. 7) likewise differ.

5.10. General Considerations

Whether the investigator is utilising PCA, DA, canonical analysis or any of the other MVA procedures, a limitation he or she always has to be aware of is the need to have adequate df. It is a problem in industry where there is a tendency—as in quality assurance—not to have replication unless there is an indication that the sample is borderline with respect to being within specifications. Samples may be re-run then or a second panel brought into play to learn if the second panel will verify the finding of the first.

The impossibility Hwang-Choi encountered (described above) when she wanted to calculate an MVA F-value for 22 terms with only 15 df illustrates the difficulty often encountered. In her case, the difficulty arose not because she failed to provide for replications. She had. The trap she fell into is a

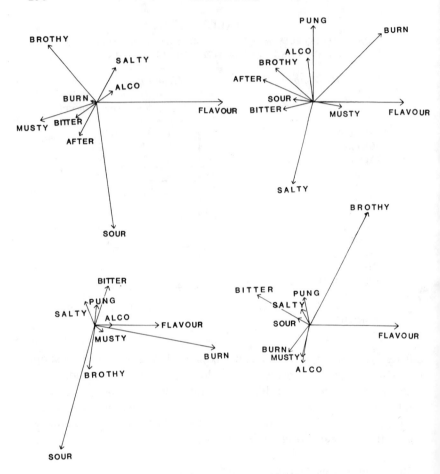

FIG. 7. Association diagrams for relations among attributes of soy sauce. Diagrams to left are for chemically hydrolysed soy sauce; at right, for fermentatively produced soy sauce. The two upper diagrams are based on results from panel composed of Chinese, Japanese and Korean nationals; the two lower ones, judgements of native Americans (excluding those of Oriental extraction). The vector lengths represent the communality or coefficient of determination of an attribute. Long vectors thus indicate an attribute possesses properties in common with other attributes; short vectors indicate poor correlation between properties of an attribute and those of other attributes. The direction of short vectors should therefore be disregarded. Abbreviated terms stand for alcoholic, burnt or caramelised, strong aftertaste and pungency. Missing terms, vectors too short to show.

common one in basic research. It is easy to say that one should make all decisions in advance about the design of the experiment and the form(s) of analysis to be employed, but one often does not know exactly what will be needed at a second stage of statistical analysis until one has the results from the first. Experience in our laboratory has shown that four replicate sensory evaluations are usually enough, and Hwang-Choi inserted five replications in her plan, but she needed nine replications to provide enough df to permit 22 variables to be analysed concurrently. The writer here is to blame, as he should have foreseen that she might need replication merely to provide sufficient df for some forms of analysis even though many replications probably would not be needed to establish statistical differences between samples.

Another illustration involves a more fortunate investigator. A dairy scientist asked us (Powers, 1979, 1981c) to examine by MVA data acquired in evaluating cheese, with trials being replicated thrice. We had told him beforehand that we thought three replications of six samples was cutting it close. We were right; the dairy scientist was lucky. At the twelfth SDA step, the 18 samples were classified into the six brands to which they truly belonged. Had 13 steps been required, classification would have been incomplete for want of df. He was lucky in a second way. He had 100% success in classifying his samples. As judged by most literature reports and our own experience, 80–90% success in classification of samples is usually the best one can do using sensory data alone. Product differences can often be resolved with as few as four chemical or instrumental variables (Powers, 1981c, 1982c), but, of course, there is always the question: Are the chemical data sufficiently correlated with sensory evaluation to serve in lieu of it? The answer almost invariably is: no! Chemical information may complement sensory information; it can rarely supplant it. Powers (1982a) has discussed the matter of pattern recognition and has pointed out that the problem is not solely one of pattern recognition but of pattern reconciliation between sensory and chemical data sets. Berglund et al. (1982) observed similarities between patterns for air inside and outside buildings as judged by odour or by GLC profiles, but between odour and GLC profiles, the patterns were different. The patterns were generated by PCA.

A problem caused by the lack of concord between sensory and chemical patterns was pointed out by Wilkin et al. (1983). Their company wanted to determine on a production basis whether its neutral spirits were organoleptically within grade. It had endeavoured to utilise GLC analysis for quality-assurance purposes, but odours important to quality were detected at points where peaks were absent. It did not want to rely on the sensory judgement of a single expert but felt rather that the mean scores provided by several well-trained assessors would be more representative of practical quality levels than the judgement of an expert who might be overly-sensitive to a particular component or biased in some way. It considered that if enough well-trained assessors and the proper descriptive terms were used, panel results would be indicative of quality differences and that quantitation would permit quality-assurance standards to be set.

The company put into practice many of the principles and procedures described in this chapter and especially those described by Piggott and Canaway (1981). The latter originated the descriptive terms, sought out suitable reference compounds and recommended methods for training and for multivariate analysis of the results. The company excluded from the panel production personnel, shift employees and those who had a defeatist attitude toward being able to use sensory analysis on an 'on-line' basis. The spirits were evaluated only by 'nosing'. Neutral spirits for blending are not devoid of odour, nor are they intended to be. Their minor odour notes, when of the proper type, are exactly what are wanted. Products of a well-run company are not likely to be grossly deficient in desirable attributes nor contain obvious amounts of undesirable compounds. As Powers (1979, 1981c) pointed out for other applications, within the range of commercial acceptability, differences in grade or quality are often exceedingly minor. Wilkin et al. (1983) reported that as soon as the 12 assessors had evaluated the intensity of each component, a composite 'score' could be calculated, permitting a decision as to whether the product met the company's specifications or those of its customers. To calculate the score, coefficients previously established by DA were used to weigh the importance of each descriptor as a determinant of quality.

Wilkin et al. (1983) did not belittle chemical analysis. They stated that the company did not have the means of correlating multiple-peak GLC patterns with sensory quality. If computing facilities are available, chemical analysis may be used to buttress sensory evaluation in decision-making, or chemical data may be used as a substitute for sensory examination, for example on night shifts when assessors might not be available. With or without chemical analysis and computing facilities, the final arbiter of sensory quality, other than acceptance in the market place, is sound sensory examination.

Sometimes for a given task either DA or PCA may be appropriate depending upon other knowledge which exists or assumptions made—DA is appropriate to examine properties of materials to ascertain to which class a sample belongs, whereas PCA is appropriate to learn whether different classes exist *at all*. Piggott and Jardine (1979) used DA to classify the nine brands of Scotch whisky and the one brand of Irish whiskey they studied. They also applied PCA. Even though they knew the brands (classes), in effect they were asking the sensory properties of the whiskies to tell them whether any of the whiskies were close to each other in character in spite of being marketed under different brand names. When PCA is applied to examine the performance of assessors, it is being used for the same purpose: to learn whether one group exists or the assessors belong to different groups because of failure to respond alike (Powers, 1984, 1986; Powers *et al.*, 1984, 1985).

This chapter is not intended to be an exposition of the application of MVA procedures to sensory results. Rather, its chief purpose is to explain the intricacies of descriptive analysis. The companion book, *Statistical Procedures in Food Research* (Piggott, 1986), takes care of the exposition of MVA procedures very well. Though not covered exhaustively here, MVA procedures employed most frequently in conjunction with QSP need to be discussed within the same chapter to illustrate their uses and limitations. For that reason, CA, DA, FA and PCA were discussed somewhat in detail. There are other MVA methods which merit mention. Stepwise regression analysis (SRA) and multiple regression analysis were not discussed in particular because many of the features of these methods are analogous to DA, at least with respect to application. Stepwise discriminant analysis is appropriate if the products or treatments fit into classes (discontinuous data) whereas SRA applies if the data are continuous. Related to regression is partial least squares regression. Martens and her co-workers have been in the forefront in explaining and applying that procedure (Martens *et al.*, 1983, 1985; Martens and Burg, 1985; Martens, 1985, 1986*a,b*). Martens and Martens (1986) have set forth the basic principles of partial least squares regression analysis. A little more attention has been devoted to it as a regression technique because it is less widely understood and used than most other regression procedures; yet it is one which has special application for the analysis of QSP data.

Correspondence analysis is described by Hoffman and Franke (1986). Canonical analysis is frequently applied to sensory data. Basically it is a correlation method between sets of variables rather than individual variables. Noble *et al.* (1982) applied it to the differentiation of wines. Stungis (1976), Harries (1973) and in particular Vuataz and co-workers (Vuataz *et al.*, 1975; Vuataz, 1977, 1981) have commented upon various features or limitations of multivariate methods. Romero (1983) described, in a simple but sound manner, the functions and distinctions among various MVA methods. Without their use the benefits of QSP would not be nearly as fruitful as they are.

6. CONCLUSIONS

The present state of descriptive sensory profiling is a result of the confluence of various forces, each adding new power to the procedure, much as the Danube becomes an even grander river once the Sava joins it. The idea and philosophy of profiling, the attaching of strict quantitation, replication and statistical analysis to profiling and the development of computers have made possible current applications. If computers and especially if easy-to-use analysis packages did not exist, multivariate descriptive analysis might still be a minor part of sensory analysis. For any given data set, there may be one best method to analyse the data, but there is no method of descriptive or statistical analysis most suitable for all sensory applications. The practitioner of QSP needs to be familiar with the scope and limitations of the various methods. There is a fourth force at work, more potent than the three tools, profiling, quantitation and computer aid, mentioned above. That is the experience and knowledge being acquired as applications are made. Both failure and success contribute to that knowledge. Indeed, failure may be more productive than success, for it often set ideas scintillating more vigorously in the mind. Out of the refining of methods in making application or in the devising of new ones, there is no doubt that even greater benefits will be realised from QSP than have been obtained so far, substantial as are present benefits.

ACKNOWLEDGEMENTS

Appreciation is expressed to the following for comments or the furnishing of materials or services: Dr Jean Caul and Mr Frederick Sullivan; Dr Morten Meilgaard, Ms Jacqueline Pearce, Ms Katherine Zook; Mrs Elaine Skinner; and Mrs Cathy B. Gober, Mrs Kathleen Shinholser and Mrs Silvia King.

REFERENCES

Amerine, M. A., Pangborn, R. M. and Roessler, E. B. (1965). *Principles of Sensory Evaluation of Food*, Academic Press, London, 602 pp.
Anderson, H. T. and Hartmann, A. O. (1971). A fifth modality of taste. *Acta Physiol. Scand.*, **82**, 447–52.
Andersson, Y. and Lundgren, B. (1981). Sensory texture/consistency descriptors of beef patties containing protein products. *J. Texture Studies*, **12**, 217–41.

Appel, C. E. (1985). Taste Flavor. *Chemtech.*, **15**(7), 420–3.
Arnold, G. M. and Williams, A. A. (1986). The use of generalized procrustes techniques in sensory analysis. In: *Statistical Procedures in Food Research*, J. R. Piggott (Ed.), Elsevier Applied Science Publishers, London. pp. 233–253.
Asher, G. (1976). An Englishman's perspective of California and her wines. *Bull. Soc. Med. Friends Wine*, **18**(1), 3–7.
ASTM Committee E-18. (1968). *Manual on Sensory Testing Methods*, STP 434, American Society for Testing and Materials, Philadelphia, p. 36.
ASTM Committee E-18. (1981). *Guidelines for the Selection and Training of Sensory Panel Members*, STP 758, American Society for Testing and Materials, Philadelphia, 35 pp.
Baines, E. (1976). Evaluation of flavours in dental creams. *J. Soc. Cosmet. Chem.*, **27**, 271–87.
Banfield, C. F. and Harries, J. M. (1975). A technique for comparing judges' performance in sensory tests. *J. Food Technol.*, **10**, 1–10.
Bartels, J. H. M., Burlingame, G. A. and Suffett, I. H. (1986). Flavor profile analysis: taste and odour control of the future. *Am. Water Works Ass. J.*, **78**, 50–5.
Barylko-Pikielna, M. and Metelski, K. (1964). Determination of contribution coefficients in sensory scoring of overall quality. *J. Food Sci.*, **29**, 109–11.
Bengtsson, K. (1943). Provsmakning som analysmetod statistisk behandling av resulaten. *Svenska Bryggareforeningens manadsblad*, **58**(3), 102–11; 149–57.
Berglund, B., Berglund, U., Lindvall, T. and Nicander-Bredberg, H. (1982). Olfactory and chemical characterization of indoor air, towards a psychophysical model for air quality. *Envir. Internat.*, **8**, 327–32.
Beyer, W. H. (1968). *CRC Handbook of Tables for Probability and Statistics*, 2nd edn, CRC Press, Boca Raton, Florida.
Böhler, G., Escher, F. and Solms, J. (1986). Evaluation of cooking quality of potatoes using sensory and instrumental methods. 1. Sensory evaluation. *Lebensm.-Wiss. u. Technol.*, **19**, 338–343.
Bone, B. (1987). The importance of consumer language in developing product concepts. *Food Technol.*, **41**(11), 58–60, 86.
Brandt, M. A., Skinner, E. Z. and Coleman, J. A. (1963). Texture profile method. *J. Food Sci.*, **28**, 409–9.
Brown, D. G. W. and Clappterton, J. F. (1978). Flavour terminology of beer. A study of terms used to describe ale flavours. *J. Inst. Brew.*, **84**, 324–6.
Cairncross, S. E. and Sjöström, L. B. (1949). Flavor profiles—a new approach to flavour problems. Paper presented at annual meeting, Institute of Food Technologists, San Francisco.
Cairncross, S. E. and Sjöström, L. B. (1950). Flavor profiles—a new approach to flavor problems. *Food Technol.*, **4**, 308–11.
Caul, J. F. (1957). The profile method of flavor analysis. *Adv. Food Res.*, **7**, 1–40.
Caul, J. F. and Vaden, A. G. (1972). Flavor of white bread as it ages. *Bakers Digest*, **46**, 42–60.
Caul, J. F., Cairncross, S. E. and Sjöström, L. B. (1958). The flavor profile in review. In: *Flavor Research and Food Acceptance*, Arthur D. Little (Ed.), Reinhold Publishing Co., New York, pp. 65–75.
Chatfield, C. and Collins, A. J. (1980). *Introduction to Multivariate Analysis*. Chapman and Hall, London, p. 89.

Civille, G. V. (1979). Descriptive analysis. In: *Sensory Evaluation Methods for the Practicing Food Technologists*, M. R. Johnson (Ed.), Institute of Food Technologists, Chicago, pp. 6-1 to 6-28.
Civille, G. V. and Lawless, H. J. (1986). The importance of language in describing perceptions. *J. Sens. Studies*, **1**, 203–215.
Clapperton, J. F. (1973). Derivation of a profile method for sensory analysis of beer flavour. *J. Inst. Brew.*, **79**, 495–508.
Clapperton, J. F. (1979). Sensory characterization of the flavour of beef. In: *Progress in Flavour Research*, D. G. Land and H. E. Nursten (Eds), Applied Science Publishers Ltd., London, pp. 1–14.
Clapperton, J. F. and Piggott, J. R. (1979a). Flavour characterization by trained and untrained assessors. *J. Inst. Brew.*, **85**, 275–7.
Clapperton, J. F. and Piggott, J. R. (1979b). Differentiation of ale and lager flavours by principal components analysis of flavour characterization data. *J. Inst. Brew.*, **85**, 271–4.
Clapperton, J. F., Dalgliesh, C. E. and Meilgaard, M. C. (1975). Progress towards an international system of beer flavour terminology. *Tech. Quart. Master Brewers Ass. Am.*, **12**, 273–80.
Clapperton, J. F., Dalgliesh, C. E. and Meilgaard, M. C. (1976). Propuesta para la redacción de un vocabulario internacional referente al gusto de la cerveza. *Cerveza y Malta* **XII** (50), 18–28.
Cross, H. R., Moen, R. and Stanfield, M. S. (1978). Training and testing of judges for sensory analysis of meat quality. *Food Technol.*, **32**(7), 48–54.
Cuřín, J. (1976). Mezinarodni system popisnych terminu objektivniho senzorickeho hodnoceni piva. *Kvasny Prumysl*, **22**, 217–22.
Daget, N. (1977). Sensory evaluation or sensory measurement? *Nestlé Research News 1976/77*, Nestec Ltd, Vevey, Switzerland, pp. 43–56.
De Graaf, C. and Frijters, E. R. (1987). Sweetness intensity of a binary sugar mixture lies between intensities of its components, when each is tasted alone and at the same total molarity as the mixture. *Chem. Senses*, **12**, 113–29.
Dravnieks, A. (1975). Evaluation of human body odors: methods and interpretation, *J. Soc. Cosmet. Chem.*, **26**, 551–71.
Dravnieks, A. (1982). Odour quality: semantically generated multidimensional profiles are stable. *Science*, **218**, 799–801.
Dravnieks, A. J. (1985). *Atlas of Odor Character Profiles*. Data Series DS 61, American Society for Testing and Materials, Philadelphia.
Dravnieks, A., Krotoszynski, B. K., Lieb, W. E. and Jungermann, E. (1968). Influence of an antibacterial soap on various effluents from axillae. *J. Soc. Cosmet. Chem.*, **19**, 611–26.
Dravnieks, A., Reilich, H. G., Whitfield, J. and Watson, C. A. (1973). Classification of corn odor by statistical analysis of gas chromatographic patterns of headspace volatiles. *J. Food Sci.*, **38**, 34–9.
Dravnieks, A., Keith, L., Krotoszynski, B. K. and Shah, J. (1974). Vaginal odors: GLC assay method for evaluating odor changes. *J. Pharmaceut. Sci.*, **63**, 36–40.
Dravnieks, A., Bock, F. C., Powers, J. J., Tibbetts, M. and Ford, M. (1978). Comparison of odors directly and through profiling. *Chem. Senses Flav.*, **3**, 191–225.

Dravnieks, A., McDaniel, H. C. and Powers, J. J. (1979). Comparison by twelve laboratories of the odor qualities of nine chemicals sniffed from the bottle and as gas-liquid chromatographic effluents. *J. Agric. Food Chem.*, **27**, 336–46.

Dürr, P. (1979). Development of an odour profile to describe apple juice essences. *Lebensm.-Wiss. u. Technol.*, **12**, 23–6.

Fischman, E. I., Shinholser, K. J. and Powers, J. J. (1987). Examining methods to test factor patterns for concordance. *J. Food Sci.*, **52**, 448–50, 454.

Forss, D. A. (1981). Sensory characterization. In: *Flavour Research—Recent Advances*, R. Teranishni, R. A. Flath and H. Sugisawa (Eds), Marcel Dekker, Basle, pp. 125–74.

Fjeldsenden, B., Martens, M. and Russwurm, H. (1981). Sensory quality criteria of carrots, swedes and cauliflower. *Lebensm.-Wiss. u. Technol.*, **14**, 234–41.

Frank, R. A. and Archambo, G. (1986). Intensity and hedonic judgments of taste mixtures: an information integration analysis. *Chem. Senses*, **11**, 427–38.

Frijters, J. E. R. (1976). Evaluation of a texture profile for cooked chicken breast meat by principal component analysis. *Poultry Sci.*, **55**, 229–34.

Gacula, M. C. Jr., Parker, L. A., Kubala, J. J. and Reaume, J. (1974). Data analysis: a variable sequential test for selection of sensory panels. *J. Food Sci.*, **39**, 62–3.

Gacula, M. C. Jr. and Washam, R. W. (1986). Scaling word anchors for measuring off flavor. *J. Food Quality*, **9**, 57–65.

Galt, A. M. and MacLeod, G. (1982). Sensory and instrumental methods in meat analysis. In: *Sensory Quality in Foods and Beverages: Its Definition, Measurement and Control*. Proc. Chem. Soc. Ind./LARS Symp., Bristol 4–8 April 1982.

Gaydou, E. M., Bianchini, J.-P. and Randriamiharisoa, R. P. (1987). Orange and mandarin peel oils differentiation using polymethoxylated flavone composition. *J. Agric. Food Chem.*, **35**, 525–9.

Godwin, D. R. (1984). Relationship between sensory response and chemical composition of tea. PhD dissertation, University of Georgia, USA, 201 pp.

Godwin, D. R., Bargmann, R. E. and Powers, J. J. (1978). Use of cluster analysis to evaluate sensory-objective relations of processed green beans. *J. Food Sci.*, **43**, 1229–34.

Golovnja, R. V., Jakovleva, V. N., Cesnokova, A. E., Matveeva, L. V. and Borisov, J. A. (1981). A method of selecting panel for hedonic assessment of new food products. *Die Nahrung*, **25**, 53–8.

Golovnja, R. V., Symonva, L. A., Yakovleva, V. N. and Enikeeva, N. G. (1986). List of chemical substances and uniforn procedure for selection of panelists by their ability to recognize odours. *Die Nahrung*, **30**, 111–18.

Gormley, T. R., Arnold, G., Buret, M. and Decallonne, J. (1986). Tomato fruit flavour—an interlaboratory taste panel study. *Lebensm.-Wiss. u. Technol.*, **19**, 144–6.

Goulden, C. H. (1952). *Methods of Statistical Analysis*, J. Wiley & Sons, New York, pp. 378–93.

Gower, J. C. (1966). Some distance properties of latent root and vector methods used in multivariate analysis. *Biometrika*, **53**(3/4), 325–38.

Gower, J. C. (1971). Statistical methods of comparison of different multivariate analyses of the same data. In: *Mathematics in the Archaeological and Historical Sciences*, R. R. Hodson *et al.* (Eds), Edinburgh University Press, pp. 138–49.

Gower, J. C. (1975). Generalized Procrustes analysis. *Psychometrika*, **40**, 33–52.
Hall, R. L. (1984). Flavor technology—methodology and formulation. In: *Role of Chemistry in the Quality of Processed Food*, O. Fennema, H. Chang, C-Y Li (Eds), Food and Nutrition Press, Westport, Connecticut, pp. 158–73.
Hanson, J. E., Kendall, D. A. and Smith, N. F. (1983). The missing link: correlation of consumer and professional sensory descriptions. *Beverage World*, **102**, 108–16.
Harper, R. (1975). Some chemicals representing particular odour qualities. *Chem. Senses Flav.*, **1**, 353–7.
Harper, R., Bate-Smith, E. C., Land, D. G. and Grifiths, N. M. (1968a). A glossary of odour stimuli and their qualities. *Perfumery Essent. Oil Record*, **59**, 22–37.
Harper, R., Land, D. G., Griffiths, N. M. and Bate-Smith, E. C. (1968b). Odour qualities: a glossary of usage. *Brit. J. Psychol.*, **59**, 231–52.
Harries, J. M. (1973). Complex sensory assessment. *J. Sci. Food Agric.*, **24**, 1571–81.
Harries, J. M., Rhodes, D. N. and Chrystall, B. B. (1972). Meat texture. 1. Subjective assessment of the texture of cooked beef. *J. Texture Studies*, **3**, 101–14.
Hashimoto, N. and Eshima, T. (1980). Multivariate analysis of the sensory pattern of beef flavor. *Report Res. Lab. Kirin Brewery Co.*, No. 23, pp. 19–31.
Hellemann, U., Powers, J. J., Salovaara, H., Shinholser, K. and Ellilä, M. (1986). Unpublished data.
Henning, H. (1916). Die qualitatenreihe des geschmacks. *Z. Psychol.*, **74**, 203–19.
Hoffman, D. L. and Franke, G. R. (1986). Correspondence analysis: graphical representation of categorical data in marketing research. *J. Market. Res.*, **23**, 213–27.
Horsfield, S. and Taylor, L. J. (1976). Exploring the relationship between sensory data and acceptability of meat. *J. Sci. Food Agric.*, **27**, 1044–56.
Hwang-Choi, I. K. (1982). A case study of the use of standard methods to analyze multivariate sensory data. PhD dissertation, University of Georgia, USA, 162 pp.
Hyvönen, I., Kurkela, R., Koivistoinen, P. and Ratilainen, A. (1978). Sweetening of coffee and tea with fructose-saccharin mixtures. *J. Food Sci.*, **43**, 1577–89.
Ingate, M. R. and Christensen, C. M. (1981). Perceived textural dimensions of fruit based beverages. *J. Texture Studies*, **12**, 121–32.
Ishii, R. and O'Mahony, M. (1987). Taste sorting and naming: can taste concepts be misrepresented by traditional psychophysical labelling systems? *Chem. Senses*, **12**, 37–51.
ISO (1987a). Sensory analysis: General methodology guidelines for the choosing, training and monitoring of selected assessors and experts. ISO/TC 34/SC 12 Document 217E. ISO Secretariat, Association Francaise de Normalisation, Paris.
ISO (1987b). Sensory analysis methodology-sensitivity of taste determination of detection thresholds. ISO/TC 34/SC 12 Document 216E. ISO Secretariat, Association Francaise de Normalisation.
Jackson, M. G., Timberlake, C. F., Bridle, P. and Vallis, L. (1978). Red wine quality: correlations between colour, aroma and flavour and pigment and other parameters of young beaujolais. *J. Sci. Food Agric.*, **29**, 715–27.

Jellinek, G. (1964). Introduction to and critical review of modern methods of sensory analysis (odour, taste and flavour evaluation) with special emphasis on descriptive sensory analysis (flavour profile method). *J. Nutr. Dietetics*, **1**, 219–60.

Jeltema, M. and Southwick, R. (1986). The use of odor profiling to obtain information on the underlying differences and similarities among compounds. In: *Sensory Evaluation, Twentieth Annual Symposium.* D. L. Downing (Ed.), Special Report 59, New York State Agr. Expt. Sta., Geneva, New York, pp. 4–8.

Jolson, M. A. and Rossow, G. L. (1971). The Delphi process in marketing decision-making. *J. Market. Res.*, **8**, 443.

Johnson, P. B. and Civille, G. V. (1986). A standardized lexicon of meat WOF descriptors. *J. Sens. Studies*, **1**, 99–104.

Jounela-Eriksson, P. (1981). Predictive values of sensory and analytical data for distilled beverages. In: *Flavour '81*, P. Schreier (Ed.), Walter de Gruyter, Berlin, pp. 145–64.

Jounela-Eriksson, P. (1982). Whisky aroma evaluated by magnitude estimation. *Lebensm.-Wiss. u. Technol.*, **15**, 302–7.

Kluba, R. M. (1986). Sensory evaluation of alcoholic beverages. In: *Sensory Evaluation, Twentieth Annual Symposium.* D. L. Downing (Ed)., Special Report 59, New York State Agr. Expt. Sta., Geneva, New York, pp. 25–31.

Kokini, J. I. (1987). The physical basis of liquid food texture and texture-taste interactions. *J. Food Engineering*, **6**, 51–81.

Konovalov, K. I. (1982). Sensory evaluation of components of beer (in Russian). *Ferment. Alcohol Ind.*, **4**, 42–7.

Krasner, S. W., McGuire, M. J. and Ferguson, V. B. (1985). Tastes and odours: the flavour profile method. *Am. Water Works Ass. J.*, **77**, 34–39.

Kwan, W. and Kowalski, B. R. (1980). Data analysis of sensory scores. Evaluations of panelists and wine score cards. *J. Food Sci.*, **45**, 213–16.

Laamanen, M. and Jounela-Eriksson, P. (1987). The use of descriptive analysis in the quality control of Sweet Gale extracts. *Lebensm.-Wiss. u. Technol.*, **20**(2), 86–90.

Land, D. G., Harper, R. and Griffiths, N. M. (1970). An evaluation of the odour qualities of some stimuli proposed as standards for odour research. *Flavour Ind.*, **1**, 842–6.

Langron, S. P., Williams, A. A. and Collins, A. J. (1984). A comparison of the consensus configuration from a generalized Procrustes analysis with the untransformed panel mean in sensory profile analysis. *Lebensm.-Wiss. u. Technol.*, **17**, 296–8.

Larmond, E. (1981). Panel discussion: 'How far can we trust taste panel results?' *Tech. Quart. Masters Brewers Ass. Am.*, **18**, 38.

Laslett, G. M. and Bremner, H. A. (1979). Evaluating acceptability of fish minces and fish fingers from sensory variables. *J. Food Technol.*, **14**, 389–404.

Lawless, H. T. (1986). Sensory interactions in mixtures. *J. Sens. Studies*, **1**, 259–74.

Lawless, H. J. and Malone, G. J. (1986). The discriminative effciency of common scaling methods. *J. Sens. Studies*, **1**, 85–98.

Lehrer, A. (1983). *Wine and Conversation*. Indiana University Press, Bloomington, USA.
Lehrer, A. (1985). Is semantics perception-driven or network-driven? *Austral. J. Ling.*, **5**, 197–207.
Liardon, R., Ott, U. and Daget, N. (1987). Analysis of coffee headspace profiles by multivariate statistics. *Nestlé Research News 1986/87*, Nestec Ltd, Vevey, Switzerland, pp. 183–6.
Lin, A. C. (1983). Chemical and sensory analysis of soy sauce. PhD dissertation, University of Georgia, USA. 113 pp.
Lisle, D. B., Richard, C. P. and Wardleworth, D. F. (1978). The identification of distilled alcoholic beverages. *J. Inst. Brew.*, **84**, 93–6.
Lundgren, B., Karlstrom, B. and Ljungqvist, A. (1979). Effect of time and temperature of storage after cooking on the sensory quality of Bintje potatoes. *J. Sci. Food Agric.*, **30**, 305–18.
Lyon, B. G. (1980). Sensory profiling of canned boned chicken: sensory evaluation procedures and data analysis. *J. Food Sci.*, **45**, 1341–6.
Lyon, B. G. (1987). Development of chicken flavor descriptive attribute terms aided by multivariate statistical procedures. *J. Sens. Studies*, **2**, 55–67.
Lyon, B. G. and Klose, A. A. (1980). Sensory profiling of canned boned chicken: comparisons of retail, school lunch, and military canned boned chicken. *J. Food Sci.*, **45**, 1336–40.
Macher, L. (1957). Die bewertung des bieres nach punkten und nach worten. Brauwelt, *Z. Ges. Brauwesen*, **97**, 373–83.
Madsen, B. N. (1980). Provesmagning. *Brygmesteren*, **37**, 233–45.
Malek, D. M., Schmitt, D. J. and Munroe, J. H. (1982). A rapid system for scoring and analyzing sensory data. *J. Am. Soc. Brew. Chem.*, **40**, 133–6.
Malek, D. M., Munroe, J. H., Schnitt, D. J. and Korth, B. (1986). Statistical evaluation of sensory judges. *J. Am. Soc. Brew. Chem.*, **44**(1): 23–7.
Martens, M. (1985). Sensory and chemical quality criteria for white cabbage studied by multivariate data analysis. *Lebensm.-Wiss. u. Technol.*, **18**, 100–104.
Martens, M. (1986a). Sensory and chemical quality criteria of stored versus not-stored frozen peas studied by multivariate data analysis. In: *The Shelf Life of Food and Beverages*, G. Charalambous (Ed.), Elsevier, Amsterdam, pp. 775–90.
Martens, M. (1986b). Sensory and chemical/physical quality criteria of frozen peas studied by multivariate data analysis. *J. Food Sci.*, **51**, 599–603, 617.
Martens, M. and Burg, van der, E. (1985). Relating sensory and instrumental data from vegetables using different multivariate techniques. In: *Progress in Flavour Research*, J. Adda (Ed.), Elsevier, Amsterdam, pp. 131–48.
Martens, M. and Martens, H. (1986). Parial least squares regression. In: *Statistical Procedures in Food Research*, J. R. Piggott (Ed.), Elsevier Applied Science Publishers, London, pp. 293–359.
Martens, M., Martens, H. and Wold, S. (1983). Preference of cauliflower related to sensory descriptive variables by partial least squares (PLS) regression. *J. Sci. Food Agric.*, **34**, 715–24.
Martens, M., Rosenfeld, H. J. and Russwurm, H., Jr. (1985). Predicting sensory quality of carrots from chemical, physical and agronomical variables: a multivariate study. *Acta Agric. Scand.*, **35**, 407–20.

McBurney, D. H. and Gent, J. F. (1979). On the nature of taste qualities. *Psychol. Bull.*, **86**, 151–67.
McDaniel, M. R. and Sawyer, F. M. (1981a). Descriptive analysis of whisky sour formulations: magnitude estimations versus a 9-point category scale. *J. Food Sci.*, **46**, 178–81, 189.
McDaniel, M. R. and Sawyer, F. M. (1981b). Preference testing of whisky sour formulations: magnitude estimation versus the 9-point hedonic. *J. Food Sci.*, **46**, 182–5.
McDaniel, M. R. Henderson, L. A., Watson, B. T. Jr and Heatherbell, D. (1987). Sensory panel training and screening for descriptive analysis of the aroma of Pinot Noir wine fermented by several strains of malolactic bacteria. *J. Sens. Studies*, **2**, 149–67.
McLellan, M. R., Lind, L. R. and Kime, R. W. (1984). Determination sensory components accounting for intervarietal variation in applesauce and slices using factor analysis. *J. Food Sci.*, **49**, 751–5.
Mccredy, J. M., Sonnemann, J. C. and Lehmann, S. J. (1974). Sensory profiling of beer by a modified QDA method. *Food Technol.*, **29**, 36–41.
Meilgaard, M. C. (1981). Beer flavour. Doctoral dissertation, Technical University of Denmark. Document 8126165. Microfilm Int. Ann Arbor, USA.
Meilgaard, M. C. (1982). Prediction of flavor differences between beers from their chemical composition. *J. Agric. Food Chem.*, **30**, 1009–17.
Meilgaard, M. C., Dalgliesh, C. E. and Clapperton, J. F. (1979a). Beer flavour terminology. *J. Am. Soc. Bew. Chem.*, **37**, 47–52.
Meilgaard, M. C. Dalgliesh, C. E. and Clapperton, J. F. (1979b). Beer flavour terminology. *J. Inst. Brew.*, 85, 38–42.
Meilgaard, M. C., Dalgliesh, C. E. and Clapperton, J. F. (1979c). Terminología sobre el flavor de la cerveza. *Cerveza y Malta*, **XV1**(62), 18–32.
Meilgaard, M. C., Reid, D. S. and Wyborski, K. A. (1982). Reference standards for beef flavour technology system. *J. Am. Soc. Brew. Chem.*, **40**, 119–28.
Meilgaard, M. C., Gail, V. and Carr, B. T. (1987). *Sensory Evaluation Techniques*, Vol. II, CRC Press, Boca Raton, Florida.
Miller, I. (1978). Statistical treatment of flavor data. In *Flavor: Its Chemical Behavioral, and Commercial Aspects*, C. M. Apt (Ed.), Westview Press, Boulder, Colorado, pp. 149–62.
Milutinović, L., Bargmann, R. E., Chang, K., Chastain, M. and Powers, J. J. (1970). Comparison of flavor and volatiles of tomato products and of peanuts. *J. Food Sci.*, **34**, 224–8.
Moncrieff, R. W. (1967). *The Chemical Senses*, CRC Press, West Palm Beach, Florida.
Moskowitz, H. R. (1977). Magnitude estimation: notes on what, how, when and why to use it. *J. Food Quality*, **3**, 195–227.
Moskowitz, H. R. (1979). Odor psychophysics and sensory engineering. *Chem. Senses Flav.*, **4**, 163–74.
Moskowitz, H. R. (1981). Changing the carbohydrate sweetness sensation. *Lebensm.-Wiss. u. Technol.*, **14**, 47–51.
Moskowitz, H. R. and Howard, R. (1982). Sensory analysis of 'thickness". *Cosmetics and Toiletries*, **97**, 34–45.

Muñoz, A. M. (1986). Development and application of texture reference scales. *J. Sens. Studies*, **1**, 55–83.

Noble, A. C. (1978). Sensory and instrumental evaluation of wine aroma. In: *Analysis of Food and Beverage*, G. Charalambous (Ed.), Academic Press, London, pp. 203–28.

Noble, A. C. (1981). Use of principal component analysis of wine headspace volatiles in varietal classification. *Vini D'Italia*, **23**, 325–40.

Noble, A. C. and Shannon, M. (1987). Profiling Zinfandel wines by sensory and chemical analysis. *Am. J. Enol. Vitic.*, **38**, 1–5.

Noble, A. C. Williams, A. A. and Langron, S. P. (1982). Descriptive analysis and quality of Bordeaux wines. *Proceedings Soc. Chem. Ind. Symposium*, Long Ashton, Bristol, April 1982.

Noble, A. C., Arnold, R. A., Buechsenstein, J., Leach, E. J., Schmidt, J. O. and Stern, P. M. (1987). Modification of a standardized system of wine aroma terminology. *Am. J. Enol. Vitic.*, **38**, 143–51.

Novais, A., Hanson, S. and Ryley, J. (1982). The texture of mashed potatoes in catering. 1. The background. The contribution of hedonic texture to overall preference. *Lebensm.-Wiss. u. Technol.*, **15**, 295–302.

Olshausen, J. J. (1957). The psychology and interpretation of taste tests. *Brewers Digest*, **32**(6), 61–6: **32**(7), 57–62.

Olshausen, J. J. (1971). *Taste Panel Presentation and a Flavour Profile*, Pub. No. 47, J. E. Siebel Sons' Company, Chicago, 26 pp.

O'Mahoney, M. and Tsang, T. (1978). Further examination of the taste of sodium carbonate using two languages English and Cantonese. *IRCS Med. Sci: Dentistry Oral Biol; Psychol. Psychiat.*, **6**, 243.

Palmer, D. H. (1974). Multivariate analysis of flavour terms used by experts and non-experts for describing tea. *J. Sci. Food Agric.*, **24**, 153–64.

Paule, C. (1986). Sensory and chemical examination of aromatic and non-aromatic rices. PhD dissertation, University of Georgia, USA, 88 pp.

Pearce, J. H., Korth, B. and Warren, C. B. (1986). Evaluation of three scaling methods for hedonics. *J. Sens. Studies*, **1**, 27–46.

Persson, I. and von Sydow, E. (1972). A quality comparison of frozen and refrigerated cooked sliced beef. 2. Relationships between gas chromatographic data and flavour scores. *J. Food Sci.*, **37**, 234–9.

Peppard, T. L. (1985). Flavours in beer. *J. Inst. Brew.*, **91**, 364–9.

Petró-Turza, M., Szárföldi-Szalma, I., Madarassy-Mersich, E., Teleky-Vámossy, Gy. and Füzesi-Kardos, K. (1986). Correlation between chemical composition and sensory quality of natural apple aroma condensates. *Die Nahrung*, **30**, 765–74.

Piggott, J. R. (1986). *Statistical Procedures in Food Research*, Elsevier Applied Science, London, 415 pp.

Piggott, J. R. and Canaway, P. R. (1981). Finding the words for it—methods and uses of descriptive sensory analysis. In: *Flavour '81*, P. Schreier (Ed.), Walter de Gruyter, Berlin, pp. 33–46.

Piggott, J. R. and Jardine, S. P. (1979). Descriptive sensory analysis or whisky flavour. *J. Inst. Brew.*, **85**, 82–5.

Piggott, J. R. and Sharman, K. (1986). Methods to aid interpretation of multidimensional data. In: *Statistical Procedures in Food Research*, J. R. Piggott (Ed.)., Elsevier Applied Science Publishers, London, p. 182.

Piggott, J. R., Hose, L. P. and Robertson, S. E. (1981). Possible primary odours. In: *Criteria of Food Acceptance*, J. Solms and R. L. Hall (Eds), Forster Verlag, Zurich, pp. 259–67.
Pino, J., Torricella, R. and Örsi, F. (1986a). Correlation between sensory and gas-chromatographic measurements on grapefruit juice volatiles. *Acta Alimentaria*, **15**(3), 237–46.
Pino, J. A., Torricella, P. G., Orsi, F. and Figureas, L. (1986b). Application of multivariate statistics for the quality classification of single-strength grapefruit juice. *J. Food Quality*, **9**, 205–16.
Pokorný, J. and Davídek, J. (1986). Application of hedonic sensory profiles for the characterization of food quality. *Die Nahrung*, **30**, 757–63.
Pokorný, J., Karnet, J. and Davídek, J. (1981). Sensory profiles of a Fernet appetizer. *Die Nahrung*, **25**, 553–9.
Powers, J. J. (1970). Measurement of relative contribution of constituents to flavour composition. *Proceedings Third International Congress of Food Science and Technology, Washington, DC*, Institute of Food Technologists, Chicago, pp. 394–402.
Powers, J. J. (1975). Validation of sensory and instrumental analyses for flavour. *Proceedings Fourth International Congress of Food Science and Technology, Madrid 1974*, Instituto de Agroquimica y Tecnologia de Alimentos, Valencia, Spain, Vol. 2, pp. 173–82.
Powers, J. J. (1976). Experiences with subjective/objective correlation. In: *Correlating Sensory Objective Measurements—New Method for Answering Old Problems*, J. J. Powers and H. R. Moskowitz (Eds), STP 594, American Society for Testing and Materials, Philadelphia, pp. 111–22.
Powers, J. J. (1979). Correlation of sensory evaluation data with objective test data. In: *Sensory Evaluation Methods for the Practicing Food Technologists, IFT Short Course 1979–1981*, Institute of Food Technologists, Chicago, pp. 9-1 to 9-35.
Powers, J. J. (1981a). Sensory-instrumental correlations—Review and appraisal. Part 1. *Lebensm.-Technol.*, **14**(6), 9–14.
Powers, J. J. (1981b). Multivariate procedures in sensory research: scope and limitations. *Tech. Quart. Master Brewers Ass. Am.*, **18**, 11–21.
Powers, J. J. (1981c). Perception and analysis: a perspective view of attempts to find causal relations between sensory and objective data. In: *Flavour '81*, P. Schreier (Ed.), Walter de Gruyter, Berlin, pp. 105–31.
Powers, J. J. (1982a). Sensory-instrumental correlations—Review and appraisal. Part 3. *Lebensm.-Technol.*, **15**(2), 2–6.
Powers, J. J. (1982b). Sensory-instrumental correlations—Review and appraisal. Part 2. *Lebensm.-Technol.*, **15**(1), 6–11.
Powers, J. J. (1984). Using general statistical programs to evaluate sensory data. *Food Technol.*, **38**, 74–82, 84.
Powers, J. J. (1986). Aplicación de métodos estadísticos de múltiplies variables a las necesidades de al industria alimentaria. *Tec. Aliment. (Mexico)*, **22**(2), 3–10.
Powers, J. J. and Quinlan, M. C. (1973). Subjective-objective evaluation of model odor systems. *Lebensm.-Wiss. u. Technol.*, **6**, 209–14.
Powers, J. J. and Quinlan, M. C. (1974). Refining of methods for subjective-objective evaluation of flavor. *J. Agric. Food Chem.*, **22**, 744–9.

Powers, J. J. and Rao, V. N. M. (1985). Computerization of the quality assurance program. *Food Technol.*, **39**(11), 136–42.

Powers, J. J. and Ware, Glen, O. (1976). Analysis of multiple paired comparisons with and without ties. *Lebensm-Wiss u. Technol.*, **13**, 143–247.

Powers, J. J. and Ware, Glen, O. (1986). Discriminant Analysis. In: *Statistical Procedures in Food Research*, J. R. Piggott (Ed.), Elsevier Applied Science Publishers, London, pp. 125–80.

Powers, J. J., Godwin, D. R. and Bargmann, R. C. (1977). Relations between sensory and objective measurements for quality evaluation of green beans. In: *Flavor Quality: Objective Measurements*, R. A. Scanlan (Ed.), ACS Symposium Series No. 51, Am. Chem. Soc., Washington DC, pp. 51–70.

Powers, J. J., Cenciarelli, S. and Shinholser, K. (1984). El uso de programas estadísticos generales en la evaluacion de los resultados sensoriales. *Rev. Agroquim. Tecnol. Aliment.*, **24**, 469–84.

Powers, J. J., Shinholser, K. and Godwin, D. R. (1985). Evaluating assessors' performance and panel homogeneity using univariate and multivariate statistical analysis. In: *Progress in Flavour Research 1984*, J. Adda (Ed.), Elsevier Science Publishers, Amsterdam, pp. 193–209.

Powers, J. J., Shinholser, K. and Brett, D. (1987). Testing factor patterns for agreement. *Proceedings II World Congress of Food Technology*, Barcelona, Spain, 3–5 March.

Quinlan, M. C., Bargmann, R. E., El-Galali, Y. M. and Powers, J. J. (1974). Correlations between subjective and objective measurements applied to grape jelly. *J. Food Sci.*, **39**, 794–9.

Riskey, D. R. (1986). Use and abuses of category scales in sensory measurement. *J. Sens. Studies*, **1**, 217–36.

Robinson, J. O. (1970). The misuse of taste names by untrained observers. *Brit. J. Psychol.*, **61**, 375–8.

Rogers, N. M., Bargmann, R. E. and Powers, J. J. (1986). Component and factor analysis applied to descriptors for tea sweetened with sucrose and with saccharin. *J. Sens. Studies*, **1**, 137–48.

Romero, R. (1983). Analisis multivariante de datos en Technologia de Alimentos. Paper presented at source on 'El analisis sensorial en el control de calidad de alimentos', Instituto de Agroquimica y Tecnologia de Alimentos, Valencia, Spain, 23–24 May 1983.

SAS User's Guide (1982). Statistics, SAS Institute, Cary, North Carolina, 955 pp.

SAS User's Guide (1985). Statistics, 5th edn, SAS Institute, Cary, North Carolina, 584 pp.

Salovaara, H., Hellemann, U. and Kurkela, R. (1982). Effect of salt on bread flavour. *Lebensm.-Wiss. u. Technol.*, **15**, 270–4.

Sawyer, F. M., Cardello, A. V., Prell, P. A., Johnson, E. A., Segars, R. A., Maller, O. and Kapsalis, J. (1984). Sensory and instrumental evaluation of snapper and rockfish species. *J. Food Sci.*, **49**: 727–33.

Sawyer, F. M., Cardello, A. V. and Prell, P. A. (1988). Consumer evaluation of the sensory properties of fish. *J. Food Sci.*, **53**, 12–18, 24.

Schiffman, S. S. and Erickson, R. P. (1971). A psychophysical model for gustatory quality. *Physiol. Behav.*, **7**, 617–33.

Schiffman, S. S., Reynolds, M. L. and Young, F. W. (1981). *Introduction to Multidimensional Scaling*, Academic Press, London, 413 pp.
Shortreed, G. W., Rickard, P., Swan, J. S. and Burtles, S. M. (1979). The flavour terminology of Scotch whiskey. *Brewers' Guardian*, **11**, 55–62.
Shinholser, K., Hellemann, U., Salovaara, H., Ellilä, M. and Powers, J. J. (1987). Factor patterns yielded by subsets of panelists examining Finnish sour rye bread. *J. Sens. Studies*, **2**, 199–213.
Sidel, J. L. and Stone, H. (1976). Experimental design and analysis of sensory tests. *Food Technol.*, **31**, 32–8.
Sjöström, L. B. (1956). Flavor analysis. *Drug Cosmet. Ind.*, **78**, 28–31.
Sjöström, L. B. and Caul, J. F. (1962). 'Dilution flavor profile' reaction. *Food. Technol.*, **16**(4), 8.
Søftoft, M. (1974). A profile method for sensory analysis of beer and its use in assessing flavour stability and flavour consistency. *J. Inst. Brew.*, **80**, 570–6.
Smith, M., Smith, L. G. and Levinson, B. (1982). The use of smell in differential diagnosis. *Lancet*, 25 Dec., 1452–3.
Stewart, E. (1961). Putting beer taste impressions into words. *Comm. Master Brewers Ass. Am.*, **22**(11/12), 3–10.
Stone, H., Sidel, J., Oliver, S., Woolsey, A. and Singleton, R. C. (1974). Sensory evaluation by quantitative descriptive analysis. *Food Technol.*, **28**, 24–34.
Stone, H., Sidel, J. L. and Bloomquist, J. (1980). Quantitative descriptive analysis. *Cereal Foods World*, **25**, 624–4.
Stungis, G. E. (1976). Overview of applied multivariate analysis. In: *Correlating Sensory Objective Measurements: New Methods for Answering Old Problems*, J. J. Powers and H. R. Moskowitz (Eds), STP 594, American Society for Testing and Materials, Philadelphia, pp 81–96.
Szczesniak, A. S. (1963). Classification of textural characteristics. *J. Food Sci.*, **23**, 385–89.
Szczesniak, A. S. (1975). General Foods texture profile revisited—ten years perspective. *J. Texture Studies*, **6**, 5–17.
Szczesniak, A. S. and Skinner, E. Z. (1973). Meaning of texture words to consumers. *J. Texture Studies*, **4**, 338–44.
Szczesniak, A. S., Brandt, M. A. and Friedman, H. H. (1963). Development of standard rating scales for mechanical parameters of texture and correlation between the objective and the sensory methods of texture evaluation. *J. Food Sci.*, **28**, 397–403.
Tilgner, D. (1962). Dilution tests for odor and flavour analysis. *Food Technol.*, **16**(2), 26–9.
Toda, J., Wada, T., Yasumatsu, K. and Ishii, K. (1971). Application of principal component analysis to food texture measurements. *J. Texture Studies*, **2**, 207–19.
Tuorila, H., Pyysalo, T., Hirvi, T. and Vehviläinen, A. K. (1980). Characterization of odours in raw and tap water and their removal by ozonization. *Vatten*, **36**, 191–9.
Vedel, A., Charle, G., Charney, P. and Tourmeau, J. (1972). B-1. *Le vin—definition*. B-2. *Caracteres organoleptiques*. Instit. Nat. des Appellations d'Origine des Vins et Eaux-de-vie, Macon, France.

von Sydow, E., Moskowitz, H., Jacobs, H. and Meiselman, H. (1974). Odor-taste interactions in fruit juices. *Lebensm.-Wiss. u. Technol.*, **7**, 18–24.

Vuataz, L. (1977). Some points of methodology in multidimensional data analysis as applied to sensory evaluation. *Nestlé Research News 1976/77*, Nestec Ltd, Vevey, Switzerland, pp. 57–71.

Vuataz, L. (1981). Information about products and individuals in multi-criteria description of food products. In: *Criteria of Food Acceptance*, J. Solms and R. L. Hall (Eds), Forster Verlag, Zurich, pp. 429–46.

Vuataz, L. Sotek, J. and Rahim, H. M. (1975). Profile analysis and classification. *Proceedings Fourth International Congress of Food Science and Technology, Madrid 1974*, Instituto de Agroquimica y Tecnologia de Alimentos, Valencia, Spain, Vol. la, p. 25.

Werkhoven, J. (1974). Tea processing. *FAO Agric. Ser. Bull.*, No. 26, FAO, Rome, pp. 195.

Whitear, A. L., Carr, B. L., Crabb, D. and Jacques, D. (1979). The challenge of flavour stability. *European Brewers Convention, Proceedings 175th Congress, Berlin*, pp. 13–25.

Wilkin, G. D., Webber, M. A. and Lafferty, E. A. (1983). Appraisal of industrial continuous still products. In: *Flavour of Distilled Beverages: Origin and Development*, J. R. Piggott (Ed.), Ellis Horwood, Chichester, pp. 154–65.

Williams, A. A. (1975). The development of a vocabulary and profile assessment method for evaluating flavour contribution of cider and berry aroma constitutents. *J. Sci. Food Agric.*, **26**, 567–82.

Williams, A. A. (1979). The evaluation of flavour quality in fruits and fruit products. In: *Progress in Flavour Research*, D. G. Land and W. H. E. Nursten (Eds), Applied Science Publishers, London, pp. 1–14.

Williams, A. A. (1982). Recent developments in the field of wine flavour research. *J. Inst. Brew.*, **88**, 43–63.

Williams, A. A. and Arnold, G. M. (1985). A comparison of the aromas of six coffees characterized by conventional profiling, free-choice profiling and similarity scaling methods. *J. Sci. Food Agric.*, **36**, 204–14.

Williams, A. A. and Carter, C. S. (1977). A language and procedure for the sensory assessment of Cox's orange pippin apples. *J. Sci. Food Agric.*, **28**, 1090–104.

Williams, A. A. and Tucknott, O. G. (1977). Misleading information from odour evaluation of gas chromatographic effluents in aroma analysis. *Chem. Ind.*, **24**, 124.

Williams, A. A., Langron, S. P., Timberlake, C. F. and Bakker, J. (1984). Effect of colour on the assessment of ports. *J. Food Technol.*, **19**, 659–71.

Williams, A. A., Baines, C. R. and Finnie, M. S. (1986). Optimization of colour in commercial port blends. *J. Food Technol.*, **21**, 451–61.

Word, K. M. and Gress, H. S. (1981). Selection, training and motivation of sensory panel judges. *Tech. Quart. Master Brewers Ass. Am.*, **18**, 22–5.

Wren, J. J. (1972). Taste-testing of beer for quality control. *J. Inst. Brew*, **78**, 69–76.

Wu, L. S., Bargmann, R. E. and Powers, J. J. (1977). Factor analysis applied to wine descriptors. *J. Food Sci.*, **42**, 944–52.

Yoshida, M. (1963). Similarity among different kinds of tastes near the threshold concentration. *Jap. J. Psychol.*, **34**, 25–35.

Zook, K. and Wessman, C. (1977). The selection and use of judges for descriptive panels. *Food Technol.*, **31**(11), 56–61.

Chapter 8

CONSUMER STUDIES OF FOOD HABITS

H. L. MEISELMAN

US Army Research Development and Engineering Center,
Natick, Massachusetts, USA

1. FOOD HABITS

The study of human eating behaviour, its antecedents, determinants, descriptions and analyses of the behaviour itself as well as its short-term and long-term outcomes, has been approached in a wide variety of disciplines and styles. The methods of history, psychology, sociology, anthropology, biology, physiology, nutrition, business, marketing, and others have been applied to human eating behaviour. As food science, food technology, and related food disciplines have moved the food industry from a non-technical to a more technical field, studies of food-related behaviour in humans have struggled to keep pace with the increasing technology of food research, the food product industry, and the food service business. Only within the last decade or two have there developed the beginnings of a field of consumer food habits, a field dedicated to the understanding of the various facets of human food-related behaviour.

Food habits is of necessity a multidisciplinary field. Viewed most broadly, it analyses the human (or other animal) from birth to death, from attitude toward food to the outcome of food intake. Within this complexity, it is difficult to propose an organisation of food-related behaviour which is meaningful to a broad variety of interests. The model which I propose as an outline of this chapter is a chronological model of food habits which applies to each food exposure situation. This model traces the chronological steps from food attitudes which are with us all of the time, to the food selections which we make in a particular eating environment, to the sensory evaluation and food acceptability decisions which we make when we contact food, to the food consumed, and finally, to the food not eaten or wasted (Table 1). This sequence represents a

TABLE 1
CHRONOLOGICAL OUTLINE OF FOOD HABITS

Food attitude
Food selection
Food acceptance/sensory evaluation
Food intake
Food waste

chronological description of food exposure and resulting behaviour which is not associated with any discipline or method, but rather is descriptive. The purpose of this chapter will be to outline the methods used to study these levels of food-related behaviour, and to present selected data resulting from certain methods.

Reviews of the various fields within food habits have appeared within the past several years, although no extensive review has combined all of the areas from food attitudes through food waste. Food attitudes were reviewed by Foley et al. (1979), food preferences and selections were reviewed by Khan (1981), food acceptability by Schaefer (1979) and Larmond (1977), food intake papers were collected by Burk and Pao (1976) and by Krantzler et al. (1982a), food waste by Comstock et al. (1979). In addition, several papers have sought to organise the food habits literature. Wilson (1973) divided food habits for her annotated bibliography as shown in Table 2. Grivetti and Pangborn (1973) described different approaches to food habits research as shown in Table 3. They recommended more interdisciplinary approaches combining these various elements.

Although the descriptive model presented above (Table 1) will be

TABLE 2
ORGANISATION OF FOOD HABITS
(From Wilson, 1973)

Cultural and environmental factors (nutritional anthropology)
    Economic
    Social
    Ethnic

Food selection
    Food beliefs, taboos and prejudices
    Food preferences, aversions and cravings
    Psychological aspects
    Food ways

Implications

TABLE 3
ORGANISATIONS OF FOOD HABITS APPROACHES
(From Grivetti and Pangborn, 1973)

Environmentalism
Cultural ecology
Regionalism
Cultural history
Functionalism
Quantitative methods
Clinical approaches

clarified as we work through the methods and data of each section, let us take some time at the outset to further describe and define each area. Food attitudes are the opinions about food which people develop through childhood and continue to modify throughout life. These attitudes are present in the absence of food itself, although the presence of food or food-cues might arouse or modify attitudes. Attitudes can be easily measured, although attitude data are not always easily interpreted. Below we will deal with several commonly measured food attitudes, especially the measurement of people's food likes and dislikes. In addition, we will review the methods of attitude measurement including questionnaire and interview design.

Food selection is a measure which is often confused with food intake. Food selection refers to the act of taking an offered food item, in a food market or in a food-service environment (home, institution, etc.). It does not connote ingestion of the food, but merely that the food is of sufficient attitudinal or sensory weight that it is taken by the person. The selection of some foods over others is therefore a potentially useful measure of acceptability or potential intake. Also, food intake is usually difficult, expensive and intrusive to measure. Hence, it is not always advisable or acceptable to measure food intake. If a good approximation of food intake can be obtained with food selection, then the latter can be helpful.

Sensory evaluation of food, strictly speaking, begins whenever a person is in a food environment, a food market or food-service establishment. The aroma of food and the sight of food surround us in these environments, and probably contribute to food acceptability and to human judgements of food quality. When food is actually selected and sampled, it is then subject to more complete sensory evaluation of all sensory attributes (temperature, texture, sound, etc.), and subjected to a more complete acceptability assessment.

Based on what foods are selected, their acceptability, and the internal state of the person, a certain amount of different foods are consumed. This food intake is of special interest to medically oriented research concerning nutrition, etc. The measurement of food intake, while seemingly simple to the layperson, is an interesting example of complexity with a large literature of conflicting data and opinion as to which method is appropriate under which condition. Food intake can be measured by report or weighing, by the individual, by the group, by the food item, or by the total food consumed. Each measure is relevant in different situations.

Finally, food waste is what is selected but not consumed. This is a measure which is often overlooked in studies, but which contains potentially useful information concerning acceptance, food habits, nutrition and economics. In addition, the total amount of food wasted in the USA is large by most estimates, as much as 30% of total food purchased. As with food intake, measurement of food waste varies with the choice of individual or group data, and individual food or aggregated food data.

Each of the areas just previewed—food attitudes, food selections, food acceptance, food intake and food waste—represent separate areas of food habits research which are only just beginning to be seen as the elements of an overall field of consumer food habits. It is still true that investigators and practitioners in food acceptability or sensory evaluation do not consider, in general, whether the food they are studying or modelling will actually be eaten, and those interested in the nutrition of food intake see the foods as nutrients, and not as acceptable or unacceptable products. Hopefully, one outcome of chapters such as this one will be to focus on the interrelationships among all levels of food habits.

2. FOOD ATTITUDES

The broad area of food attitudes includes a wide range of topics and techniques. Included are areas of traditional interest and method such as food preferences, and less understood areas such as food aversions. Food attitudes involve some of the most easily measured food habits, and less researched areas such as food aversions (Rozin and Fallon, 1981), because opinions are less expensive to measure than other forms of behaviour. Further, food attitudes extend from the entirely subjective ('How much do you like pizza?'), to questions which approach behaviour ('How often would you like to eat pizza?'), to questions which approximate behaviour

('How often do you think you eat pizza?'). As with other food habits, food attitude data are no more or less than the methods used to collect them, and our analysis of food attitudes begins with an overview of the methods available.

2.1. Questionnaires and Interviews

There are two basically different methods for collection of attitudinal data: the written questionnaire and the spoken interview. The term 'survey' is used here to refer to either method. Measurement of behaviour can be used to assess attitude; for example, the number of people selecting a food item is an indication of their attitude toward that item. The questionnaire and the interview possess different strengths and weaknesses, and the two are often linked in attitude research.

The spoken interview is often the first step in the attitude measurement process. If one knows little about a subject matter—which is a good assumption in most cases!—then a probing interview is a good starting place. By its very nature, the interview process has the following characteristics:

1. It is relatively labour-intensive because it is either one-on-one (one interviewer with one interviewee) or one interviewer to a small group.
2. It is relatively time-consuming, both in the data collection process, and in the data reduction and data analysis processes.
3. It can be collected in person or by telephone.
4. It requires relatively highly trained personnel, who are familiar with the topic and with sound interviewing procedures.
5. It is relatively expensive because of the intensive labour, time and professional staff.
6. It provides the (only) opportunity to probe the individual as to his/her motivations, feelings, past behaviour, future expectations, etc., allowing for question and follow-up. No follow-up is possible in most questionnaires; whatever comes back on the paper is what you have for data.
7. It provides a major opportunity to influence the individual. This can take many forms: interesting the individual in your study, leading the individual to attend in detail to an aspect of behaviour normally overlooked, etc.

The written questionnaire requires more knowledge about the subject matter than the interview; just ask a colleague in another field to write intelligent survey questions about your field, and you might be amazed at how naive they sound. Only the novice questionnaire writer is totally pleased with his or her survey efforts; the more experienced person sees room for constant improvement in survey design. The questionnaire, in general, has the following characteristics:

1. It may require substantial leadtime for development of a good instrument.
2. It can be administered to individuals, and to small or large groups.
3. It can be collected in person or by mail.
4. It possesses large opportunities for losing data due to lack of following instructions, lack of cooperation and lack of survey returns, etc.
5. It can be very brief (one question) or very lengthy, depending on the time and cooperation of individuals.
6. Although a questionnaire must usually be written by a trained professional it can be administered by someone with less training; therefore repetitive field work involving substantial labour can employ less highly trained and paid individuals.
7. Because it is administered in groups by less trained staff it can be a less expensive alternative to interviewing.
8. Because a lot of data can be collected quickly, the questionnaire can be used to collect a lot of information of questionable quality.

2.1.1. Structured and Unstructured Questions

Both interviews and questionnaires can utilise structured or unstructured questions. The totally unstructured interview involves a more casual question and answer, covering a range of topics to various levels of detail. What can emerge, on the positive side, is a deep understanding of how a person feels and what he reports he has done in a specific area of behaviour (e.g. 'Tell me how you feel about food and eating.' or 'Tell me how your eating habits have changed over the past ten years'). Such interviews are the basis for understanding potential issues for further study; they are generally not the answer to any question. Even the totally unstructured interview should be phrased in an unbiased way rather than by using a biased phrasing, e.g. 'What do you *dislike* about eating any food?'. The unstructured questionnaire item is similar except that the spacing within the survey will determine how much the respondent should write. Asking a broad question and then leaving 2 or 20 lines for the answer will draw different amounts of information.

Structured interview and questionnaire items are different due to the differences in setting. Within the interview, there is less opportunity for visual aids (although they can be used) and hence most items depend on verbal description of the possible answers. In general, use of multiple-choice answers is limited to up to three possibilities (e.g. 'Is your appetite small, moderate, or large?'). Because of the natural tendency for some people to avoid 'extreme' answers, the central answer is most probable, independent of its actual answer content. Because the questionnaire is easier and cheaper to use and is ideally suited to multiple-choice items, these items are not advised in an interview unless an interview is needed for other information and no questionnaire is planned. The interview is also ideal for multiple-choice items in which there are only two or three logical possibilities (e.g. 'Have you ever been pregnant?').

2.2. Questionnaire Items

Questionnaire items can be presented in a larger number of different formats. Some of the most common are the following:

1. Checklists, in which the respondent is presented with a list of alternatives and generally asked to check all that apply (e.g. 'Check all the foods you have eaten, check all the diseases you have had'). Checklists usually encourage more answers than other forms of the same question. Therefore the results of checklists must be viewed conservatively.
2. Multiple choice, in which the respondent must select the one best answer from a number of alternatives. If one answer is permitted, the alternatives must be mutually exclusive so that two answers are not possible (e.g. 'What is your highest level of education: high school, college, graduate education?').
3. Rating scales, in which a subject or issue is answered according to degrees of some attribute (degree of liking, favourableness, anxiety, etc.). Generally one should use previously developed and tested scales. Rules for scale design are presented in Table 4.

Fortunately, the questionnaire designer can utilise the results of research studies which have mathematically scaled the terms typically used in multiple choice and scaled surveys. In fact, the survey researcher should develop his/her own scales only rarely! It is preferable and safer to use scales which you have used previously with demonstrated success (defined statistically) or which have been used and demonstrated by others. The most well-known scale in food research is the nine-point hedonic scale

TABLE 4
RULES FOR SCALE DESIGN

a. An entire scale should use one root word, e.g. *like–like/dislike*. Do not shift from *prefer* to *dislike*.
b. Every scale point should be modified with modifiers of the root such as *very*, *slightly*, etc.; wording can affect the perceived scale length.
c. There should be the same number of scale values above and below neutral, if neutral is used, or above and below the midpoint. Do not use e.g. five levels of *like* and three of *dislike*. Many statistical techniques assume equal intervals between scale points.
d. Carefully consider the use of the neutral points (e.g. *neither like nor dislike*); use one if it is logically necessary.
e. Use an adequate number of scale points; remember that respondents avoid extremes, more so on longer scales. Shorter rating scales reduce unequal differences between scale points; longer rating scales produce better scale discrimination.
f. Higher scale numbers should mean more of the rated quality.

(Table 5) developed by the US Army in the 1940s. It is interesting to note that this scale satisfies the six points mentioned above: it is adequately long (nine points), it possesses a neutral point, it uses one root word (*like–dislike*), and uses the same modifiers above and below neutral (*slightly, moderately, very* and *extremely*).

Bass *et al.* (1974) have scaled verbal descriptors of frequency (Table 6) and amount (Table 7), both key concepts in food attitude research. These data provide guidelines for selecting four-nine point category scales for these concepts. The four-point scale of frequency (never, sometimes, often and always) is commonly used. For some situations one should question

TABLE 5
THE NINE-POINT HEDONIC SCALE USED FOR FOOD ACCEPTANCE AND FOOD PREFERENCE

9 Like extremely
8 Like very much
7 Like moderately
6 Like slightly
5 Neither like nor dislike
4 Dislike slightly
3 Dislike moderately
2 Dislike very much
1 Dislike extremely

TABLE 6
SCALES OF FREQUENCY
(Adapted from Bass et al., 1974)

9	8	7	6	5	4
Always	Always	Always	Always	Always	Always
Continually	Continually	Constantly	Frequently if not always	Very often	Often
Very often	Very often	Often	Quite often	Fairly many times	Sometimes
Quite often	Rather frequently	Fairly many times	Sometimes	Occasionally	Never
Fairly many times	Sometimes	Sometimes	Once in a while	Never	
Sometimes	Now and then	Once in a while	Never		
Occasionally	Not often	Never			
Not very often	Never				
Never					

TABLE 7
SCALES OF AMOUNT

9	8	7	6	5	4
All	All	All	All	All	All
An exhaustive amount of	Almost entirely	An extraordinary amount of	Almost completely	An extreme amount of	A great amount of
An extreme amount of	An extreme amount of	A great amount of	Very much	Quite a bit of	A moderate amount of
A great deal of	A lot of	Quite a bit of	Fairly much	Some	None
Quite a bit of	Fairly much	A moderate amount of	To some degree	None	
An adequate amount of	Some	Somewhat	None		
Some	A limited amount of	None			
A little	None				
None					

whether the frequency terms 'never' and 'always' and the amount terms 'none' and 'all' are too demanding.

Moser and Kalton (1972) have provided a particularly good and general list of concerns in the wording of questions for interviews and surveys:

1. Questions insufficiently specific. Although general questions can be appropriate ('What do you like and dislike about this food?'), sometimes a specific problem is translated into a more general question ('How would you rate the food in your cafeteria?' becomes 'How would you rate your cafeteria?'). More general questions are harder to interpret.

2. Simple language. Use language which is appropriate to your audience. When in doubt, use the simpler language. In many cases, the particular slang or jargon of the audience can be used for better communication and to win over the audience. This jargon can be picked up during the initial interview process.

3. Ambiguous wording. If you ask whether a person eats more when depressed, you are assuming that the person gets depressed, that he changes his eating level, and that he remembers both.

4. Double-barrelled questions. Asking, 'Do you like apple pie and cherry pie?' is difficult to answer for the person who feels differently about the two products. Either ask two questions, or eliminate one item from the survey.

5. Leading questions. This refers both to the form of questions ('Which breakfast cereal do you eat') and to the specifics involved in the question ('Which breakfast cereal, such as Cheerios or Rice Krispies, do you eat?'). The question should not make a particular answer more probable.

6. Presuming question. The example in point (5) not only leads to an answer, but also presumes information about the respondent.

7. Hypothetical or conditional questions. Asking about an unknown is risky ('If frozen eggs were available would you buy them?') because so many factors might affect the consumer's decision (price, etc.). Nevertheless, when viewed cautiously such questions can be useful where more direct questions are difficult, for example in studies of disgust and aversion (Rozin et al., 1986).

8. Self-reference. If one asks, 'Should school lunch serve hamburgers?' or 'Would you like hamburgers in school lunch?' or 'Would your child like hamburgers in school lunch?' there is different information involved. Many people will recommend things for others, but will know that they, themselves, would not like them.

9. Sensitive topics. Food topics do not contain as many of these topics as family planning, drugs and other areas. The food area still can embarrass with questions of obesity and diet, with questions of how much money is

spent on food, and whether a parent prepares wholesome meals for the children. Special care is needed when such things are asked, as well as anytime the respondent perceives that he might lose or gain something as a result of the survey.

10. Periodical behaviour. Asking people how often they eat out of doors, how often they prefer to drink cold beverages, etc., might involve behaviour which varies with the season.

11. Memory. When the surveyor is depending on the memory of the respondent, it is best to assume that memory will not be as good as we would like to think it is. In many cases it has been demonstrated that memory is very poor for seemingly simple information. Beware.

By careful design, surveys can be designed for different age, education, and socioeconomic groups. By combining speaking and reading of materials, young children's food attitudes can be studied (Fauslow et al., 1982). Specialised surveys have also been designed for elderly persons (Axelson and Peufield, 1983).

2.3. Food Preference

The nine-point hedonic scale of food preference has been the one most commonly used. It was developed by the Quartermaster Food and Container Institute (QMFCI) in Chicago in the late 1940s (see Peryam and Pilgrim, 1957).

The development of this scale involved more research than other food preference measures. A rating scale was selected rather than a paired-comparison method, in which pairs of items are used rather than lists of individual items. It was determined that the rating scale approach and the paired-comparison approach yielded relative preferences in good agreement.

The two basic questions addressed were the number of scale points and their labelling. The nine-point scale had already been used in laboratory food acceptance testing, and the researchers compared it in a food preference testing (using 50, 100 and 150 item lists) to a seven-point scale (eliminating the 'like extremely' and 'dislike extremely' categories) and a five-point scale (eliminating the 'extremely' and 'slightly' categories). The three survey lengths showed no difference in test–retest reliability and the nine-point category showed the highest test–retest reliability (0·96), so the longer list length and the longer scale length were both adopted.

The naming of the scale points was the next step in the development of the scale. Ideally, the names should be chosen by scaling their meaning, so that the distance between 'extremely' and 'very much' should be the same as that between 'very much' and 'moderately'.

One subtle point involved in using rating scales is positioning on the page. In their survey the rating scale followed a list of foods presented on the left-hand side of the paper. The question raised was 'Should the scale begin with the 'dislike' or 'like' end of the rating scale?'. They found, in a test on a list of 45 food items, that the proportion of answers in each of the nine categories was almost identical (correlation coefficient = 0·96). However there were significant differences between the form of the scale in which 'dislike extremely' was placed on the extreme left and the one in which 'like extremely' was placed in that position. Beginning the scale with 'dislike extremely' led to a significantly greater frequency of the 'dislike' categories. Beginning the scale with 'like extremely' did not produce the analogous increased frequency for 'like' categories. In practical terms the effects are very small. The correlation between the 45 pairs of food means was 0·997, and it is the mean which is used for predictive purposes with these data. The researchers suggested that the scale should begin with 'like extremely' but hastened to add that no clear problem resulted from the reverse.

The issue of preference frequency has been another focus of food preference scaling and has been phrased in a variety of ways: 'How often would you like to eat the menu items?', 'How often would you be willing to eat the items?', 'How often would you like to see the food offered?'..., 'How often would you like to eat the item?'.

Preference frequency scales have been of two types, one using verbal categories of frequency and the other using quantitative categories (Table 8). Almost all frequency scales used have had four or nine categories. The verbal-based scales have depended heavily on the existing temporal system of day, week and month. Two scales have used the term 'often' in addition to day, week or month referrents (Leverton, 1944; Schuck, 1961), although this could represent difficulties in trying to translate into actual temporal units. Benson (1958) also used a four-category scale but stuck to temporal terms (once a day, week, month, year). Hartmuller (1971) and Knickrehm et al. (1967) used identical nine-category scales from 'twice a day' to 'once a month' (plus 'never want'). The QMFCI research on frequency scales also used a nine-category scale which overlapped greatly with the one just cited. In some administrations it was extended to 'every three months' and to 'once a year'. The question which arises then is: 'What is the most appropriate time frame for the preference frequency scale?' This question has not been directly addressed, most scales using the month as the unit.

TABLE 8
SCALES OF PREFERRED FREQUENCY

	Knickrehm et al. (1967)	Hartmuller (1971)	QMFCI	Benson (1958)	Schuck (1961)	Leverton (1944)
Often					*	*
Twice a day	*	*	*			
Once a day	*	*	*	*		
Every other day	*	*				
Several times per week, 15 months			*			
Twice a week	*	*	*			
Once a week	*	*	*	*	*	*
Every other week	*	*	*			
Once a month	*	*	*	*		
Every 3 months			*			
Once a year			*	*		
Never/unwilling to eat	*	*	*			*
Not familiar	*	*	*		*	*

For most purposes it would appear that items consumed only once per year would be insignificant, unless very specialised food services (class A restaurants, catering) were of interest.

The QMFCI scale also listed the frequency per month of all verbal scale categories. The category 'every three months' was rated 0·3 and the category 'once a year' was rated 0·1. This reinforces the use of the month as the unit. It also provides both the test respondent and the researcher with a quantified scale for analysis and prediction. In some cases (e.g. Knickrehm et al., 1967) subjects responded on the frequency scale by listing the number of the verbal category. For example, twice a week was coded as 4. The potential problem here is that the best respondent is not using the frequency statement in his answer, whereas in other scales he is.

A preference scale (Fig. 1) developed more recently for the military used a quantitative preference frequency scale based on the week and month (Meiselman et al, 1972). The subject was asked how often he would like an item in terms of days per week (answer 1, 2, 3, 4, 5, 6 or 7) and weeks per month (answer 1, 2, 3 or 4). While this does directly ask the preference frequency question in quantitative terms, it forces the subject into a week–month system. If he wants squash 13 times per month, he cannot so indicate. Further it assumes that the weekly pattern is repeated. This is also

	BREAKFAST		MID-DAY		EVENING		NEVER
	days per week	weeks per month	days per week	weeks per month	days per week	weeks per MEAL month	
102. Lemonade	⊙②③④⑤⑥⑦	①②③④	①②③④⑤⑥⑦	①②③④	①②③④⑤⑥⑦	①②③④	○
103. Carrot Salad	①②③④⑤⑥⑦	①②③④	①②③④⑤⑥⑦	①②③④	①②③④⑤⑥⑦	①②③④	○
104. Tomato Vegetable w/Noodle Soup	①②③④⑤⑥⑦	①②③④	①②③④⑤⑥⑦	①②③④	①②③④⑤⑥⑦	①②③④	○
105. Cheeseburger	①②③④⑤⑥⑦	①②③④	①②③④⑤⑥⑦	①②③④	①②③④⑤⑥⑦	①②③④	○
106. Grapefruit & Pineapple Juice	①②③④⑤⑥⑦	①②③④	①②③④⑤⑥⑦	①②③④	①②③④⑤⑥⑦	①②③④	○
107. Blackberry Pie	①②③④⑤⑥⑦	①②③④	①②③④⑤⑥⑦	①②③④	①②③④⑤⑥⑦	①②③④	○
108. Brownies	①②③④⑤⑥⑦	①②③④	①②③④⑤⑥⑦	①②③④	①②③④⑤⑥⑦	①②③④	○
109. Honeydew Melon	①②③④⑤⑥⑦	①②③④	①②③④⑤⑥⑦	①②③④	①②③④⑤⑥⑦	①②③④	○
110. Chow Mein	①②③④⑤⑥⑦	①②③④	①②③④⑤⑥⑦	①②③④	①②③④⑤⑥⑦	①②③④	○
111. Grapeade	①②③④⑤⑥⑦	①②③④	①②③④⑤⑥⑦	①②③④	①②③④⑤⑥⑦	①②③④	○
112. Raisin Bread	①②③④⑤⑥⑦	①②③④	①②③④⑤⑥⑦	①②③④	①②③④⑤⑥⑦	①②③④	○
113. Yellow Squash	①②③④⑤⑥⑦	①②③④	①②③④⑤⑥⑦	①②③④	①②③④⑤⑥⑦	①②③④	○
114. Macaroni Salad	①②③④⑤⑥⑦	①②③④	①②③④⑤⑥⑦	①②③④	①②③④⑤⑥⑦	①②③④	○
115. Yellow Cake	①②③④⑤⑥⑦	①②③④	①②③④⑤⑥⑦	①②③④	①②③④⑤⑥⑦	①②③④	○
116. Fruit Cocktail (Canned)	①②③④⑤⑥⑦	①②③④	①②③④⑤⑥⑦	①②③④	①②③④⑤⑥⑦	①②③④	○
117. Eggnog	①②③④⑤⑥⑦	①②③④	①②③④⑤⑥⑦	①②③④	①②③④⑤⑥⑦	①②③④	○
118. Pineapple (Canned)	①②③④⑤⑥⑦	①②③④	①②③④⑤⑥⑦	①②③④	①②③④⑤⑥⑦	①②③④	○
119. Carrot, Raisin & Celery Salad	①②③④⑤⑥⑦	①②③④	①②③④⑤⑥⑦	①②③④	①②③④⑤⑥⑦	①②③④	○
120. Apples (Canned)	①②③④⑤⑥⑦	①②③④	①②③④⑤⑥⑦	①②③④	①②③④⑤⑥⑦	①②③④	○
121. Fish	①②③④⑤⑥⑦	①②③④	①②③④⑤⑥⑦	①②③④	①②③④⑤⑥⑦	①②③④	○
122. Imitation Lemon Beverage	①②③④⑤⑥⑦	①②③④	①②③④⑤⑥⑦	①②③④	①②③④⑤⑥⑦	①②③④	○

FIG. 1. Format of food preference survey (Meiselman *et al.*, 1972).

the case in some verbal categories scales. A more recently developed survey (Fig. 2) (Meiselman and Waterman, 1978) avoids weekly units and asks for preference frequency per month using a scale which permits coding of any number from 0 to 31 (actually 39 is possible) days per month. Note again that the monthly unit was the unit of choice.

The numerical and verbal scales possibly reduce to the same thing when the subject is using numbers in the numerical scale and using the verbal categories in the non-numerical scale. When the subject uses numerical codes for the verbal scale categories, problems can arise. The focus of his attention is then on a number which does not directly represent frequency. He then begins to use the category scale of numbers without necessarily referring them to their referrent frequencies. This is similar to what can happen in the hedonic acceptance scale in which one begins to use a number without realising its referrent (extremely good, very bad, etc.).

One potential advantage of certain quantitative scales of frequency is that they can be ratio scales, that is, scales with equal intervals and a zero point. Ratio scales permit statements of ratios so that one could say x is preferred twice as often as y, etc. The frequency scale developed by US Army Natick Laboratories (Meiselman and Waterman, 1978) is such a scale (from 0 to 39). Both the old QMFCI scale and the scale used by Meiselman *et al.* (1972) are not continuous series of numbers; hence the subject is selecting categories rather than dealing in ratios.

The scales discussed so far have been either hedonic scales or preference frequency scales. Schutz (1965) developed a food action rating scale (FACT Scale), by scaling 18 action statements representing affective attitudes towards foods. Nine were selected to give equal intervals. The standard deviation and mean of the FACT scale and the nine-point hedonic scale are very similar; the two scales correlate 0·97 for food means. The overall tendency for the FACT means to be lower than the hedonic means apparently results from slightly lower FACT ratings for desserts and semisolid and liquid foods.

Van Riter (1956) used a scale based on home use of foods (specifically vegetables) including scale categories: 'never served at home', 'one or more of my family dislike the food', and 'prepared differently at home'. These categories are indicators of factors that are possibly important in food preference determination. Whether they are good measures of the preferences themselves is unclear without a more complete evaluation.

2.4. Examples of Food Preference Data

Although a large amount of food preference data is collected by various institutions and commercial organisations, little of it reaches the open

1	2	3	4	5	6	7	8	9
dislike extremely	dislike very much	dislike moderately	dislike slightly	neither like nor dislike	like slightly	like moderately	like very much	like extremely

	NEVER TRIED	HOW MUCH you like or dislike the food (1-9)		HOW OFTEN you want to eat the food in days per month (01-30)	
001 Tea	○	1 2 3 4 5 6 7 8 9		0 1 2 3	0 1 2 3 4 5 6 7 8 9
002 Roast Turkey	○	1 2 3 4 5 6 7 8 9		0 1 2 3	0 1 2 3 4 5 6 7 8 9
003 Strawberry Shortcake	○	1 2 3 4 5 6 7 8 9		0 1 2 3	0 1 2 3 4 5 6 7 8 9
004 Celery & Carrot Sticks	○	1 2 3 4 5 6 7 8 9		0 1 2 3	0 1 2 3 4 5 6 7 8 9
005 Boiled Pigs' Feet	○	1 2 3 4 5 6 7 8 9		0 1 2 3	0 1 2 3 4 5 6 7 8 9
006 Hot Turkey Sandwich with Gravy	○	1 2 3 4 5 6 7 8 9		0 1 2 3	0 1 2 3 4 5 6 7 8 9
007 Sliced Tomato Salad	○	1 2 3 4 5 6 7 8 9		0 1 2 3	0 1 2 3 4 5 6 7 8 9
008 Braised Liver with Onions	○	1 2 3 4 5 6 7 8 9		0 1 2 3	0 1 2 3 4 5 6 7 8 9
009 Tomato Juice	○	1 2 3 4 5 6 7 8 9		0 1 2 3	0 1 2 3 4 5 6 7 8 9
010 Peas	○	1 2 3 4 5 6 7 8 9		0 1 2 3	0 1 2 3 4 5 6 7 8 9
011 Cooked Turnips	○	1 2 3 4 5 6 7 8 9		0 1 2 3	0 1 2 3 4 5 6 7 8 9
012 Fried Rice	○	1 2 3 4 5 6 7 8 9		0 1 2 3	0 1 2 3 4 5 6 7 8 9
013 Corned Beef	○	1 2 3 4 5 6 7 8 9		0 1 2 3	0 1 2 3 4 5 6 7 8 9
014 Jellied Fruit Salad	○	1 2 3 4 5 6 7 8 9		0 1 2 3	0 1 2 3 4 5 6 7 8 9
015 Apricot Pie	○	1 2 3 4 5 6 7 8 9		0 1 2 3	0 1 2 3 4 5 6 7 8 9

FIG. 2. Format of food preference survey (Meiselman *et al.*, 1974; Meiselman and Waterman, 1978).

TABLE 9
FOOD PREFERENCES OF COLLEGE STUDENTS ARRANGED BY DECREASING PREFERENCE WITHIN FOOD CLASS (Einstein and Hornstein, 1970)

Breakfast
  Doughnuts
  Hot cakes
  French toast
  Scrambled eggs
  Blueberry pancakes
  Frosted cornflakes
  Fried eggs
  Ham omelet
  Oatmeal
  Cream of wheat
  Bran flakes
  Soft-cooked eggs
  Grits
  Spanish omelet

Appetisers
  Orange juice
  Fruit cup (fresh)
  Grape juice
  Banana (fresh)
  Apple juice
  Cantaloupe
  Grapefruit juice
  Tomato juice
  Pineapple juice
  V-8 juice (vegetable)
  Stewed prunes
  Stewed rhubarb

Soups
  Chicken noodle
  Vegetable
  Tomato
  Cream of mushroom
  Clam chowder
  Beef barley
  Navy bean

Salads
  Tossed green
  Lettuce and tomato
  Head lettuce
  Chef's
  Chef's salad bowl
  Potato
  Cole slaw
  Assorted relishes

  Waldorf
  Jellied peach and banana
  Cottage cheese
  Cinnamon pear
  Jellied bing cherry
  Orange coconut
  Peach and cranberry
  Cucumber and onion
  Pickled beets
  Carrot raisin
  Vegetable aspic (fresh)

Sandwiches
  Hamburger
  Cheeseburger
  Hot roast beef
  Bacon, lettuce and
    tomato
  Ham
  Grilled ham-cheese
  Barbecue beef
  Grilled cheese
  Tuna salad
  Chicken salad
  Submarine (Hoagie)
  Hot roast pork
  Bologna and lettuce
  Egg salad
  Cream cheese and jelly
  Grilled reuben

Entrées
  Beef steak
  Roast turkey
  Roast beef
  Fried chicken
  Spaghetti (Italian)
  Canadian bacon
  Bacon strips (crisp-grilled)
  Swiss steak
  Grilled pork chops
  Baked smoked ham
  Frankfurters
  Fried shrimp
  Salisbury steak
  Beef stew
  Pizza
  Ham-cheese platter (cold)
  Roast pork

  Chicken and dumplings
  Barbecued spareribs
  Grilled ham steak
  Macaroni and cheese
    (baked)
  Sausage (grilled)
  Meat loaf
  Roast veal
  Chicken pot pie
  Lasagna
  Ravioli
  Chicken a la king
  Cold cuts (platter)
  Chili con carne
  Fried flounder
  Beef turnover
  Fruit-cottage cheese
    platter
  Roast leg of lamb
  Fried scallops
  Veal cutlet parmigiano
  Creamed chipped beef
  Ham loaf (baked)
  Tuna-noodle casserole
  Fried haddock
  Turkey croquettes
  Grilled stuffed frank
  Chicken chow mein
  Corned beef
  Deviled crab
  Corned beef hash
  Shrimp creole
  Chicken cacciatore
  Veal fricassee
  Fried perch
  Stuffed tomato cold
    platter
  Baked swordfish
  Baked halibut
  Veal scaloppini
  Hungarian goulash
  Stuffed peppers
  Corn fritters and bacon
  Salmon loaf (baked)
  Sautéed liver
  Welsh rarebit
  Lamb stew
  Cheese blintzes
  Sautéed chicken livers

TABLE 9—contd.

Vegetables		
French-fried potatoes	Baked squash	Cherry cobbler
Whole kernel corn	Stewed tomatoes	Apple crisp
Baked potato (Idaho)	Succotash	Boston cream pie
Corn on the cob	Fried eggplant (French)	Peach cobbler
Whipped potatoes	Turnips	Pumpkin pie
Green beans (buttered)	Kale (buttered)	Peanut butter cookies
Oven-browned potatoes		Baked apple
Green peas (buttered)	*Breads*	Pineapple upside-down
Parsley potatoes (buttered)	Soft rolls	cake
Potatoes au gratin	Hot biscuits	Strawberry chiffon pie
Baked beans	Cornbread	Butterscotch pudding
Noodles (buttered)		Raspberry gelatin
Rice (steamed)	*Desserts*	Chocolate bavarian cream
Sweetpotatoes	Ice cream	Cheese cake
Spanish rice	Apple pie	Tapioca
Lima beans	Ice cream sundae	Baked custard
Broccoli (buttered)	Strawberry shortcake	Rice pudding
Spinach	Sliced peaches	Lemon snow pudding
Asparagus (buttered)	Chocolate-chip cookies	Plum cobbler (purple)
Brussels sprouts	Brownies	Bread pudding
Glazed carrots	Orange sherbet	
Sauerkraut	Chocolate-nut sundae	*Beverages*
Cauliflower	Devil's-food cake	Milk
Smothered onions	Lemon meringue pie	Lemonade
Harvard beets	Angel-food cake	Hot chocolate
Black-eyed peas	Gingerbread	Iced tea
Cabbage (steamed)	Blueberry pie	Hot tea
	Chocolate pudding	Coffee

literature. However, there is a growing body of data for the investigator to tap so that many food preference decisions need not be made intuitively. One of the largest of available data bases is that of the US Armed Forces which have been collecting food preference data for almost 40 years. These data are available as technical reports (e.g. Meiselman *et al.*, 1974) and as journal articles (e.g. Meiselman and Waterman, 1978). The US military data contain a large number of foods, some of which overlap with the cuisines of different cultures.

Presentation of food preference data has been handled in a number of ways. The most straightforward is a listing of each food item, organised alphabetically or within food classes, and presenting various descriptive statistics (mean rating, standard deviation, number of respondents). These data presentations are essential for the person who will use the data to make decisions and wishes to make various calculations and comparisons with the data base.

Food preference data, like all attitudinal data, are best understood when

TABLE 10
US ARMED FORCES FOOD PREFERENCES FROM PERIODIC SURVEYS

Food preferences of US military personnel were tested on surveys using the nine-point hedonic scale (H), and a preferred frequency scale (F) (by Peryam et al. (1960), Kamen et al. (1967), Moskowitz et al. (1972), Meiselman and Waterman (1974), Meiselman and Wyant (1979), and in 1984 by Meiselman and Bell (unpublished))

	1984		1978		1974		1967	1963	1960
	H	F	H	F	H	F	H	H	H
Appetisers					5·81	9·99			
Fruit cocktail (canned)	6·69	10·01	6·60	10·31	6·43	10·42			
Fruit cup			6·73	11·22	5·94	8·61			
Guacomole dip	4·92	5·57	5·84	9·02	5·60	8·38			
Tomato juice	4·88	8·22	5·44	12·05	5·70	11·87			
Tomato juice	5·09	7·66	5·50	11·36	5·84	11·38			
Vegetable juice	5·38	9·09	5·51	12·09	5·37	9·30			
Shrimp cocktail							7·08		
Soups					5·33	7·13			5·99
Bean soup	4·97	4·66			5·27	7·12	5·88	6·56	5·96
Beef barley					5·41	6·84	6·43	5·66	
Beef rice					5·46	7·26		5·92	
Chicken noodle	6·83	9·07	6·73	9·23	6·57	10·25			6·71
Clam chowder	5·71	7·14	5·71	2·88	5·47	7·44	5·70	5·86	
Corn chowder					5·07	6·37			5·44
Cream of mushroom	5·04	5·59	5·34	8·22	5·11	7·12			5·11
Cream of potato					5·21	6·88			
Creole soup					4·99	6·55		5·79	
Egg drop soup					4·52	5·30			
Fish chowder			5·14	7·82	5·03	6·22			
Minestrone					5·27	6·54		6·29	
Onion soup	4·40	4·04	4·67	7·45	4·76	5·71	5·62		4·59
Split pea soup			4·81	6·93	4·58	5·75	5·71	5·54	5·67
Tomato soup	5·58	5·81	5·94	8·61	5·90	8·53	6·31		6·22
Tomato vegetable soup					5·59	7·71			
Turkey rice					5·60	7·46		6·51	
Vegetable soup	6·47	8·79	6·40	9·97	6·16	9·23			7·02
Fruit and vegetable juices					5·79	11·20			7·00
Apple juice	6·79	13·49			6·19	12·9	7·33		
Cranberry juice	4·64	5·53	4·96	9·13	4·94	6·73			
Grape juice					6·36	13·54	7·67		
Grapefruit juice	5·30	10·38			5·79	11·20			6·99
Grapefruit–orange juice					6·23	13·54			
Grapefruit–pineapple juice					5·73	10·76			
Orange juice	7·86	20·35	7·86	20·18	7·50	19·02	7·86	7·69	8·02
Pineapple juice	6·21	9·78	6·06	11·32	5·81	9·96	6·98		7·61
Prune juice			3·59	7·01	4·06	4·85			
Tomato juice	4·88	8·22	5·50	11·36	5·70	11·87	6·03		7·42
Tomato juice	5·09	7·66	5·44	12·05	5·84	11·38			
Vegetable juice	5·38	9·09	5·51	12·09	5·37	9·30			

TABLE 10—contd.

	1984		1978		1974		1967	1963	1960
	H	F	H	F	H	F	H	H	H
Fruit drinks and iced tea					6·11	11·39			
Cherry-flavoured drink					5·67	9·53		6·04	
Fruit punch	6·50	11·36			6·18	10·69	7·03	7·13	
Grape-flavoured drink					6·05	10·96		6·91	
Grape-lemonade					5·67	9·37			
Iced tea	6·93	16·14	7·10	17·62	6·91	16·87	7·26		6·97
Lemonade	6·82	12·95	7·28	15·18	6·81	13·64			7·83
Lime-flavoured drink					5·38	8·70	7·40	6·15	
Orange-flavoured drink					6·19	11·35		7·32	7·52
Hot beverages					5·88	13·68			
Freeze-dried coffee					4·85	9·47			
Fresh coffee	6·13	17·47	5·41	18·20	6·43	19·24	6·89	7·11	7·01
Hot chocolate	6·94	12·43	7·15	13·51	6·76	13·13	7·41	6·88	7·66
Instant coffee	4·47	8·56	4·42	13·37	4·85	9·82		4·79	5·16
Tea	6·32	14·03	6·66	16·59	6·50	16·74	6·03	6·05	5·98
Milk products					6·19	13·35			
Buttermilk	3·02	3·01	3·14	10·01	3·69	5·75			
Chocolate milk	6·68	13·59	6·83	14·7	6·87	16·15			7·93
Fruit flavoured yogurt	5·14	9·00			4·71	6·89			
Ice cream	7·27	14·86			7·38	17·56			
Milk	8·06	24·52	7·86	24·32	7·98	24·59		8·36	8·60
Milk shake	7·23	11·64	7·22	11·82	7·38	14·73			
Milk shake	7·23	11·64			7·19	14·31			
Skimmed milk/low fat milk	5·19	13·17	4·37	14·95	3·84	7·22			
Soft serve ice cream	6·95	12·08			6·71	12·93			
Carbonated beverages					5·80	11·26			
Cherry soda					5·82	9·95			
Cola	6·73	15·70	6·60	16·40	6·80	16·72			
Ginger ale	5·99	9·48			5·99	10·73			
Grape soda					5·95	10·83			
Lemon-lime soda					5·79	11·35			
Lo-cal. soda	3·66	6·25	3·93	14·20	3·85	6·08			
Orange soda					6·22	12·34			
Pepper soda					5·37	10·77			
Root beer					6·42	12·59			
Beer			5·71	14·81	7·26	19·78			
Hot breads and doughnuts			6·92	10·90	6·30	10·57			
Baking powder biscuits					6·21	11·20	7·95		8·33
Coffee cake	5·91	7·51	6·27	8·80	5·78	8·66	6·71	6·71	7·13
Corn bread	6·30	8·07	6·76	10·97	6·30	10·13	6·59		7·07
Danish pastry					6·40	10·22	7·70		
Doughnuts	6·67	10·71	6·92	10·90	6·85	13·51	7·5	7·33	7·55
English muffins					6·42	10·88		7·18	
Sweet rolls	6·80	11·48	6·94	11·59	6·58	12·02	7·70		

(continued)

TABLE 10—contd.

	1984		1978		1974		1967	1963	1960
	H	F	H	F	H	F	H	H	H
Hot cross buns								7·69	
Blueberry muffins					6·51	10·57			
Plain muffins					5·66	7·89	7·23		6·84
Breakfast cereals					5·60	9·83			
Cold cereals	6·48	13·20	6·34	13·39	6·06	12·54		6·98	6·53
Hominy grits	5·05	8·23	5·69	11·35	5·36	9·02			
Hot oatmeal	5·47	7·44			5·64	9·57	6·59	6·02	6·13
Hot whole wheat cereal					5·35	8·18	5·94	6·44	
Cream of wheat								5·80	5·80
Griddle cakes			6·87	11·07	6·49	11·59			
French toast					6·53	12·45			7·24
Griddle cakes	6·70	11·25			6·40	11·14	6·94		7·05
Waffles					6·54	11·18		7·10	
Eggs					6·98	16·56			
Eggs to order	7·74	21·41	7·18	17·28	7·51	21·02		7·82	
Omelette	7·26	15·16	6·86	13·33	6·44	12·11			6·16
Scrambled eggs								7·02	6·71
Hard-cooked eggs									6·65
Devilled eggs			6·54	9·80					6·93
Fried eggs								7·43	7·71
Breakfast meats					6·29	10·97			
Bacon	7·12	15·96	7·19	14·51	7·33	17·29		7·62	7·45
Canadian bacon					6·93	13·04			
Creamed chipped beef					5·83	8·97	5·41	6·48	6·08
Creamed ground beef	5·58	8·79	5·39	9·31	5·75	9·20	5·60	5·88	
Grilled bologna	5·06	4·95			5·01	6·39			
Ham			7·23	10·55	7·08	11·50			
Ham			7·15	10·51	6·89	11·04			
Pork sausage patties	6·55	13·81			6·32	11·60	7·16		6·91
Sausage links			6·47	11·45	6·71	13·69			6·89
Scrapple					5·08	7·00			
Fish and seafood			5·44	9·45	6·20	9·20			
Baked fish	6·27	7·51			5·71	8·14			4·58
Baked tuna and noodles	5·62	5·93	5·93	7·84	5·72	7·66		6·03	5·88
French fried fish sticks					6·35	9·13			
French fried scallops					6·20	9·28		6·07	
French fried shrimp	7·55	12·14	7·44	12·85	7·13	11·98		7·09	
Fried fish	6·69	8·73	6·75	9·74	6·39	9·41		6·48	5·42
Fried oysters					5·65	8·48		5·99	
Lobster					6·91	11·37			
Salmon	5·35	5·23			5·77	7·55			6·06
Seafood platter					6·75	10·30	6·68	6·44	
Shrimp creole					6·28	9·64	6·51		

TABLE 10—contd.

	1984		1978		1974		1967	1963	1960
	H	F	H	F	H	F	H	H	H
Meats					6·24	9·03			
Baked chicken	6·87	8·38	6·78	9·57	6·74	10·27			
Baked ham	6·91	8·79			6·82	10·39		7·42	7·68
Baked stuffed pork chops	7·00	8·56	7·19	10·03	6·58	9·36			7·46
Barbequed beef cubes					6·13	7·52		6·43	7·65
Barbequed spareribs	7·13	9·10	7·29	10·67	6·84	10·51	7·53		6·89
Boiled pigs' feet			4·42	6·89	4·32	5·01			
Braised liver and onions	4·17	3·51	4·75	7·24	4·64	5·66	4·55		6·11
Breaded veal steaks					6·41	9·69		7·75	7·60
Chitterlings			5·20	9·73	4·67	6·79			
Corned beef	5·29	4·79	5·17	6·47	5·19	6·34	5·92	6·46	6·08
Fried chicken	7·36	10·03	7·45	10·83	7·38	12·32	7·59	7·80	8·24
Grilled ham	6·99	8·74			6·75	10·41			7·60
Grilled lamb chops	6·12	6·77	6·04	8·46	5·86	7·89	6·55		6·71
Grilled minute steaks					6·80	10·55			
Grilled steak	7·94	12·87	7·95	13·23	7·76	14·43		8·30	8·31
Hot roast beef sand. w/gravy					6·94	10·68	7·68	7·76	
Hot turkey sand. w/gravy	6·85	6·94	6·77	7·52	6·91	9·86	7·68		
Italian sausage					6·59	10·22			
Pepper steak					6·59	9·61			
Pickled pigs' feet					4·79	6·58			
Polish sausage					6·23	8·70			
Pork hocks	5·17	5·84	5·23	8·29	5·25	7·14			
Pot roast	7·08	8·64			6·67	9·75			7·74
Roast beef	7·49	10·64	7·41	11·13	7·04	11·54		7·79	8·02
Roast lamb					5·84	8·52	6·43	6·35	6·13
Roast pork	6·63	7·51	6·37	8·22	6·49	9·25		7·37	7·72
Roast turkey	6·98	5·88	6·95	6·72	6·96	8·20			
Roast veal					6·07	8·26	7·06		7·34
Sauerbraten					5·38	6·30			
Spareribs w/sauerkraut					5·74	7·64			
Swiss steak	6·99	8·63	6·93	8·94	6·94	11·12	7·22		7·92
Veal parmesean	6·55	7·38	6·34	7·32	6·28	8·62			
Stews and extended meats					6·25	8·86			
Baked tuna and noodles	5·63	5·93	5·93	7·84	5·72	7·76		6·03	5·88
Beef stew	6·98	8·83	6·81	9·38	6·68	10·08		6·02	6·73
Beef stroganoff			6·58	8·07	6·33	8·70			
Chicken cacciatore					5·97	7·82			
Chili con carne	6·22	6·62	6·53	8·49	6·54	9·41	6·80	6·79	6·61
Chili macaroni					5·86	6·79			
Corned beef hash	5·15	5·40	5·12	6·77	5·16	6·51	5·80		5·76
Enchiladas	6·04	6·57			6·19	8·65			
Ham loaf					5·94	7·99			7·15
Lasagna	7·43	9·83	7·08	9·76	6·61	9·51			
Meat loaf	6·64	7·78	6·57	8·81	6·69	9·94	6·96		7·20

(*continued*)

TABLE 10—contd.

	1984		1978		1974		1967	1963	1960
	H	F	H	F	H	F	H	H	H
Pizza	7·51	10·26	7·28	9·67	7·04	11·64	7·46		
Pork chop suey	5·74	5·88	5·97	7·72	5·69	7·61		5·88	6·36
Ravioli					6·45	9·33			
Salisbury steak					6·72	10·07		7·29	
Shrimp creole					6·28	9·64			
Spaghetti w/meat sauce	7·40	10·08			7·27	11·74		7·31	
Spaghetti w/meatballs	7·40	9·74	7·38	9·99	7·27	11·70	7·54	7·47	7·47
Stuffed cabbage					5·37	6·89			
Stuffed green peppers	5·38	5·22	5·56	7·44	5·64	7·41		6·13	6·47
Sukiyaki					5·83	7·56	6·23		
Swedish meat balls					6·58	9·57			6·95
Sweet and sour pork					5·77	7·50	6·41		
Turkey pot pie					6·29	8·88		7·20	
Veal burger					5·72	5·99	6·23		6·74
Chicken a la king								6·56	7·02
Chow mein							6·36	5·95	5·71
Short order, Sandwiches					6·34	9·61			
Bacon, lettuce and tomato	7·01	9·38	7·31	10·88	7·24	12·47			
Baked bean sand.					4·39	5·30			
Bologna sand.					5·68	7·96		6·34	6·25
Burritos	6·41	7·57	6·45	9·10	6·27	9·17			
Cheeseburger	6·97	11·37	7·01	11·95	7·00	12·44	7·50		7·52
Egg salad sand.					6·14	8·73			
Fishwich	6·28	7·37	6·30	8·34	6·00	8·35			
Frankfurter, cheese and bacon					6·08	8·67	7·29		
Frankfurter	6·23	7·59	6·34	8·98	6·27	9·33		6·50	6·52
Grilled cheese sand.					6·70	10·90	7·29		7·32
Grilled ham and cheese	6·92	9·38	6·98	1·96	6·84	11·02	7·47		
Ham sand.					6·77	10·38			
Hamburger	6·83	10·01	7·07	11·01	6·92	11·90	7·53	7·52	
Hot pastrami			6·12	7·54	5·82	7·54			
Hot reuben					5·49	7·65			
Hot tamales	5·74	6·17	6·03	8·39	6·14	8·79			
Meatball sub	6·50	7·37			6·58	9·39			
Peanutbutter and jelly sand.	5·90	7·01	5·95	9·19	5·87	9·12			
Pizza	7·51	10·26	7·28	9·69	7·04	11·64	7·46	7·17	
Salami sand.					5·59	7·46			5·98
Sloppy Joe	6·77	8·43	6·67	9·12	6·74	10·59			
Submarine sand.	6·85	8·96	6·48	8·72	6·60	9·58			
Tacos	6·83	8·79	6·86	10·10	6·53	10·06			
Tuna salad sand.	6·48	8·08	6·70	9·96	6·39	9·74			
Turkey club sand.					6·63	9·77			
Western sand.	6·56	7·78	6·38	9·14	6·49	9·53			
Luncheon meat								6·47	

TABLE 10—contd.

	1984		1978		1974		1967	1963	1960
	H	F	H	F	H	F	H	H	H
Potato and potato substitutes					5·99	9·03			
Baked macaroni and cheese	6·85	9·21	6·94	10·52	6·49	9·68	6·86	7·16	6·65
Baked potato	6·95	9·00	7·10	10·36	6·74	10·74	7·13	6·77	6·73
Boiled navy beans					4·99	6·12			
Boston baked beans					6·12	8·26	5·16	5·71	6·14
Buttered noodles	5·99	7·17	5·95	8·22	5·81	8·37	6·30		6·18
Corn bread stuffing					5·61	7·24			
French fried potatoes	7·27	11·99	7·28	12·03	7·42	14·55	7·92		8·17
Fried rice	6·19	7·81	6·46	9·43	5·85	8·52	5·87	5·61	
Giblet stuffing					5·84	7·40			
Hashed brown potatoes	7·06	13·60	7·14	12·64	7·03	14·13	7·12		7·11
Hot potato salad	6·54	9·02	7·00	10·67	5·50	7·44			6·38
Mashed potatoes	6·93	11·01	6·95	12·01	7·00	13·90	7·57	7·52	7·81
Pork and beans	6·07	6·65	6·15	8·15	6·21	9·14			
Potato chips	6·22	9·67	6·59	10·75	6·77	12·67			7·45
Refried beans	5·10	4·94	5·77	7·33	5·17	6·55			
Rice pilaf					5·33	7·15			
Sausage stuffing					5·17	6·44			
Savory bread stuffing	5·74	6·17	6·00	7·37	5·42	6·54			
Scalloped potates					5·98	8·46	6·46		6·65
Spanish rice					5·99	8·82	6·21		5·92
Steamed rice	5·90	8·57	6·21	10·23	5·71	9·09		5·85	5·52
Sweet potatoes	5·60	6·40			5·55	7·53	5·72	6·59	6·53
Green vegetables					5·22	7·41			
Asparagus	5·09	6·86	5·21	9·21	5·12	7·33	5·00		4·86
Broccoli	5·86	8·08	5·91	10·99	5·19	7·16	5·63	5·58	4·56
Brussel sprouts	5·08	6·19	5·35	9·14	5·02	6·57	5·20		4·83
Buttered mixed veg.	6·33	11·10	6·23	12·35	5·95	10·40	6·56		
Buttered peas and carrots	5·52	6·52	5·58	8·90	5·39	7·39	6·47		
Buttered zucchini	4·70	4·67			4·76	6·04			
Cabbage	5·52	6·85	5·77	9·24	5·30	7·30			5·48
Canned green beans			6·98	12·07	5·79	9·18			6·80
Canned lima beans			5·10	8·23	4·86	6·19			6·81
Canned peas			6·03	10·57	5·55	7·91	7·00	6·78	6·99
Collard greens	5·36	6·74	6·18	10·54	5·39	8·29			
Creamed frozen peas					4·92	6·83			
Fried cabbage					4·85	6·00	5·10		
Fried okra	4·78	5·98	5·47	9·44	5·01	7·21	7·50		5·35
Frozen green beans	6·83	11·25			5·60	8·26			
Frozen lima beans	4·68	5·45			4·69	6·27			
Frozen peas	5·98	9·02			5·38	7·34			
Mustard greens					4·87	6·56			
Spinach	5·27	7·33	5·64	10·56	5·05	7·46		4·53	4·83
Turnip greens					4·79	6·60			5·40

(continued)

TABLE 10—contd.

	1984		1978		1974		1967	1963	1960
	H	F	H	F	H	F	H	H	H
Yellow vegetables					5·67	8·50			
Baked yellow squash	4·25	4·00	4·68	7·23	4·31	4·97			4.52
Buttered carrots	5·62	7·07	5·56	9·16	5·23	7·38		5·75	6·12
Buttered succotash	5·73	7·26			5·04	6·27		5·82	5·51
Buttered wax beans					5·33	7·16			5·85
Buttered whole kernel corn	7·51	12·49	7·55	13·19	7·19	13·12			7·20
Corn-on-the-cob					7·50	13·36	7·82		8·03
Corn fritters	5·65	6·12	5·97	8·53	5·79	7·62			
Cream-style corn					6·65	11·23		7·18	6·89
French fried carrots					4·03	4·61	4·54		
Other vegetables					4·76	6·04			
Buttered cauliflower	5·18	5·97	5·34	9·88	4·84	6·14			4·68
Creamed onions	3·74	2·77	4·06	6·49	4·29	4·90			
French fried cauliflower					4·28	4·97	4·35		
French fried onion rings	6·73	9·83	6·64	9·99	6·74	11·21		6·54	5·77
Fried eggplant	4·77	4·51	4·90	7·09	4·86	6·17		5·65	5·35
Fried parsnips					3·85	4·10	4·41		4·47
Harvard beets	4·25	3·92	4·72	7·50	4·75	5·56		5·33	5·46
Mashed rhutabagas					3·93	4·35	4·72		4·74
Sauerkraut	4·62	4·50			5·14	6·86			5·68
Stewed tomatoes	4·57	4·57	4·80	7·78	4·91	6·13		5·45	5·67
Fruit salads					5·45	7·67			
Banana salad			6·02	8·59	5·40	6·73			
Cottage cheese and fruit salad	5·04	6·98	5·40	10·43	5·04	7·40	5·82		6·44
Jellied fruit salad	5·44	6·46	5·86	8·62	5·55	8·29			6·99
Mixed fruit salad	6·53	9·99			6·28	9·94	7·33	7·02	
Pineapple cheese salad					4·89	6·08		6·20	6·70
Slice orange salad					5·77	7·96			
Waldorf salad	4·91	5·58	5·89	9·59	5·23	7·30		6·33	
Vegetable salads					5·05	7·18			
Carrott, raisin and celery	4·19	4·41	5·02	8·74	4·38	5·40			5·04
Celery and carrot sticks			6·20	14·45	5·68	11·49			
Cole slaw	5·83	7·90	6·02	9·82	6·23	10·37	6·53	6·43	6·76
Cucumber and onion salad					5·35	7·70			6·27
Frijole salad			5·22	8·27	5·27	7·08			
Garden cottage cheese salad					4·96	6·74		5·45	
Jellied vegetable salad					4·78	5·67			
Kidney bean salad					4·29	4·74	4·87		4·87
Pickled beets and onion salad					4·47	5·47			5·28

TABLE 10—contd.

	1984		1978		1974		1967	1963	1960
	H	F	H	F	H	F	H	H	H
Tossed green salads					6·57	13·37			
Chef's salad	7·21	12·99	7·28	12·16	6·57	11·81			
Lettuce salad					6·63	13·60		6·81	6·69
Sliced tomato salad	5·56	8·58	6·28	11·62	6·29	12·13			
Tossed green salad	7·44	18·98	7·67	19·46	7·03	17·18		7·17	
Tossed vegetable salad					6·31	12·13		7·39	6·97
Cucumber and tomato salad									6·49
Salad dressings					5·59	10·14			
Blue cheese					4·93	8·16			
Caesar					5·34	8·92			
French					6·36	12·64			6·52
Italian					6·02	12·82			
Russian					5·46	8·85			
Sour cream					4·68	6·49			
Thousand island					6·47	13·39			
Vinegar and oil					5·50	9·86			
Mayonnaise									6·68
Fresh fruit					6·76	12·61			7·28
Apples	7·25	16·88			7·12	15·34			7·80
Bananas	7·15	13·82	7·14	12·87	6·86	13·18	7·56	7·59	
Cantaloupe	6·71	10·85	7·06	13·15	6·85	12·68		7·38	7·75
Fruit cup			6·73	11·22	5·94	8·61	7·56		
Grapefruit half	5·64	9·61	6·20	12·46	6·14	12·31	6·36		7·10
Grapes					6·89	13·10	7·64		7·66
Honeydew melon					6·46	8·72			7·17
Oranges	7·45	16·61	7·49	16·12	7·14	15·37		7·53	7·74
Peaches	7·31	13·15	7·54	13·88	7·16	13·49			8·05
Pears	6·93	11·87	7·14	12·83	6·84	12·33		7·67	
Plum					6·22	11·18			7·03
Tangerines					6·90	13·22	7·65		7·40
Watermelon	6·98	10·01	7·45	11·34	7·08	11·82		7·54	7·99
Canned fruits					5·68	8·41			6·60
Apple sauce	6·68	9·47	6·76	10·26	6·45	10·47	6·96		7·19
Apricots					5·19	6·99			5·50
Figs					4·33	4·83			5·60
Fruit cocktail	6·69	10·01	6·60	10·31	6·43	10·42	7·34		7·88
Grapefruit sections					5·64	8·82		6·49	6·68
Peaches	6·80	11·00	6·70	10·69	6·65	11·60		7·55	7·92
Pears	6·68	10·35			6·47	10·37			7·63
Pineapple	6·60	9·75			6·03	9·05			7·69
Plums					5·13	6·58	5·69		6·26
Stewed prunes	3·04	2·27	3·72	7·15	4·08	4·62	4·97		5·32
Sweet cherries					5·64	7·38		6·52	7·04

(continued)

TABLE 10—*contd.*

	1984		1978		1974		1967	1963	1960
	H	F	H	F	H	F	H	H	H
Cookies and brownies					5·67	7·94			7·08
Brownies	6·69	9·00	6·88	10·12	6·65	10·81	7·31	7·27	6·99
Butterscotch brownies					5·36	7·01			
Chocolate chip cookies	6·90	10·13	7·01	11·03	6·18	9·13	7·64		7·61
Chocolate cookies					5·99	8·89		6·98	
Coconut raisin cookies	5·16	5·29	5·36	7·60	5·12	6·61			
Fruit bars					5·14	6·47	6·16		
Lemon cookies					5·46	7·24	6·26		
Molasses cookies					5·37	7·10			6·46
Nut bars					5·16	6·16			
Nut cookies			5·84	7·74	5·57	7·34			
Oatmeal cookies	6·30	8·09			6·16	9·92			6·97
Peanut butter cookies	6·11	7·63	6·51	9·58	5·94	8·89			6·73
Raisin cookies					5·36	7·34		6·73	
Sugar cookies					5·75	7·98		6·72	7·35
Vanilla wafers					5·81	8·19			7·18
Cakes					6·19	8·45			7·26
Angel food					6·20	8·65			
Banana	6·23	7·48	6·41	8·40	6·34	9·61			7·84
Boston cream pie	6·37	7·07			6·34	8·91	6·87		7·71
Cheesecake	6·47	8·42	6·36	9·43	5·85	7·90			
Cherry upside down cake					6·11	8·51			
Chocolate cream cake					6·04	8·24	7·47	7·43	7·67
Devil's food	6·55	7·43	6·81	9·00	6·48	9·45		7·28	
Ginger bread					5·64	7·72	6·47	6·91	6·78
Marble cake					5·97	7·69		7·03	
Peach shortcake					6·24	8·60	6·72		7·76
Peanut butter cake					5·41	7·07	5·98		
Pineapple upside down					6·41	9·21			7·69
Pound cake	6·26	7·07	6·51	8·20	5·93	7·78	6·63		
Raspberry shortcake					6·09	9·12		7·47	
Spice cake					5·77	7·41		6·72	7·16
Strawberry shortcake	6·92	7·71	7·15	8·97	7·43	11·88			8·32
White cake	5·85	6·08	5·92	7·45	5·83	7·48			
Yellow cake					5·61	6·87			
Pies					5·90	8·34			7·18
Apple	7·01	9·60	7·25	10·78	6·98	11·67	7·45	7·85	7·81
Apricot	4·16	3·32	4·64	5·76	4·77	5·80	5·81	6·13	6·16
Banana cream					6·45	9·85			8·03
Blackberry					6·04	8·56		7·22	7·43
Blueberry					6·17	8·85		7·13	
Butterscotch cream					5·66	7·50			7·26
Cherry	6·24	7·16	6·62	8·94	6·40	9·40			7·61
Chocolate cream pie	6·23	7·37			6·37	9·42			7·55
Coconut custard					5·74	7·87		7·28	

TABLE 10—contd.

	1984 H	1984 F	1978 H	1978 F	1974 H	1974 F	1967 H	1963 H	1960 H
Fried pie (fruit)					5·67	8·01			
Lemon chiffon					6·02	8·22			
Lemon meringue	6·19	6·86	6·72	9·23	6·50	9·67		7·59	7·48
Peach					6·00	8·13		7·19	7·69
Pineapple cream					5·60	7·47		7·08	
Pineapple					5·36	6·88			7·47
Pumpkin					6·34	9·06	6·95		7·28
Raisin			4·34	6·35	4·49	5·12			6·00
Strawberry chiffon	6·00	6·33	6·44	8·56	6·39	9·69			
Sweet potato pie	5·19	6·15			5·10	7·20			
Boysenberry								5·95	
Mincemeat								6·08	6·25
Puddings and other desserts					5·71	7·97			6·89
Apple crisp	6·38	7·48	6·78	8·68	6·50	9·61	6·27		6·74
Banana cream pudding					6·33	9·42			
Bread pudding					5·20	6·39	5·93	6·43	6·42
Butterscotch pudding					5·62	7·55			7·17
Cherry cake pudding					5·60	7·84			
Chocolate cake pudding					5·88	8·05			
Chocolate pudding	6·26	7·36	6·21	9·06	6·17	9·21	7·16		7·16
Coconut cream pudding					5·73	8·06			
Fruit flavoured yogurt	5·14	9·00	4·86	11·08	4·71	6·89			
Rice pudding	4·75	4·54	5·33	7·54	5·21	6·73	6·09		6·10
Strawberry gelatin	5·63	7·32	5·86	9·43	5·73	8·03	6·66	6·61	
Vanilla cream pudding					5·86	7·90			7·04
Ice cream and sherbet					6·78	12·14			7·63
Banana split			6·95	9·45	7·02	11·57			
Butterscotch sundae					6·14	8·95			7·66
Hot fudge sundae	6·95	9·53	6·88	9·67	7·01	11·59			7·88
Ice cream	7·27	14·86	7·52	15·25	7·38	17·56		7·70	8·26
Milk shake	7·23	11·64			7·38	14·73			
Pineapple sundae					5·93	8·64			
Sherbet	6·33	8·73	6·61	10·21	6·36	10·46	7·13		6·50
Soft serve ice cream	6·95	12·08	7·05	12·62	6·71	12·93			
Strawberry sundae					6·66	10·63			
Ice cream sundae								7·78	8·24

large amounts of data are available from repeated surveys so that trends are more easily seen. The US military food preferences from several surveys are gathered in Table 10. Earlier surveys used nine-point hedonic scale (H) and later surveys added a preferred frequency scale (S). Sample sizes on each survey are usually in the thousands. Such data permit more precise specification of item preference as well as examination of temporal trends.

Various approaches have been used to summarise the mass of data in such tables. One approach has been to list the food items within each food class (by increasing or decreasing preference). Einstein and Hornstein (1970) have used this approach to quickly summarise their college student data (Table 9). This will provide the decision maker with the answer as to whether one item is better than another. In addition, Meiselman and Waterman (1978) further analysed their preference data to produce a list of foods which were significantly lower or higher than the class means. This produced a list (Table 11) of 'winners' and 'losers' for use in decision making. It underscores that one should not compare different food classes when comparing preferences, since, for example, many cooked vegetables will be lower than many fresh fruits or desserts. However, some cooked vegetables are preferred more than others.

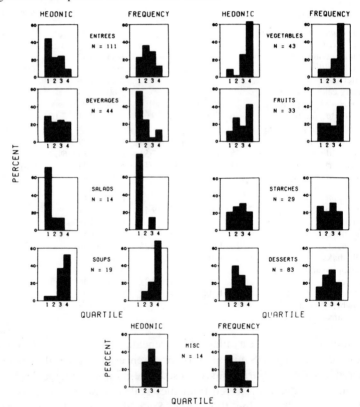

FIG. 3. Quartile distribution of food item ranks for hedonic and frequency scales. Food within each class is expressed as a percentage of the total number of foods in that class.

TABLE 11
FOOD PREFERENCES OF US MILITARY PERSONNEL WHICH WERE SIGNIFICANTLY HIGHER OR LOWER THAN THE FOOD CLASS MEAN

Food class	High-preference	Low-preference
1. Appetisers		
2. Soups	Tomato vegetable noodle soup	Corn chowder
		Fish chowder
	Tomato soup	Split pea soup
	Chicken noodle soup	Egg drop soup
		Onion soup
		Creole soup
3. Fruit and vegetable juices	Orange juice	Cranberry juice
	Grape juice	Prune juice
4. Fruit drinks and ice tea	Lemonade	Grape lemonade
	Iced tea	Lime-flavoured drink
		Cherry-flavoured drink
5. Hot beverages		Instant coffee
		Freeze-dried coffee
6. Milk products	Milk	Skim milk
	Ice cream	Buttermilk
		Fruit-flavoured yogurt
7. Carbonated beverages	Cola	Low-calorie soda
9. Hot bread and doughnuts	Doughnuts	Plain muffins
	Sweet breads	
10. Breakfast cereals	Cold cereal	
11. Griddle cakes	Griddle cakes	
12. Eggs		
13. Breakfast meats	Bacon	Grilled bologna
	Canadian bacon	Scrapple
14. Fish and seafood	French fried shrimp	Baked fish
	Seafood platter	Salmon
	Lobster	Baked tuna and noodles
15. Meats	Roast beef	Grilled lamb chops
	Swiss steak	Spareribs with sauerkraut
	Pot roast	Corned beef
	Grilled steak	Pork hocks
	Grilled minute steak	Pickled pigs' feet
	Barbecued spareribs	Sauerbraten
	Grilled ham	
	Baked ham	
	Italian sausage	
	Fried chicken	
	Baked chicken	
	Hot turkey sandwich with gravy	
	Hot roast beef sandwich with gravy	
16. Stews and extended meats	Lasagna	Chicken cacciatore
	Pizza	Chili macroni
	Spaghetti with meat sauce	Ham loaf

(continued)

TABLE 11—contd.

Food class	High-preference	Low-preference
16. Stews and extended meats—contd.	Spaghetti and meat balls Meatloaf Swedish meat balls Salisbury steak Beef stew	Vealburger Stuffed cabbage Corn beef hash Stuffed green peppers Pork chop suey Sweet and sour pork Sukiyaki Baked tuna and noodles
17. Short order, sandwiches	Hamburger Cheeseburger Ham sandwich Bacon, lettuce and tomato sandwich Grilled cheese sandwich Grilled ham and cheese sandwich Sloppy Joe Pizza	Frankfurter, cheese, and bacon Salami sandwich Bologna sandwich Hot Reuben sandwich Hot pastrami sandwich Fishwich
18. Potato and potato substitutes	French fried potatoes Baked potatoes Hashed brown potatoes Mashed potatoes Potato chips	Sweet potatoes Hot potato salad Boiled navy beans Refried beans Rice pilaf Cornbread stuffing Savoury bread stuffing Sausage stuffing
19. Green vegetables	Canned green beans Frozen green beans Canned peas Collard greens Buttered mixed vegetables	Frozen Lima beans Canned Lima beans Creamed frozen peas Fried cabbage Brussels sprouts Mustard greens Turnip greens Buttered zuchini squash
20. Yellow vegetables	Cream-style corn Corn-on-the cob Buttered whole kernel corn	Baked yellow squash French fried carrots
21. Other vegetables	French fried onion rings	Mashed rutabagas Fried parsnips
22. Fruit salads	Mixed fruit salad	Pineapple cheese salad
23. Vegetable salads	Cole slaw Celery and carrot sticks	Pickled beet and onion salad Carrot, raisin, and celery salad Kidney bean salad
24. Tossed green salads		
25. Salad dressings	Thousand Island dressing French dressing	Sour cream dressing Blue cheese

TABLE 11—contd.

Food class	High-preference	Low-preference
26. Fresh fruit	Oranges Apples	Plums Honeydew melon Fruit cup
27. Canned fruits	Peaches Pears Applesauce Fruit cocktail	Plums Apricots Figs Stewed prunes
28. Cookies and brownies	Chocolate chip cookies Peanut butter cookies Chocolate cookies Oatmeal cookies Brownies	Molasses cookies Coconut raisin Fruit bars Nut bars Butterscotch brownies
29. Cakes	Strawberry shortcake Pineapple upside down cake Devil's food cake Banana cake	Spice cake White cake Peanut butter cake Yellow cake Cheesecake Gingerbread
30. Pies	Cherry pie Apple pie Pumpkin pie Strawberry chiffon pie Banana cream pie Lemon meringue pie	Raisin pie Pineapple pie Apricot pie Pineapple cream pie Sweet potato pie
31. Pudding and other deserts	Chocolate pudding Banana cream pudding Apple crisp	Bread pudding Rice pudding Fruit-flavoured yogurt
32. Ice cream and sherbet	Ice cream Milk shake	Butterscotch sundae Pineapple sundae

The relative rankings of food items within their classes can be graphically depicted (Fig. 3). This provides a quick summary of how the preference for different items is distributed over rankings. Figure 3 shows that, for example, salads are generally popular while many main dishes are disliked by this population.

Although it is dangerous and difficult to compare food preferences from study to study when different methods are employed, such analyses can be interesting. Meiselman (1979) showed that data collected on a nine-point scale on military personnel and on a three-point scale on college students yielded similar rankings for soups. An even more ambitious comparison of groups, one from the USA, and the other from the UK (Piggott, 1979), is presented in Table 12. The relatively low overall rankings for Brussels sprouts, turnip and, perhaps, cabbage are worth noting. The differences in

TABLE 12
COMPARISON OF PREFERENCE OF US MILITARY PERSONNEL AND UNITED KINGDOM RESIDENTS

	Results for UK residents (Piggott, 1979)	Results for US military personnel (Meiselman et al., 1974)
Baked beans	6·20 (1·92)	6·12 (2·12)
Brussels sprouts	6·13 (2·49)	5·02 (2·67)
Cabbage	6·12 (2·35)	5·30 (2·57)
Carrots	6·60 (1·93)	5·23 (2·40)
Cauliflower	6·56 (2·36)	4·84 (2·78)
Lamb (roast)	7·29 (1·83)	5·84 (2·48)
Milk	7·23 (1·93)	7·98 (1·74)
Oranges	7·33 (1·53)	7·14 (1·78)
Peaches	7·47 (1·76)	7·16 (1·82)
Rice (steamed)	6·73 (1·92)	5·71 (2·38)
Tea	7·00 (2·01)	6·50 (2·37)

Values in parentheses are the mean standard deviations.

ratings in many cases reflect genuine differences in American and British cuisines, e.g. the results for lamb. That these differences are not always as marked as the stereotype would suggest is shown by the results for tea.

3. FOOD ACCEPTANCE

Once a food is selected, it is evaluated on sensory and non-sensory dimensions. The food is judged acceptable or unacceptable to some degree. Based on the hunger state, the portion size available, situational and other factors, the food is then consumed or wasted. The sensory aspects of food evaluation are broader than assumed by many people. Food is evaluated on the basis of taste, smell and vision, but also pain, texture, sound, temperature and perhaps others. While each sense is understood to varying degrees, the relationship of sensory processes to food acceptance is not well understood especially in the context of complex foods.

The non-sensory factors contributing to food acceptance are also not well understood. Experience, situation, hunger and taboos contribute to acceptability but the mechanisms of interaction and the degree of control are not well known. Rozin and colleagues have begun to explore the area of taboos and aversions (Rozin and Fallon, 1981; Rozin et al., 1986).

Meiselman and Hirsch (1988) have begun to explore how the situational factors of eating environments contribute to the acceptance and consumption of foods. They have concluded that acceptability of groups of items or entire feeding systems is better measured using food consumption rather than the traditional methods appropriate to individual food items (Meiselman *et al.*, 1988).

Traditionally, sensory methods of evaluation are divided into analytical and affective methods (Table 13). Analytical methods try to use people as

TABLE 13
ANALYTICAL AND AFFECTIVE METHODS FOR EVALUATING FOODS
(Adapted from the Institute of Food Technologists, 1981)

Classification of methods by function	Appropriate methods
Analytical: Evaluate differences or similarity, quality and/or quantity of sensory characteristics of a product with a screened, trained panel of 5–10 members.	
1. Discriminate:	
a. Difference: Measures simply whether samples are different	Paired comparison Duo-trio Triangle Ranking Rating difference/scalar difference from control
b. Sensitivity: Measures ability of individuals to detect sensory characteristic(s)	Threshold Dilution

2. Descriptive: Measures qualitative and/or quantitative characteristics	Attribute rating
	  Category scaling
	  Ratio scaling (magnitude estimation)
	Descriptive analysis
	  Flavour profile analysis
	  Texture profile analysis
	  Quantitative descriptive analysis

Affective: Evaluate preference and/or acceptance and/or opinions of product with a large number of panelists representative of target populations	Paired-preference
	Ranking
	Rating
	  Hedonic (verbal or facial) scale
	  Food action scale

machines, not as consumers, to describe products in an accurate and repeatable manner or to discriminate among real differences in products. Affective methods try to measure the evaluative component of consumers' responses.

Food quality, or from the consumer viewpoint, food acceptance, is the most critical aspect of food as measured in institutional settings and probably in home settings as well. In studies within US military food service systems this has been shown repeatedly (Meiselman, 1979). Only when food quality is brought to an acceptable level, do other food service variables achieve high import.

The collection of food acceptance data is a key ingredient in studies of product development, quality control, food product acceptance in the market place, and food service evaluation. During the product development cycle, the food technologist will usually want periodic assessment of the item being developed. This is done with either a consumer panel or an expert trained panel.

The collection of food acceptance data is a key ingredient in studies of product development, quality control, food product acceptance in the market place, and food service evaluation. During the product development cycle, the food technologist will usually want periodic assessment of the item being developed. This is done with either a consumer panel or an expert trained panel.

The expert trained panel is selected and trained in such a way that it is supposed to bring the human being to the level of a measurement machine with high reliability of judgements independent of psychological factors such as motivation and individual experience. The techniques of selecting and training expert judges are covered in other references (ASTM, 1981) and are not within the subject matter of consumer studies. The expert is not viewed as representing the consumer. The role of the expert in product development is to determine flaws in the development process (too salty, poor texture), and possibly to attribute these flaws to specific processing steps (burnt note). This information can be used by the food technologist to improve processing. What is relevant in the context of this chapter is that the relationship between expert judgements and consumer judgements is not well researched and understood. We need to understand how to predict consumer opinion from expert judgement; until we can do that expert judgements remain difficult to interpret in terms of product acceptance.

When the product is beyond this stage of development, it can be submitted to a consumer panel. This panel must represent the ultimate consumer to be maximally effective. The people who develop food products, or the people who market food products, are not necessarily the best representatives of the consumers who will ultimately consume the product. Many food developers utilise in-house panels of 'consumers' and in addition some of them use field tests of consumers in homes and institutions. Let us briefly examine these three options: in-house laboratory testing, home testing and institutional testing.

3.1. Laboratory Acceptance Tests
In-house laboratory acceptance testing represents the most controlled environment in which to conduct acceptance tests. However, as is the case with all laboratory research as compared with field research, realism is sacrificed for control. Within the laboratory testing area, one can control a number of environmental variables (odour, light, temperature and humidity, etc), a number of stimulus variables (serving temperature, portion size, etc.), and the degree of social interaction (isolation booths, group table settings, etc.). However, the realism of the test kitchen is never that of a real eating environment, even when attempts are made to make the test area more natural and home-like. In-house testing utilises either laboratory personnel or consumers brought in for these tests. In the former case, the approximation of in-house personnel to consumers is usually not exact, and in the latter case the people who are available for such testing are not usually representative of the full range of consumers.

Several publications have presented detailed discussions of how to conduct food product tests (Larmond, 1977; Schaefer, 1979), and this detailed field will only be skimmed here to give the reader a feel for the number of decisions which have to be reached in designing and conducting laboratory food tests of acceptability (Table 14).

TABLE 14
ISSUES INVOLVED IN DESIGNING AND CONDUCTING A LABORATORY FOOD ACCEPTANCE TEST

1. Testing environment
2. Panel selection
3. Panel training or instruction
4. Food sample selection
5. Rating form selection
6. Food sample preparation
7. Food sample presentation
8. Reduction and analysis of acceptance data

The testing environment should be a convenient and attractive area. Since people have to get to the facility, it is important that it be conveniently located to facilitate people coming for tests. Any laboratory in which food will be served should be attractive. Generally, in facilities doing large-scale testing, a waiting area is needed for people to wait for their tests, and perhaps to spend a social moment after testing. The acceptance testing area should be a pleasant area in which people like to spend time. The details of construction and design of a testing area can be found in ASTM (1968), Sidel and Stone (1976) and Larmond (1977).

In general, most large research facilities utilise as large a number of their personnel as possible for a consumer panel. At the US Army Natick RD and E Center approximately 500 employees out of a total of 1100 serve on the acceptance panel. Where a large population of employees is not available, it would be dangerous to limit consumer testing to small groups of in-house personnel. This sampling can become highly biased, either toward supporting in-house development or toward being critical of in-house development based on knowledge or rumour.

One general issue which faces all acceptance testing is the utilisation of users and non-users. If one is testing a new coffee formulation, should one use coffee drinkers only? In general, I believe the answer is to involve only users, except where there is a compelling reason to the contrary. For example, with entirely new products, there is no user group established. Do we begin with users of related products or a totally random group? If we believe that a new formulation of a product will attract previous non-users, then we should utilise users and non-users, but their results should be statistically compared.

Once selected for inclusion in a panel, the personnel are usually exposed to standard laboratory procedures either in a pretraining or by attending their first test. It is important that they become thoroughly instructed in what they are supposed to be doing, including philosophy, methods, etc. They should be assured of confidentiality, and that their honest opinion as a consumer (not a professional, etc.) is what is needed. Wherever possible the new panelist might be utilised on several practice tests for experience before embarking on periodic testing.

The selection of the food samples which will be tested is a critical element in most acceptance (and sensory) tests, and one which is sometimes overlooked in methodological discussions. First, a decision is made whether a standard is to be used, in the general sense. Will some item be included against which the other samples will be compared? When testing will be repeated over several sessions or over a long period of time (a food storage test), a standard is often helpful for anchoring the judgements. As discussed below, a standard avoids the difficult problem of absolute ratings. When a standard is to be used, the standard should be selected by the technologist working with the acceptance expert. The technologist might have a tendency to select a standard to make the test items look better or worse. The standard should reflect what the consumer usually sees, not the best or worst on the market. When a standard is used, its selection is critical, because all later judgements will be related to it. If you are left out of the selection of the standard, you might as well be left out of the acceptance measures.

The selection of test samples beyond the standard should cover the full range of product. This again can be handled by the technologist and measurement person working together. The total number of samples depends on the length of the session, size of the test panel, and number of sessions possible. Presenting a large number of samples in one session is generally discouraged, especially when the samples must be tasted (rather than just observed). One must decide whether to present a large number of samples in several sessions with the resulting changes in people, or to present them in one longer session with the changes in people within the longer session.

The selection of a method for the measurement of food acceptability has been the subject of much research and discussion. The interested reader is referred to the publication by the Institute of Food Technologists (1981) for further detailed discussion and references. The methods identified by the IFT are shown in Table 15. The application of these methods to specific product development and product evaluation problems is shown in Table 16. As noted above, the methods can be divided between affective methods, usually a ranking or rating procedure, and analytical methods. The analytical methods include a number of methods to test discrimination and several descriptive methods. In addition to these methods, newer statistical methods are available which lie between strict sensory evaluation and market testing. These methods are presented in the new volume edited by Thomson (1988).

The preparation of food samples is critical in the maintenance of control within the food test situation. The food item must be prepared according to a strict recipe or method, held for serving under similar levels of control, and finally presented to the respondent under controls. Basic variables for scrutiny are: recipe formulation, standardisation of ingredients, duration and temperature of cooking, duration and temperature of holding after cooking until serving, portion size of serving, method of serving (cheese,

TABLE 15
METHODS FOR SENSORY EVALUATION AND ACCEPTANCE TESTING OF FOODS
(Adapted from the Institute of Food Technologists 1981)

Method	Number of samples per test
1. Paired-comparison (or paired-preference)	2
2. Duo-trio	3 (2 identical, 1 different)
3. Triangle	3 (2 identical, 1 different)
4. Ranking	2–7
5. Rating difference/scalar difference from control	1–18 (the larger number only if mild-flavoured or rated for texture only)
6. Threshold	5–15
7. Dilution	5–15
8. Attribute rating (category scaling: and ratio scaling or magnitude estimation)	1–18 (the larger number only if mild-flavoured or rated for texture only)
9. Flavour profile analysis	1
10. Texture profile analysis	1–5
11. Quantitative descriptive analysis	1–5
12. Hedonic (verbal or facial) scale rating	1–18 (the larger number only if mild-flavoured or rated for texture only)
13. Food action scale rating	1–18 (the larger number only if mild-flavoured or rated for texture only)

TABLE 16
APPLICATION OF SENSORY EVALUATION AND ACCEPTANCE TESTING METHODS
(Adapted from the Institute of Food Technologists, 1981)

Type of application	Appropriate test methods listed in Table 15
New product development	1, 2, 3, 4, 5, 8, 9, 10, 11, 12, 13
Product matching	1, 2, 3, 5, 8, 9, 10, 11, 12, 13
Product improvement	1, 2, 3, 5, 8, 9, 10, 11, 12, 13
Process change	1, 2, 3, 5, 8, 9, 10, 11, 12, 13
Cost reduction and/or selection of a new source of supply	1, 2, 3, 5, 8, 9, 10, 11, 12, 13
Quality control	1, 2, 3, 5, 8, 9, 10, 11
Storage stability	1, 2, 3, 4, 5, 8, 9, 10, 11, 12, 13
Product grading or rating	8
Consumer acceptance and/or opinions	1, 4, 12
Panelist selection and training	1, 2, 3, 4, 5, 6, 7, 8
Correlation of sensory with chemical and physical measurements	5, 8, 9, 10, 11

plain or on bread, etc.), and dish in which the food is served. Every aspect of the test should be intentionally, not accidentally, selected, noted and reported to make replication possible.

Finally, the reduction and analysis of acceptance data is a topic which is not within the scope of this paper, and is covered elsewhere in detail (Sidel and Stone, 1976; Larmond, 1977; O'Mahony, 1986).

3.2. Home Testing

Food testing by consumers in the home does not represent any special theoretical issue in food habits, but it does present many special practical problems. Most, if not all, of the process of data collection is removed from the supervision of the investigator, who is unable to directly assess the validity of the procedure and resulting data. This has made many investigators uncomfortable with the data generated in the home. However, it is possible to do acceptance research in the home, and this is probably necessary for realistic testing of infants (Ashbrook and Doyle, 1985) and others. It is true that much data within the food industry and other industries are gathered in the home, usually from consumer panels: groups of individuals who are carefully selected and repeatedly tested. This area has been reviewed by Sudman and Ferber (1979).

The selection and maintenance of a consumer panel is a key problem in panel home testing, keeping in mind that a panel can be composed of over 10 000 households. The cooperation rate from consumer home panels is approximately 50%, although estimates range from 90% for closely monitored government panels to under 5% for some loosely monitored contact-by-mail panels. According to Sudman and Ferber, cooperation is best in households (1) with more than two members, (2) with a younger wife, (3) with more education. Conversely, it is difficult to obtain cooperative panel members from (1) single person households and (2) older households. New panel members must be continually recruited to compensate for the ageing of a selected panel over the course of several years of testing, and to offset dropouts.

The recruitment of the consumer panel shares many of the problems of data collection for attitude data in general or food intake data. The choice is between more expensive, but more effective methods of direct person-to-person interview, or the less expensive methods of telephoning and mail surveys. For consumer home panel recruitment, direct interviewing produces far better results if the cost is manageable.

Sudman and Ferber have listed the major panels operating in the USA, Western European countries, Japan, Brazil, and Australia, and have provided information on the size of their panels, the items which they survey, survey frequency, and the method of survey employed (Table 17). The most frequent method is mail survey, making the techniques of questionnaire design discussed above applicable.

3.3. Field Food Acceptance

The institutional food service setting provides an excellent opportunity to collect food acceptance information. Many institutions, including the military, schools, hospitals, and factories, are interested in the acceptability of their food service systems and products. Of course, one method for determining food acceptance in these situations is to survey the personnel outside of the eating environment, either at the institution or at home by interview or questionnaire. The data generated would be recall of foods eaten at the institution. It is sometimes preferable to collect the food acceptance ratings from the consumers as they are eating, or just after they have completed eating. These type of data are difficult to collect in a home setting but are easy to collect in an institutional setting. Collection of direct consumer acceptance ratings also provides the researcher or manager with an opportunity to observe the food system in operation and to interact with customers. One spin-off from this effort is the feeling on the part of the consumers that the management cares about their opinions. This does place on management the burden to correct problems which do arise.

The direct interviews procedure is the simplest here, and works quite well. In advance of the interviews, the interviewer should become familiar with what is on the menu, and should look over the dining area to develop a sampling strategy. This is necessary in order to avoid the bias involved in talking only to people on the right-hand side of the room(s), or to people seated at large tables, etc. Ideally, a sampling plan will make every seat (or person) equally likely to be interviewed. The interview process can begin as soon as people have received their food and have sampled a portion of each item. Just as people are finishing their meal is a good time to approach them. In most situations it is proper to ask permission for the interview in a casual and friendly manner, 'May I ask you about your food at this meal?'. Perhaps 1 or 2% of respondents will refuse at this point. Accept their refusal, or briefly explain why the information is needed and then accept their refusal and approach someone else. If approval is granted, then either stand or sit with the person, and present the interview.

The interview procedure can be limited to only food acceptance ratings. In this case either a pure interview format can be used (e.g. 'Please rate your food as acceptable, neutral or unacceptable'), or a longer scale of answers can be shown to the respondent on a plastic wrapped card with the instruction to 'Pick the rating which best fits how you feel about your food item' (name the item). Ratings can be obtained for one item, several items, or all the items. In addition, an overall rating can be obtained for the entire meal on the same scale, and ratings of other meal characteristics can be obtained (e.g. 'How hot is the food?').

In general it is probably better to interview one person per table, and then move on to another table, perhaps returning to the original table when some of the customers change. This avoids the problem of one person's response affecting the responses of others.

This type of one-on-one food acceptance data collection is comparable to individual food intake measurement or individual food waste measurement and represents the ideal type of food acceptance data collection. However, other techniques are available which make the data collection process less labour-intensive. These techniques fall under what might be called feedback systems. They comprise a variety of forms which are either handed out to consumers or can be picked up by consumers and returned later. Avoidance of the interview represents a substantial labour saving.

Feedback forms can address food acceptance and/or other aspects of food service. Several factors should guide development of such feedback instruments. First, the instrument should be brief and clear; it should take only 1–2 min to fill out. Secondly, the format of questions should provide information which can be acted upon. For example, asking whether parts of the meal were unacceptable could produce the answer 'yes' which cannot easily be a basis for corrections. Even knowing which items were unacceptable does not tell you what was wrong with the item (too cold, too small, too tough, etc.).

Although many food service or food quality feedback forms are seen in restaurants, cafeterias, etc., very few appear to have been scientifically evaluated. Meiselman (1982) has compared the performance characteristics of five different feedback forms (Fig. 4) designed for evaluation of food quality in military dining halls. The simplest card contained separate scaled ratings of different food characteristics. These separate scales were located within boxed areas of the card. Data collected on a number of food products showed that the card with different scales for different food attributes worked the best. Most interesting was the test done with an intentionally small cake portion. The consumers rated portion size small but did not otherwise downrate the cake (Table 18). This was a measure of the validity of the cards.

More recently, Bell and Meiselman (unpublished work) have compared

TABLE 17
WORLDWIDE MAJOR PANEL OPERATORS IN THE USA AND THE WORLD
(Adapted from Sudman and Ferber, 1979)

Country/Firm	Size	Items covered	Frequency	Method
A. Australia				
1. Roy Morgan Research	—	—	—	—
B. Austria				
2. GFK	2 000	Grocery, health and personal care, textiles	Weekly	Mail
C. Belgium				
3. A.C. Nielsen	—	Consumer goods	Weekly	Mail
4. Aspemar S.A.	1 800	Grocery, health and personal care, textiles	Weekly	Mail
D. Brazil				
5. A.C. Nielsen	2 000	Grocery, health and personal care	Monthly	Dustbin
E. Canada				
6. International Surveys Limited	—	—	—	—
F. Denmark				
7. Observa	1 500	Grocery, textiles, media	Weekly	Mail
G. France				
8. Attwood	4 000	Grocery, health and personal care	—	Mail
9. Secodip	9 000	Grocery, personal care, textiles	Weekly	Mail
H. Germany				
10. Attwood	4 100	Grocery, health and personal care	—	Mail
11. GFM	5 000	Grocery, health and personal care, textiles	Monthly	Mail
12. GFK	11 000	Grocery, personal care, textiles	Weekly	Mail

CONSUMER STUDIES OF FOOD HABITS

	Sample	Product categories	Frequency	Method
I. Great Britain				
13. Attwood	5 600	Grocery, health and personal care	Weekly	Mail
14. AGB	5 800	Grocery, health and personal care	Weekly	Dustbin
J. Ireland				
15. Attwood	750	Grocery, health and personal care	—	Mail
K. Italy				
16. Attwood	1 520	Grocery, health and personal care	—	Mail
17. A.C. Nielsen	4 000	Grocery, health and personal care	Monthly	Dustbin
L. Japan				
18. Marketing Center	2 700	Pharmaceuticals	Monthly or bi-monthly	Face-to-face
		Hospitals and physicians		
M. Netherlands				
19. Dentsu	—	—	—	—
20. Attwood	2 000	Grocery, health and personal care	—	Mail
21. Intromart	3 600	Grocery, personal care, textiles	Weekly	Mail
N. Spain				
22. Dympanel	3 000	—	—	—
O. Sweden				
23. Swedish Consumer Panel AB	1 500	Grocery, health and personal care	—	Mail
P. Switzerland				
24. IHA	2 000	Consumer goods, textiles	Weekly	Mail
Q. USA				
25. Market Research Corporation of America (MRCA)	7 500	Grocery, health and personal care, textiles	Weekly	Mail
26. National Purchase Diary (NPD)	13 000	Grocery, personal care, toys, eating out, textiles	Monthly	Mail
27. Marketing Information Center	6 100	Pet food and care, beer, tobacco	Bi-weekly	Mail
28. A.C. Nielsen	1 200	Television viewing	Bi-weekly	Meter
29. Home Testing Institute	Custom	Custom	Custom	Mail and phone
30. Slade Research	Custom	Custom	Weekly	Phone

FIG. 4. Five different feedback forms tested by Meiselman (1982).

CONSUMER STUDIES OF FOOD HABITS 313

```
DID YOU SELECT  WHITE CAKE  ?     ☐ YES (continue answering)   C
                                  ☐ NO  (STOP & return card)
```

After you have tasted the WHITE CAKE , please rate it for each of the following characteristics by checking one box in each category.

Temperature	Flavor	Portion Size	Texture
Too Hot ☐	Good Flavor ☐	Too Big ☐	Bad Texture ☐
Slightly Too Hot ☐	Slightly Good Flavor ☐	Slightly Too Big ☐	Slightly Bad Texture ☐
Just Right ☐	Neutral Flavor ☐	Just Right ☐	Neutral Texture ☐
Slightly Too Cold ☐	Slightly Bad Flavor ☐	Slightly Too Small ☐	Slightly Good Texture ☐
Too Cold ☐	Bad Flavor ☐	Too Small ☐	Good Texture ☐

Considering everything, how was the WHITE CAKE ?

Good ☐   Slightly Good ☐   Neutral ☐   Slightly Bad ☐   Bad ☐

COMMENTS:

TSA Form 938¹ 30 Oct 75

```
DID YOU SELECT  WHITE CAKE  ?     ☐ YES (continue answering)   C
                                  ☐ NO  (STOP & return card)
```

After you have tasted the WHITE CAKE , please rate it for each of the following characteristics by checking one box in each category.

Temperature	Flavor	Portion Size	Texture
Much Too Hot ☐	Very Good Flavor ☐	Much Too Big ☐	Very Bad Texture ☐
Too Hot ☐	Good Flavor ☐	Too Big ☐	Bad Texture ☐
Slightly Too Hot ☐	Slightly Good Flavor ☐	Slightly Too Big ☐	Slightly Bad Texture ☐
Just Right ☐	Neutral Flavor ☐	Just Right ☐	Neutral Texture ☐
Slightly Too Cold ☐	Slightly Bad Flavor ☐	Slightly Too Small ☐	Slightly Good Texture ☐
Too Cold ☐	Bad Flavor ☐	Too Small ☐	Good Texture ☐
Much Too Cold ☐	Very Bad Flavor ☐	Much Too Small ☐	Very Good Texture ☐

Considering everything, how was the WHITE CAKE ?

Good ☐   Slightly Good ☐   Neutral ☐   Slightly Bad ☐   Bad ☐

COMMENTS:

TSA Form 938² 30 Oct 75

FIG. 4—contd.

TABLE 18
VALIDITY TEST FOR FOUR FEEDBACK CARDS
(From Meiselman, 1982)

Attributes	Card 1 Mean	Card 1 SD	Card 1 N	Card 2 Mean	Card 2 SD	Card 2 N	Card 3 Mean	Card 3 SD	Card 3 N	Card 4 Mean	Card 4 SD	Card 4 N	Card 5 Mean	Card 5 SD	Card 5 N
Normally like	1·33	0·47	40												
Like to eat here	1·33	0·47	40												
Appearance	4·04	1·21	49												
Temperature	3·93	1·30	44				4·04			4·04	1·32	56	5·37[a]	2·28	43
Texture	3·69	1·47	45				3·32			3·32	1·41	59	4·41[a]	1·82	44
Flavour	3·82	1·51	45				3·69			3·69	1·34	59	5·15[a]	1·69	48
Moisture	3·57	1·52	44												
Cook's preparation	3·93	1·44	43												
Portion size[b]	3·07	1·68	44				2·62			2·62	1·61	58	3·09[a]	2·04	47
Goes with meal	3·95	1·48	42				3·86			3·86	1·29	56	3·84	1·27	43
Overall				5·12[a]	1·94	49									

[a] Responses to this question range from 1 (very bad) to 4 (neutral) to 7 (very good).
Responses to all other questions range from 1 (very bad) to 3 (neutral) to 5 (very good).
[b] The validity test consisted of serving a very small portion of cake to determine which product attributes would be affected.
SD: standard deviation.

the use of direct interviews with the use of self-administered questionnaires for customer feedback of food quality. This measured the role of the interviewer in the process. In ratings of food quality by young male military consumers there was no difference between the two methods in one test. Before the non-interviewer approach is widely adopted, more testing is suggested.

4. FOOD SELECTION AND FOOD INTAKE

The measurement of food intake comprises probably the largest area of food habits research. This is because food intake is related to so many different issues. Nutrition, medicine, psychology, anthropology, and many other disciplines are interested in knowing what people eat. People's food intakes are used as a measure of preference, of nutrient intake, of economic status, etc. Finally, the simple question, 'What did you eat yesterday?' has turned out to be a very complex situation, without a clear method emerging as the best.

Throughout this section we will deal with both food selection and food intake measures. I have tried to maintain the distinction between them throughout this chapter because they really represent different behaviours. A food can be selected and not tasted or eaten (leaving waste). Intake brings in the added issues of food acceptability, hunger state, and portion size, etc. And although we are often interested in determining actual intake, many measures of intake are measures of selection in our terminology. Still, the methods for determining selection and intake are so closely linked we will review the two areas together.

4.1. Variables in Diet Measurement

There are several basic variables involved in determining methodology in food intake. First is the level of data collection: group, household, or individual (Burk and Pao, 1976). Group or aggregate data are most easily collected. For example, one can determine the total amount of milk served in a cafeteria, then subtract the amount of milk returned in cartons, and divide by the number of people selecting milk to determine the average intake of milk per individual. Although this type of aggregate data is of limited use in making statements about individual food behaviour (preference, nutrition, etc.), it is widely used as an index of individual behaviour. It provides no measure of variability of the average so that one cannot determine what percentage of the group lies above or below the average, or whether two groups differ statistically in their aggregate intake.

Another aggregate measure commonly used is the household measure of food intake. Here, one person, usually the food preparer, keeps track of amount prepared, amount consumed, and the number of people served, and then an average amount consumed can be determined. These data can be interesting in comparing households as individual entries, but they are not useful for comparing the individuals within households (children vs adults) etc. If one were studying the effect of household income on aggregated household food intake or studying household food intake in different ethnic areas, then a household measure of intake would be a simple means of dealing with the intake measure.

When one requires one person to measure the intake of another related person, then the question of bias also arises. Would the homemaker correctly report the food intake of children over whom she has responsibility, or might there be a tendency for the reports to be shifted in the direction of what she believes would be better nutrition?

If resources and respondent cooperation were always available, probably most studies would choose to determine individual food intakes. From individual data, one can collapse data into households, families or any other subgroups of interest. In addition, the variability, both within subgroups (children vs adults) and between subgroups (family vs family), can be compared statistically. Only with individual data can valid statements be made about the nutrition of the individual. Determining that a group consumes 10 ounces of milk per day on average is very different from determining that individual milk intakes average 10 ounces with a standard deviation of 2 ounces and with 10% of the sample falling at zero intake and 10% of the sample exceeding 50 ounces. However, as we will see below, the measurement of individual intakes presents the maximum intrusion into the individual's life, including his eating, requires the maximum in resource by the researcher, and for all this investment depends on a selection of methods of uncertain validity and reliability.

Another basic question in determining food intake is whether to deal with historical data recalled by the respondent or with weighed intake taken at the time of food consumption. Again, here, the ideal is clear: the measurement of each item in a meal both before and after eating (with no food trading permitted among people) with the difference attributed to intake. This approach is not only very expensive but in many situations is very intrusive; the individual's intake becomes a main focus, perhaps the main focus, of the meal.

Different approaches have been used in the collection of historical data and actual (weighed) intakes. Historical data can be recalled by the respondent, for example in a 24-h recall of what was eaten the day before. More generally, people can be asked to complete a food frequency list; this is not the list of preferred frequency noted above in which the respondent indicates how frequently a food is preferred, but requests the respondent to indicate how often a food is eaten during a particular time period. The food frequency checklist presented in Fig. 5 was used in comparing college students' responses to the frequency checklist and actual food intake (Mullen et al., 1984). A similar frequency scale based on the day, week, month, year was used on a very brief 28-item questionnaire self-administered by food store shoppers (Fjeld et al., 1984). Both the recall and frequency list approaches have a number of method choices which will be noted below.

Determination of intake is obtained through a dietary record in which a person is asked to record the amounts and types of foods as they are consumed. There is reason to believe however that the recording might be done historically, once per day or less often, making this an historical method. The main intake method is weighed intake in which each meal item is weighed before and after the meal for each person. If portion sizes are controlled, the pre-weighing step can be eliminated.

In many studies of food intake measurement, the data are immediately converted to nutrients through a conversion process and thereafter the results are presented as nutrient intakes. People eat foods not nutrients. People report food intakes not nutrient intakes. Lastly, the conversion process of going from foods to nutrients is itself the subject of controversy. Hence, the focus should be on what foods people eat.

Another general measurement issue is the degree of precision requested of the respondent. Asking a layperson the number of sandwiches he/she consumed is a relatively easy task. Asking them the number of slices of cheese in each sandwich might be acceptable also. But asking the number of ounces/grams of sliced beef consumed cannot be assumed to be within the capability of the layperson unless this is demonstrated. Some procedures try to use training and food models for amount estimation.

Many of the food selection and intake measures can be repeated to provide additional data. Although many studies still use one 24-h recall, the recommendation is increasingly to use multiple recalls to obtain a more representative picture of average intake (Nesheim, 1981). Axelson (1984) has demonstrated that using only one 24-h recall to establish a baseline against which a treatment effect will be compared can be misleading because of recall-to-recall variability. Similarly, longer dietary records not

I.D. _____ (1–5) ☐ 1 ☐ (6)

Food	Usual amount eaten each time	How many times a day	How many times a week	How many times a month	How many times a year	Rarely or never	For computer use only
Main dishes							
Meat and vegetable stew	_____ cup(s)						
Chicken pot pie	_____ how many?						
Meatloaf	_____ slice(s)						
Meatballs	_____ how many?						
Chili con carne	_____ cup(s)						
Corned beef hash	_____ cup(s)						
Spaghetti	_____ cup(s)						
Macaroni and cheese (packaged dry)	_____ cup(s)						
Lasagne	_____ serving(s)						
Beef macaroni casserole	_____ serving(s)						
Tuna noodle casserole	_____ serving(s)						
Tamale	_____ how many?						
Enchilada	_____ how many?						
Pizza	_____ slice(s)						
Tuna or salmon salad	_____ cup(s)						
Egg salad	_____ cup(s)						
French toast	_____ slice(s)						

Chow mein .... _____ cup(s) .... | | | | | | |
Other: _____ | | | | | | |
                                                                     | | | | | | |
Soups
Bean, split pea, lentil .... _____ cup(s) .... | | | | | | |
Cream of mushroom .... _____ cup(s) .... | | | | | | |
Meat or vegetable soup, any kind
  (include onion, chicken noodle,
  minestrone, etc) .... _____ cup(s) .... | | | | | | |
Beef bouillon or broth .... _____ (cup)s .... | | | | | | |
Chicken bouillon or broth .... _____ (cup)s .... | | | | | | |
Other _____ | | | | | | |
                                                                     | | | | | | |

Meat, Fish and Poultry
Beef
  Hamburger .... _____ patties .... | | | | | | |
  Lean cut, such as round steak (include
  sirloin tip, top round roast, etc.) .... _____ ounces* .... | | | | | | |

* Please estimate as best you can, using the following as a guide: 3 ounces = size of the palm of your hand; 5 ounces = size of your whole hand; 1 lb = 16 ounces.

FIG. 5. Food frequency list tested by Mullen et al. (1984).

only provide more representative data, they may possess higher validity in certain settings (Krantzler et al., 1982b). The frequency checklist and diet history are usually administered only once.

4.2. Methods for Determining Selection and Intake

Although any data used to measure intake can be used to measure selection, the converse is obviously not true. The more common methods of recall, record, weighed intake, and frequency list can be used to develop a list of foods selected at a meal, daily, weekly, etc. In each of these methods, one can also ask how much was consumed, either in more general terms (one portion, two portions), in semi-quantitative terms (three slices of cheese), or in quantitative terms (8 ounces milk). Hence, whether one asks for intake or selection data really depends on the reasons the data are being collected, the opportunity to instruct and train respondents, the resources available, etc.

Table 19 presents some characteristics of key food selection and food intake measures. A plus sign indicates that the particular measure is relatively strong on this characteristic, and a minus sign indicates it is relatively weak or impossible (e.g. recalls can be handled by telephone, whereas frequency lists cannot). Some points in the table will receive more general agreement from readers than others.

All of the methods can be self-administered except the diet history. In practice, the recall is usually administered by a trained professional making it an expensive procedure, and weighed intake is usually conducted by non-subjects to avoid feedback and minimise subject effort.

The interview is generally used with all of the procedures except a dietary record, which is carried by and filled out by the subject. Morgan et al. (1987) compared nine food intake methods, four exclusively based on personal interviews, one each exclusively mail, or exclusively telephone, and three more which were combinations of the three methods. Some telephone methods produced good data at less cost or effort. Similarly Krantzler et al. (1982b) found that agreement between the number of foods correctly recorded and the number of foods actually observed, a measure of validity, was virtually the same for telephone diet records and interviewer based diet records. Because of increasing labour costs, telephone methods probably represent the methods of the future.

Visual estimation of dietary intake is not commonly used but has been evaluated and holds promise for providing intake data at less cost and less subject intrusion. Hirsch et al. (1984) compared weighed and visually estimated military meals consumed by soldiers on a 34-day field training

TABLE 19
CHARACTERISTICS OF DIETARY METHODS
(Krantzler, N. J., Schutz, H. G., Grivetti, L. E., Meiselman, H. L. and Mullen, B. J., pers. comm.)

Method	Self-administered	Interview	Telephone	Reliability	Validity accuracy	Validity representative	Cost	Time	Subject burden
Recall									
24 h	+	+	+	+	+	−	+	+	+
Record									
1 day	+	−	+	+	+	−	++	++	++
3 day	+	−	++	++	+	−	+++	++	++
7 day	+	−	++	+	++	++	++	−	−
14 day	+	−	+		++	++	+	−	−
Frequency list	+	+	−	−	−	+	+	+	+
Weighed intake	+	+	−	+	+	−	−	−	−
Diet history		+	−	−	−	+	+	−	+
Pantry inventory	+	+	−	+	+	+	−	−	−

+, Indicates that the particular measure is relatively strong on this characteristic; −, indicates that it is relatively weak.

exercise. They observed strong agreement between weighed and estimated values for less variable pre-packaged military rations, less so for more variable freshly prepared foods (Tables 20, 21). Furthermore, the correlations between estimated and weighed amounts appeared to improve over time (Table 21). Rose et al. (1987) confirmed the agreement of the visual estimation method in a series of military feeding studies. These studies showed a very slight underestimation with visual estimation. Probably the small overestimate measured by Hirsch et al. and the small underestimate observed by Rose et al. could be understood by methodological differences. Further, Rose et al. showed test–retest reliability for estimators in excess of 95%. These studies strongly suggest that visual estimation has more potential in estimating intake.

Reliability and validity are usually considered the essential characteristics of any measuring instruments and have been reviewed by Burk and Pao (1976) and Krantzler et al. (1982a) for dietary methods. Reliability is probably not substantially different for any of the methods with the exceptions of the frequency list and the diet history. Validity is more difficult to establish because it involves the distinction between validity for a larger group or for an individual. Group validity would generally be lower for weighed intake. The representativeness of the data, another aspect of validity, is reduced for any method which obtains only data from a short time period (one 24-h recall, 1- or 3-day record, a few weighed intakes). By their nature, the frequency list and diet history probably provide less accurate data.

The practical aspects of dietary measurement have not been addressed as often as the traditional measurement issues. However, these methods pose serious concerns for the researcher and subject alike. As Table 19 indicates,

TABLE 20
ESTIMATED AND WEIGHED NUTRIENT INTAKES IN 34-DAY MILITARY FIELD TEST
Adapted from Hirsch et al. (1984)

	Military rations		Fresh food	
	Estimated	Weighed	Estimated	Weighed
Protein (g)	76.0	74.3	105.5	104.4
Fat (g)	86.0	83.3	119.2	116.5
Carbohydrate (g)	215.3	211.1	266.1	281.8
Calories	1939.0	1891.1	2559.9	2588.0
Total Food (g)	649.6	639.7	1961.4	2029.4

TABLE 21
CORRELATION COEFFICIENTS FOR NUTRIENT INTAKES BETWEEN WEIGHED AND ESTIMATED METHODS OVER TIME
Adapted from Hirsch et al. (1984).

	Main dishes				Fruits/vegetables				Starches			
	Day 9	16	22	32	9	16	22	32	9	16	22	32
Military rations												
Protein (g)	0·82	0·95	0·96	0·94	0·84	1·00	0·98	0·99	0·69	0·89	0·96	0·80
Fat (g)	0·90	1·00	0·95	0·96	0·84	1·00	0·80	1·00	0·65	0·86	0·96	0·93
Carbohydrate (g)	0·93	0·98	0·96	0·97	0·93	1·00	0·99	0·94	0·65	0·85	0·95	0·84
Calories	0·82	0·98	0·96	0·95	0·92	1·00	0·99	0·94	0·60	0·85	0·95	0·85
Fresh food												
Protein (g)	0·46	0·77	0·84	0·58	0·51	0·84	0·71	0·70	0·69	0·70	0·49	0·76
Fat (g)	0·33	0·79	0·83	0·73	0·46	0·44	0·68	0·66	0·55	0·87	0·73	0·77
Carbohydrate (g)	0·51	0·66	0·85	0·58	0·58	0·99	0·85	0·77	0·70	0·80	0·52	0·80
Calories	0·41	0·77	0·83	0·65	0·59	0·98	0·85	0·77	0·69	0·82	0·55	0·78

the frequency list method possesses good practical qualities, given its reduced measurement qualities. It is done at one time (in person or by mail), with little or no professional time for administration. On the other end of the continuum is weighed intake which is heavy in cost, time and even subject burden (the person must eat in a particular place in most studies, and must handle the food according to rules). The other methods fall between these two, and practical problems actually encountered will depend on whether the study is staffed with all expensive professionals, whether the sample size is small or large, how many meals will be sampled and which meals (breakfast, lunch, dinner, snacks), etc. Morgan et al. (1987) included practical comparisons within comparison of nine dietary methods. The three methods using mail data collection had lowered response rates over time. The lowest respondent burden in time was for telephoned recalls and the lowest processing burden was for a recall diary contribution combined with telephoned recalls.

4.3. Example of Food Selection/Food Intake Research

Recently, a study by Krantzler et al. (1982b) brought together several of these techniques utilising the telephone to the fullest extent possible. They compared two different recall methods (6 h and 24 h), and two different recent methods (3 day and 7 day); all this was done over the telephone. These methods were compared with data obtained in a college dining hall in which the 107 students ate. These data were used to validate the information provided by the students through the different dietary methods, providing a validity check on the dietary methods of recall and record.

The percentage agreement between estimated dietary intake and observed dietary intake is shown in Fig. 6. Note that the 7-day record received the highest validity score. The validity scores were also calculated for each food group and are shown in Table 22. Overall, sugar/sweeteners and dairy foods were correctly recalled most often, and nuts/seeds, snacks/desserts, and condiments were recalled least often. Table 23 further breaks down the data to show errors as under-reports or over-reports. An under-report was an instance in which the reported diet was less than the observed diet. Under-reporting was more frequent for recalls than for records overall.

The same authors also evaluated a food frequency questionnaire (Mullen et al., 1984) with the same student population utilising the same measure of validity (i.e. agreement with foods consumed). Regression equations were calculated for each of 31 subjects using the frequency data as the

TABLE 22
VALIDITY SCORES FOR EACH FOOD GROUP
(From Krantzler et al., 1982b)

Food group	6-h recall (%)	24-h recall (%)	3-Day recall (%)	7-Day recall (%)
Meat/fish	85·5	70·4	94·4	96·0
Dairy	82·6	83·3	89·1	91·0
Fats/oils	76·9	81·1	79·5	89·9
Grains/starches	77·0	64·7	77·2	90·7
Vegetables/legumes	69·0	64·4	80·9	89·5
Nuts/seeds	53·3	61·5	77·8	76·9
Berries/fruits	63·2	59·8	63·1	86·0
Beverages	81·5	72·4	84·4	87·1
Snacks/desserts	58·3	58·0	72·9	82·2
Condiments	68·4	52·8	49·3	67·3
Sugar/sweeteners	80·0	87·5	83·3	90·5
Sandwiches	58·8	73·1	100·0	94·1
Casseroles	70·0	78·6	72·7	100·0
Soups	100·0	33·3	100·0	71·4

TABLE 23
UNDER-REPORTING AND OVER-REPORTING ERRORS FOR EACH FOOD GROUP FOR RECALLS
AND RECORDS
(From Krantzler et al., 1982b)

	Under-reported (%)		Over-reported (%)	
	Recalls	Records	Recalls	Records
Meat/fish/poultry	22·0	4·4[a]	15·6	3·7[a]
Dairy	16·8	9·7[a]	13·6	12·0
Fats/oils	20·6	19·6	36·5	23·4
Grains/starches	30·5	13·8[a]	17·3	12·0
Vegetables/legumes	33·7	13·7[a]	23·4	13·7[a]
Nuts/seeds	42·9	22·7	14·3	22·7
Berries/fruits, etc.	38·4	21·7[a]	25·8	10·0[a]
Beverages	23·4	14·0	14·7	6·6[a]
Snacks/desserts	41·8	20·9[a]	24·5	16·9
Condiments	41·7	38·1	14·5	15·1
Sugar/sweeteners	15·4	61·5[a]	61·5	60·6
Sandwiches	32·6	2·7	7·0	2·7
Casserole type	25·0	10·7	0·0	3·6
Soups	37·5	20·5	18·8	10·3

[a] Significant differences.

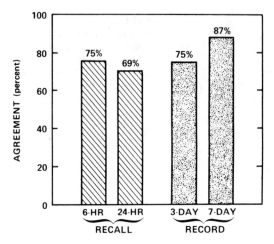

FIG. 6. Validity of dietary recalls and dietary records reported by Krantzler et al. (1982b).

independent variable and intake data as the dependent variable. Eighty-four per cent of r-values were 0·50 or better and 55% were 0·71 or better. These values suggest that the relatively inexpensive food frequency questionnaire is a possible alternative in situations in which practical considerations demand a less expensive and intrusive procedure and possible reduced accuracy of the data is acceptable.

5. FOOD WASTE

The measurement of food waste, or plate waste, has been used because of intrinsic interest in how much food is wasted, as a part of the process of measuring food intake, as a measure of other food habits such as preference, and as a measure of the efficacy of nutrition education and other interventions. When food waste is measured for its own sake the issue is generally economic; how much food is being wasted, how much protein is being wasted, etc. These often apply to an institutional food service system such as a national school lunch programme, the military, etc. Plate waste is also part of the measurement process in determining individual food intake, since plate waste must be subtracted from starting portion size to yield net amount consumed. Finally, food waste has been used as a measure of other food habits, including food attitudes and preferences, food selections, and food intake itself. As an indirect measure, waste has been used as an index of nutritional practice, sometimes to assess the impact of nutrition education programmes. People in the field have heard both jokes and serious references to 'weighing the garbage'.

The issues concerning food waste measurement are similar to those governing food intake measurement. The same ideal method exists for both, i.e. a weighed measure of each food item for each person studied. As with food intake measurement a number of different variables are involved in the choice of method. First, food waste data are collected by aggregate or by individual. Data can be aggregated over individuals and over foods. The grossest measure (in several senses!) of food waste is the total weight of all food waste from a group of people from a number of different foods. All other things being equal, this provides an index of acceptance, or intake. Rathje and Ho (1987) have measured all discarded meat and their wrappings to determine how much meat is purchased and how much separable fat is discarded. This of course relates to the clarifying nutritional image of meat and fat.

Data can be aggregated over individuals, but not over foods. This produces a total weight of food waste for each item in a meal. Alternatively one item can be weighed and compared to the total weight of all the remaining food waste (e.g. the percentage of potatoes wasted as a portion of total food wasted). Aggregating food waste over individuals is the more common procedure because of the intensive effort required to collect individual food wastes in non-clinical settings. In metabolic clinics and similar facilities, the goals of the research, and the facilities available, permit individual waste measures for individuals for each food item.

Food intake is measured either historically or by actual measure. Food waste has generally not been measured historically, i.e. asking people to recall how much of their food they left on their plate. However, in addition to direct measures of food left on plates, plate waste measures have used visual estimation of amount wasted. This has utilised both untrained consumer judges as well as trained expert judges to make the visual estimations.

5.1. Measurement of Aggregate Plate Waste

Although food waste aggregated over both individuals and foods (total garbage) has been used as a measure of acceptance and as a measure of total nutrient loss, it is of little interest to the food habits scientist or practitioner. Its main advantage is that it can be totally unobtrusive; the person need never know that his food waste is being studied either in a cafeteria or a home. We will not further discuss this technique.

Food waste aggregated over individuals but separated by foods is a technique with wide appeal and wide practice. In general, all food items (menu items) are separated into different collection containers as the plates or trays from a meal are collected. A count is kept of the plates or trays to determine the total number of diners. Decisions should be made before the overall meal, and again before each meal, as to what foods are separated and what foods are combined. For example, a hot dog would be removed from a roll, sandwich fillings would be removed from bread, and multiple sandwich fillings (ham and cheese) would be separated further. Other items would not be separated (for example, butter from bread, ingredients of pizza, frosting from cake, components of salad). Obviously, all non-food items (paper, plastic, etc.) are discarded. In addition, all non-edible components of food must be discarded (chicken bones, fruit stones, etc.). This can be done either for each dinner, or the proportion of non-edible waste can be calculated for each menu item (chicken bones on chicken) on a subsample of the trays and this value used as a basis for correcting the overall waste figure for that menu item.

Aggregate food waste has also been measured by asking homemakers to maintain 7-day records of food waste divided into 15 food categories. A parallel study (Van Garde and Woodburn, 1987) with 50 households compared these waste records with weighed food waste and found an agreement of 97%, with no significant difference for any of the 15 food categories.

5.2. Measurement of Individual Plate Waste

Individual plate waste can be measured with the same conditions used in the measurement of aggregate plate waste, except that the separated food components are weighed separately for each person rather than combined for all individuals and then weighed to give a total waste weight of a menu item. The main difference is the number of individuals whose food waste is measured. Comstock (unpublished work) has estimated that a two-person study team can handle up to 500 trays for the study of aggregated food waste at the rate of about 100 per hour, but can handle only up to 120 trays measuring individual plate waste at a rate of only up to 50 per hour and possibly far less (Table 24). When individual plate waste is measured in a home or small clinic this poses no problem. However, when individual plate waste is measured in a large institutional setting, a sampling procedure must be used to select individuals. This clearly identifies some individuals as being subjects in the study and others as not being subjects. Unless the collection and measurement of waste is done out of sight and without the

TABLE 24
CHARACTERISTICS OF PLATE WASTE METHODS
(From Comstock, unpublished work)

	Individual weights	Aggregate weights	Visual estimation
Approximate number of trays/h for two people	30	100	120
Approximate set-up and clean-up time (min)	35	45	15
Approximate maximum number of trays/meals	120	500	300

knowledge of the diners, it poses serious questions of intrusion into food habits. Interestingly, since individual plate waste is used as the standard against which other plate waste methods have been compared, there are no validity studies of this procedure, even though validity might be compromised by the intrusion and sampling problems of the procedure.

5.3. Visual Estimation of Food Waste

Researchers have repeatedly returned to visual estimation of food waste as an appealing alternative to the mess and labour-intensive procedures of separating and weighing garbage. The estimation schemes have been based on training an estimator to rate what portion of the initial portion was left as food waste. Most scales have used as end points of their visual estimation scales 'all eaten' or 'all wasted'. In between they have used standard fractions of $\frac{1}{4}$, $\frac{1}{3}$, $\frac{1}{2}$, $\frac{2}{3}$ and $\frac{3}{4}$ as well as scale values such as 'one bite eaten' or 'some eaten'. Two of the most widely used scales are those of Comstock *et al.* (1981) and LaChance (1976) (Table 25).

TABLE 25

Comstock et al. scale		LaChance scale
5	all remained	5
4	just tasted	
3	$\frac{3}{4}$ remained	4
2	$\frac{1}{2}$ remained	3
1	$\frac{1}{4}$ remained	2
0	none remained	1

Visual estimation of food waste can be done on up to 300 people per meal (Comstock, unpublished work) at a rate of 120 per hour (Table 24). This compares favourably with the food waste weighing procedure aggregating over individuals but not foods, and is substantially greater in potential numbers than individual weighings. Visual estimation can produce information on individuals if they (or their tray) are identified by name or number. The visual estimation procedure is also easily suited to a number of sampling strategies to obtain representative samples of diners measuring fewer people with less effort.

The reliability of visual estimations of plate waste between different observers has been reported at above 90% (Kirks and Wolff, 1985). This is important because some studies have used only one observer. Validity has been assessed in several studies by calculating the correlation between visual estimations of waste and actual weight of waste (Kirks and Wolff, 1985; Thompson et al., 1987). Although correlations over 0·8 and often over 0·9 were obtained, one must question what a correlation means in this context. When one is interested in food waste the correlation will only tell you if the differences observed by the two methods are in the same order. This is generally of less interest than the level of waste. Graves and Shannon (1983) report the weights visually estimated and weighed, again showing good intermethod agreement with some exceptions. When Kirks and Wolff (1985) tried to use visual estimation of waste to assess nutritional change, they found that visual estimation and weighing lead to different estimates of amount wasted and hence to different conclusions about the nutrition programme studied.

TABLE 26
MEAN PERCENTAGE WEIGHED WASTE FOR EACH VISUAL ESTIMATION CATEGORY[a]
(Adapted from Comstock et al., 1981)

No food left	$\frac{1}{4}$ left	$\frac{1}{2}$ left	$\frac{3}{4}$ left	Just tasted	Everything left
2	32	55	73	90	96

[a] Based on 6 974 total observations.

Comstock and Symington (1982) have provided further support for the visual estimation of plate waste within the school lunch setting. They determined that over half of the students produced plate waste in one of the two extreme categories, which correspond to the visual estimation categories of 'all or almost all wasted' and 'none or almost none wasted'.

These estimates are easily made. These results need replication in other food service settings before the technique of visual estimation is more generally adopted.

6. CONCLUSIONS

This chapter on consumer studies has attempted to present a model of human food habits based on the sequence of events leading up to food ingestion and food waste. This sequence of food attitude, food selection, food acceptance, food intake and food waste was reviewed, focusing on methodological issues in each area and across areas. In some cases methods were proposed for application in a variety of practical settings.

Although the chapter has focused on the description and discussion of each food habits area and its methods, the material suggests that multiple food habits be studied in most settings. It is only by studying different food habits that we will come to understand the inter-relationships among them, and thereby understand what brings the consumer to eat what he does.

REFERENCES

ASTM (1968). *Manual on Sensory Testing Methods*, American Society for Testing and Materials, Philadelphia.
ASTM (1981). *Guidelines for the Selection and Training of Sensory Panel Members*, American Society for Testing and Materials, Philadelphia.
Axelson, M. L. and Peufield, M. P. (1983). Food—and nutrition—related attitudes of elderly persons living alone. *J. Nutr. Educ.*, **15**, 23–27.
Axelson, J. M. (1984). Repeated measurements in evaluation. *J. Nutr. Educ.*, **16**, 12–14.
Bass, B. M., Cascio, W. F. and O'Connor, E. J. (1974). Magnitude estimations of expressions of frequency and amount. *J. Appl. Psychol.*, **3**, 313–20.
Benson, P. H. (1958). Relation of food frequency of serving. Final report of Project No. 7-84-15-007, Quartermaster Food and Container Institute for the Armed Forces, Chicago.
Burk, M. and Pao, E. (1976). Methodology for large-scale surveys of household and individual diets. USDA Home Economics Research Report, No. 40, Washington, DC.
Comstock, E. M. A guide to the measurement of plate waste (unpublished work).
Comstock, E. M. and Symington, L. E. (1982). Distributions of serving sizes and plate waste in school lunches. *J. Am Dietet. Assoc.*, **81**, 413–22.

Comstock, E. M., Symington, L. E., Chmielinski, H. E. and McGuire, J. S. (1979). Plate waste in school feeding programs: Individual and aggregate measures. USDA Contract No. FNS58-3198-9-30, NARADCOM Technical Report No. 81/011. Natick, Mass.

Comstock, E. M., St. Pierre, R. G. and Machiernan, Y. D. (1981). Measuring individual plate waste in school lunches. *J. Am. Dietet. Assoc.*, **79**, 290–6.

Einstein, M. A. and Hornstein, I. (1970). Food preference of college students and nutritional implications. *J. Food Sci.*, **35**, 429–37.

Fauslow, A. M., Pease, D., Gilmore, S. C. and Brun, J. K. (1982). An inventory for assessing food behaviors of elementary school children. *J. Nutr. Educ.*, **14**, 96–8.

Fjeld, C. R., Storer, J., Warholic, J., Sommer, R. and Becker, F. D. (1984). Food intake frequencies of food co-op shoppers. *J. Nutr. Educ.*, **16**, 142–4.

Foley, C., Hertzler, A. A. and Anderson, H. L. (1979). Attitudes and food habits—A review. *J. Am. Dietet. Assoc.*, **75**, 13–18.

Graves, K. and Shannon, B. (1983). Using visual plate waste measurement to assess school lunch food behavior. *J. Am. Dietet. Assoc.*, **82**, 163–5.

Grivetti, L. E. and Pangborn, R. M. (1973). Food habit research: A review of approaches and methods. *J. Nutr. Educ.*, **5**, 204–8.

Hartmuller, V. W. (1971). Development of a separation rating of menu items for the Home Economics Food Service. Master's thesis, Purdue University, USA.

Hirsch, E., Meiselman, H. L., Popper, R., Smits, G., Jezior, B., Lichton, I., Wenkam, N., Burt, J., Fox, M., McNutt, S., Thiele, M. N. and Dirige, O. (1984). The effects of prolonged feeding meal, ready-to-eat (MRE) operational rations. US Army Natick Research and Development Center Technical Report TR-85/035, Natick, Mass.

Institute of Food Technologists (1981). Sensory evaluation guide for testing food and beverage products. *Food Technology*, November, pp. 50–9.

Kamen, J. M., Peryam, D. B., Peryam, D. R. and Kroll, B. J. (1967). Analysis of US Army Food Preference Survey (1963). US Army Natick Laboratories Technical Report 67-15-PR, Natick, Mass.

Khan, M. A. (1981). Evaluation of food selection patterns and preferences. *CRC Critical Reviews in Food Science and Nutrition*, CRC Press, Boca Raton, Florida, pp. 129–53.

Kirks, B. A. and Wolff, H. K. (1985). A comparison of methods for plate waste determinations. *J. Am. Dietet. Assoc.*, **85**, 328–31.

Knickrehm, M. E., Cotner, C. and Kendrick, J. G. (1967). A study of the frequency of acceptance of menu items by residence hall students at the University of Nebraska. Department of Food and Nutrition, Report No. 1, University of Nebraska.

Krantzler, N. J., Mullen, B. J., Comstock, E. M., Holden, C. A., Schutz, H. G., Grivetti, L. E. and Meiselman, H. L. (1982a). Methods of food intake assessment—An annotated bibliography. *J. Nutr. Educ.*, **14**, 108–19.

Krantzler, N. J., Mullen, B. J., Schutz, H. G., Grivetti, L. E., Holden, C. A. and Meiselman, H. L. (1982b). Validity of telephoned diet recalls and records for assessment of individual food intake. *J. Clin. Nutr.*, **36**, 589–90.

LaChance, P. S. (1976). Simple research techniques for school foodservice. II. Measuring plate waste. *School Foodser. J.*, **30**, 66.

Larmond, E. (1977). *Laboratory Methods for Sensory Evaluation of Foods*, Canada Depart. of Agriculture, Ottawa.

Leverton, R. M. (1944). Freshman food likes. *J. Home Econ.*, **36**, 589–90.
Meiselman, H. L. (1979). Determining consumer preference in institutional food service. In: *Food Service Systems*, G. Livingston and C. Chang (Eds), Academic Press, London, pp. 127–54.
Meiselman, H. L. (1982). Design and evaluation of cards for customer feedback of food quality. *J. Food Serv. Systems*, **2**, 7–21.
Meiselman, H. L. and Bell, B. L. (Unpublished work).
Meiselman, H. L. and Waterman, D. (1978). Food preferences of enlisted personnel in the Armed Forces. *J. Am. Dietet. Assoc.*, **73**, 621–9.
Meiselman, H. L., Van Horne, W., Hasenzahl, B. and Wehrly, T. (1972). The 1971 Fort Lewis Food Preference Survey. US Army Natick Laboratories Technical Report 72–43–PR, 123 pp, Natick, Mass.
Meiselman, H. L., Waterman, D. and Symington, L. E. (1974). Armed Forces Food Preferences. US Army Natick Laboratories Technical Report 75–63–FSL, Natick, Mass.
Meiselman, H. L., Hirsch, E. S. and Popper, R. D. (1988). Sensory, hedonic, and situational factors in food acceptance and consumption. In: *Food Acceptability*, D. M. H. Thomson (Ed.), Elsevier Applied Science Publishers (in press).
Morgan, K. J., Johnson, S. R., Rizek, R. L. and Reese, R. (1987). Collection of food intake data: An evaluation of methods. *J. Am. Diet. Assoc.*, **87**, 888–96.
Moser, C. A. and Kalton, G. (1972). *Survey Methods in Social Investigation*, 2nd edn, Basic Books, New York.
Moskowitz, H. R., Nichols, T. L., Meiselman, H. L. and Sidel, J. L. (1972). Food preferences of military men, 1967. US Army Natick Laboratories Technical Report 72–70–PR, Natick, Mass.
Mullen, B. J., Krantzler, N. J., Grivetti, L. E., Schutz, H. G. and Meiselman, H. L. (1984). Validity of a food frequency questionnaire for assessment of individual food intake. *Am. J. Clin. Nutr.*, **39**, 136–43.
Nesheim, R. (1981). *Assessing Food Consumption Patterns*, National Academy Press, Washington, DC.
O'Mahony, M. (1986). *Sensory Evaluation of Food*. Marcel Dekker, New York.
Peryam, D. R. and Pilgrim, F. J. (1957). Hedonic scale of measuring food preferences. *Food Technol.*, **11** (suppl) 9–14.
Peryam, D. R., Polemis, B. W., Kamin, J. M., Eindhoven, J. and Pilgrim, F. J. (1960). Food preferences of men in the armed forces. Quartermaster Food and Container Institute for the Armed Forces, Chicago.
Piggott, J. R. (1979). Food preferences of some United Kingdom residents. *J. Human Nutr.*, **33**, 197–205.
Rathje, W. L. and Ho, E. E. (1987). Meat fat madness: Conflicting patterns of meat fat consumption and their public health implications. *J. Am. Dietet. Assoc.*, **87**, 1357–62.
Rose, M. S., Buchbinder, J. C., Dugan, T. B., Szeto, E. G. Allegretto, J. D., Rose, R. W., Carlson, D. E., Samonds, K. W. and Schnakenberg, D. D. (1987). Determination of nutrient intakes by a modified visual estimation method and computerized nutritional analysis for dietary assessments. US Army Research Institute of Environmental Medicine Technical Report T6-88.
Rozin, P. and Fallon, A. E. (1981). The acquisition of likes and dislikes for foods. In: *Criteria of Food Acceptance*, J. Solms and R. L. Hall (Eds), Forster Verlag, Zurich, pp. 35–48.

Rozin, P., Millman, L. and Nemeroff, C. (1986). Operation of the laws of sympathetic magic in disgust and other domains. *J. Person. Soc. Psychol.*, **50**, 703–12.
Schaefer, E. E. (Ed.) (1979). *ASTM Manual on Consumer Sensory Evaluation*. American Society for Testing and Materials, Philadelphia.
Schuck, C. (1961). Food preferences of South Dakota college students. *J. Am. Dietet. Assoc.*, **39**, 595–7.
Schutz, H. G. (1965). A food action rating scale for measuring food acceptance. *J. Food Sci.*, **20**, 365–74.
Sidel, J. L. and Stone, H. (1976). Experimental design and analysis of sensory tests. *Food Technol.*, **30**, 32–8.
Sudman, S. and Ferber, R. (1979). *Consumer Panels*, American Marketing Association, Chicago.
Thompson, C. H., Head, M. D. and Rodman, S. M. (1987). Factors influencing accuracy in estimating plate waste. *J. Am. Dietet. Assoc.*, **87**, 1219–1220.
Thompson, D. M. H. (1988). *Food Acceptability*. Elsevier Applied Science Publishers (in press).
Van Garde, S. J. and Woodburn, M. J. (1987). Food discard practices of householders. *J. Am. Dieter. Assoc.*, **87**, 322–29.
Van Riter, I. G. (1956). The acceptance of twenty-six vegetables by college students. *J. Home Econ.*, **48**, 771–3.
Wilson, C. S. (1973). Food Habits: A selected annotated bibliography. *J. Nutr. Educ.*, **5** (Suppl. 1, No. 1), 38–72.
Wyant, K. W. and Meiselman, H. L. (1979). The USAF food habits study: Part I, Method and Overview. US Army Natick Research and Development Command Technical Report TR-79/04, Natick, Mass.

Chapter 9

STATISTICAL ANALYSIS OF SENSORY DATA

GORDON L. SMITH

*Ministry of Agriculture, Fisheries and Food,
Torry Research Station, Aberdeen, UK*

1. INTRODUCTION

The purpose of this chapter is to explain how to perform statistical analysis of the data from sensory analysis experiments. This will consist of three stages:

1. The *presentation* and *summary* of data. Much insight can be gained from inspecting the data before performing any statistical tests, and the inspection can be assisted by sensible tabulation, graphing or summarising. In many cases this stage will help the choice of statistical model to use in the analysis, and will enable peculiarities in the data to be noticed which might affect the performance of a statistical test.
2. The *analysis* by a *suitable statistical procedure*. There is often a choice of statistical methods which appear to be suitable for a particular set of data, but they usually require certain conditions of the data, e.g. the scale of assessment should be 'equal interval', or the relationship between two types of measurement should be linear, and failure of the data to meet the appropriate criteria would render a test invalid.
3. The *inference*, or *conclusions* to be drawn from the results. The usual aim is to be able to infer something about a large set (e.g. population of consumers or a batch of fish fingers) using the analysis of results from only a sample from that set. It may be a simple conclusion such as 'one type of biscuit is preferred to another' or a numerical assertion about the shelf-life of a product. Inferring something about a population from a sample has a degree of uncertainty or

'error' attached to it, and this depends on assumptions about the distributions of scores or properties throughout the population.

There is another stage whose correct place comes before any data have been collected, and that is *experimental design*. In any investigation, the design (e.g. number of assessors, number of samples presented, questions to be asked of the assessors, codes for responses) should be planned beforehand on statistical and commonsense principles. Failure to do so may lead to insufficient data being collected, a waste of time and resources when fewer samples or assessors could have been used with little loss of efficiency, or biased or irrelevant responses from assessors. Although this stage is an essential initial one, many of the aspects depend on statistical properties introduced in the part of this chapter dealing with the analysis of data. The matter of experimental design is therefore left to a later section.

2. PRESENTATION OF DATA

It is unwise to submit data slavishly to a statistical analysis on a calculator or computer without first examining the data. Properly tabulated or graphed data make it easier for the experimenter to understand what is going on, and also aid his communication with the readers of his report (Ehrenberg, 1975).

2.1. Summarising One-Dimensional Data

It is common to summarise a set in terms of two *summary statistics*, one expressing the centre of the set and the other describing the spread about that centre. The most often used are the mean and standard deviation. The mean, $\bar{x}$ ('x bar'), and standard deviation, s, of n observations are defined by

$$\bar{x} = \frac{\sum x_i}{n}$$

and

$$s = \sqrt{\frac{\sum (x_i - \bar{x})^2}{n-1}}$$

The expression inside the square-root sign is the *variance*. Note the divisor, $n-1$, which should be used in all cases where data from a 'sample' are used to estimate parameters of a 'population'.

The expression $\sum(x_i - \bar{x})^2$ is the sum of squares of deviations about the mean and may be alternatively calculated by

$$\sum x_i^2 - \frac{(\sum x_i)^2}{n}$$

If a routine for sum of squares is introduced in a computer program, least rounding error results from using the following recursive relation. If x_i is the mean of the first i observations and SS_i the sum of squared deviations of the first i observations about $\bar{x}_i$, then

$$\bar{x}_1 = x_1; \quad SS_1 = 0;$$
$$SS_i = SS_{i-1} + (x_i - \bar{x}_{i-1})^2(1 - 1/i) \quad \text{for } 2 \leq i \leq n$$

An algorithm in BASIC is shown in Fig. 1.

```
10  REM  ***ALGORITHM FOR SUM OF SQUARES ABOUT THE MEAN***
20  REM  ***THE ARRAY X SHOULD BE DEFINED IN A DIMENSION STATEMENT***
100 SS = 0:SUM = X(1):XBAR = SUM
110    FOR I = 2 TO N
120    SS = SS + (X(I) - XBAR) ^ 2 * (1 - 1 / I)
130    SUM = SUM + X(I)
140    XBAR = SUM / I
150    NEXT I
```

FIG. 1. Algorithm for sum of squares.

The *coefficient of variation* (CV) (also known as the relative standard deviation) is the standard deviation expressed as a percentage of the mean:

$$CV = \frac{s}{\bar{x}} \times 100\%$$

The *range* is defined as the difference between the *extremes*, or highest and lowest values. Often, however, the extremes are loosely referred to in published literature as the range.

Other summary statistics are of use, particularly if the data do not appear to be normally distributed.

The *median* is the middle value when the data have been arranged in order from highest to lowest or vice versa. If data have been grouped, i.e. observed on a continuous scale but only recorded in discrete intervals, then the population median may be estimated more precisely by interpolating between the limits of the group within which the middle value falls. It may be considered that this could apply to hedonic ratings.

Other measures which describe the spread of data are the upper and lower *quartiles*. In the same way that the median divides the data set into two equal parts, so the quartiles continue the process and along with the median divide the data into four equal numbers of observations in each part.

Tukey (1977) gives a simple method of graphically presenting summarised data in a 'box and whisker' plot by means of the median, 'hinges' (similarly to the two quartiles) and the extremes.

Any *percentile* of the population distribution (i.e. value below which a given percentage of the distribution lies) can be similarly estimated from sample data. If the percentile corresponding to $P\%$ is required, it is estimated by the observation whose index is $P(n + 1)/100$.

The mean is sensitive to a value which departs quite considerably from the rest of the data. The median is rather crude in some cases as an estimate

TABLE 1
RAW DATA: HEDONIC RATING SCORES OF A BISCUIT GIVEN BY A PANEL OF ASSESSORS

Assessor	Score	Assessor	Score	Assessor	Score	Assessor	Score
1	7	14	6	27	6	39	4
2	6	15	2	28	6	40	7
3	6	16	5	29	9	41	6
4	5	17	7	30	7	42	8
5	7	18	6	31	8	43	5
6	5	19	7	32	5	44	5
7	6	20	6	33	7	45	7
8	9	21	6	34	7	46	8
9	6	22	2	35	6	47	6
10	7	23	5	36	5	48	6
11	7	24	8	37	6	49	7
12	8	25	7	38	6	50	6
13	5	26	5				

Mean = 6·18.
Median = 6·21.
Standard deviation = 1·395.
Standard error of the mean = 0·197.
95% confidence interval for the mean = 5·98 to 6·38.

Lower quartile = $4·5 + \dfrac{13 - 3 - 0·5}{10} = 5·45.$

Upper quartile = $6·5 + \dfrac{38 - 30 - 0·5}{13} = 7·08.$

95% confidence interval for the median = 5·82 to 6·62.
10% trimmed mean = 6·225.

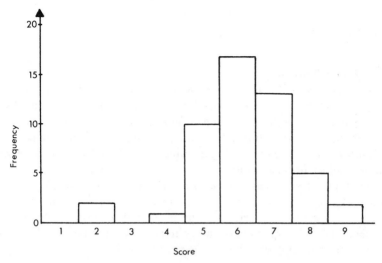

FIG. 2. Histogram of data from Table 1.

of the centre of a distribution and a class of *robust estimators* which are not susceptible to the distorting effects of 'outliers' have been studied in recent years (Andrews *et al.*, 1972). The most commonly used of these is the trimmed mean, the arithmetic mean of those observations that remain after a specified number of the highest and lowest are discarded, thus reducing the influence of extreme observations.

Summary statistics are shown in Table 1 for an example of hedonic data.

TABLE 2
FREQUENCY TABLE OF DATA FROM TABLE 1

Score, x_i	Frequency, f_i	Product, $f_i x_i$	Cumulative frequency
1	0	0	0
2	2	4	2
3	0	0	2
4	1	4	3
5	10	50	13
6	17	102	30
7	13	91	43
8	5	40	48
9	2	18	50
Sums	50	309	

The data can be presented as frequencies as in Table 2, or graphically as a histogram as in Fig. 2.

2.2 Tests for Normality

As the most suitable choice of summary statistics depends on the underlying distribution—especially if the data are normally distributed—it is useful to have a method of testing for normality. With small samples (less than about 10 observations) it is not generally possible to make any conclusions about the underlying distribution, but moderate-to-large sets can give sufficient information.

The simplest test is to use *normal probability graph paper*, where the scale on one axis corresponds with points of the normal distribution. For each observation in the data set, its value is plotted against its cumulative frequency (i.e. the proportion of values in the set less than or equal to it). If the data are normally distributed, the points will lie on or near a straight line.

Two summary statistics which allow the shape of the underlying distribution to be compared with that of the normal distribution are the *coefficients of skewness* and *kurtosis*. The coefficient of skewness, g_1, uses the averages of cubed derivations from the mean, i.e.

$$m_3 = \frac{\sum(x_i - \bar{x})^3}{n}$$

and the variance (with a divisor of n)

$$m_2 = \frac{\sum(x_i - \bar{x})^2}{n}$$

and is defined by

$$g_1 = \frac{m_3}{m_2^{3/2}}$$

When the distribution is symmetrical, the sums of cubed positive deviations and of cubed negative deviations will tend to cancel each other out and g_1 will be close to 0, but if it is skewed then one will heavily outweigh the other. If $z = g_1\sqrt{(n/6)}$ is greater in magnitude than 1·96, then it is concluded that the distribution is skewed.

If it is concluded that the distribution is skewed, it cannot be normal, but if it is symmetrical, then it may or may not be normal. The coefficient of kurtosis, g_2, is defined by

$$g_2 = \frac{m_4}{m_2^2} - 3$$

where

$$m_4 = \frac{\sum(x_i - \bar{x})^4}{n}$$

and m_2 is as defined above.

If the distribution is normal, g_2 will be close to 0; if $z = g_2\sqrt{(n/24)}$ is positive and greater than 1·96 it may be concluded that the distribution is more concentrated towards the centre than would be expected for the normal distribution; if z is negative and greater in magnitude than 1·96, then it may be concluded that the data come from a distribution with a heavier concentration at the extremes.

For the data of Table 1, the coefficient of skewness is $-0·69$, and $z = -1·96$, so there is a suspicion of skewness in the distribution. The effect may have been mainly caused by two extreme values of 2, and the interpretation of a high value for skewness has to be carefully considered in relation to the sample.

3. INTERVAL ESTIMATION

The main purpose of obtaining a mean freshness score of, for example, five cod sampled from a consignment is to estimate the mean score for the whole consignment or 'population'. The sample standard deviation which indicates the amount of variation of scores in the population is used to give an interval within which we are fairly confident the population mean will lie, usually the 95% *confidence interval*.

If the mean of n observations from a normally distributed population is $\bar{x}$ and the standard deviation is s, the 95% confidence interval for the population mean is from

$$\bar{x} - t^*_{(n-1)}\frac{s}{\sqrt{n}} \quad \text{to} \quad \bar{x} + t^*_{(n-1)}\frac{s}{\sqrt{n}}$$

where $t^*_{(n-1)}$ is the value of the t-distribution with $n-1$ degrees of freedom corresponding to 0·975 cumulative probability (see Appendix 1). This cumulative probability value is chosen since the probability that t lies between $-t^*$ and $+t^*$ is therefore 0·95.

The extremes of the interval are called the 95% *confidence limits for the mean*. The correct value of t^* to use for a given number of degrees of freedom can be found from statistical tables, but for samples of more than about 10 observations it is convenient to consider t^* as being approximately 2.

The interval is called the 95% confidence interval because if an interval was calculated as shown for each of a very large number of samples of size n from the population, 95% of these intervals would contain the true population mean.

Intervals corresponding to other degrees of confidence (e.g. 90%, 99%) could be chosen with t^* corresponding to a suitable cumulative probability (0·95, 0·995 respectively) but unless there are compelling reasons to do otherwise, it is advisable to follow the usual convention of using 95%.

The expression $s/\sqrt{n}$ in the confidence limits is the *standard error of the mean*, and represents the standard deviation of the distribution of means of a large number of samples, each of n observations, from the population.

Corresponding expressions exist for the median. If the population is normally distributed, the *standard error of the median* is $1 \cdot 253 \, (s/\sqrt{n})$ where s is the standard deviation of scores in the sample and the 95% *confidence interval for the median*, x_{med}, is from

$$x_{med} - 1 \cdot 253 t^*_{(n-1)} \frac{s}{\sqrt{n}} \quad \text{to} \quad x_{med} + 1 \cdot 253 t^*_{(n-1)} \frac{s}{\sqrt{n}}$$

If, however, the distribution is clearly not normal, and in particular is markedly skewed, it is better to base the 95% confidence interval on order statistics. This 95% confidence interval for the median lies between the values of the observations numbered $((n+1)/2) \pm 0 \cdot 98 \sqrt{n}$ (to the nearest whole number) when the data are ordered.

Another statistic sometimes encountered is the *tolerance interval*. Used mostly in quality control, it is constructed to contain a fixed portion of a population distribution with a specified confidence.

4. COMPARING RESULTS FROM TWO SETS

In comparing results from two sets, we are most often interested in answering questions like 'Do the samples come from different distributions?' or more specifically 'Do the samples come from populations whose distributions have different means?' This question is often put in the form 'Are the (sample) means significantly different?'

The rationale behind these, and many other statistical tests, is to test the strength of evidence in the *sample* data *against* a *hypothesis* that the *population* means were the same, the so-called 'null hypothesis'. If the difference observed (or a more extreme one) is improbable under that hypothesis, then we would decide to reject the null hypothesis, and accept an alternative hypothesis that the population means were different. The probability that we reject the null hypothesis when it is actually true is the significance level of the test.

It is usually the case that a two-sided test is performed; that is the experimenter concludes that the two means are different without having a particular interest in the direction of the difference.

If the aim of the experiment is to investigate whether the mean score for a particular product is greater than the mean for the other (e.g. if product B, a new formulation, is more acceptable than product A, the standard formulation) the experimenter will be interested only in the outcome of the test in one particular direction. In such a situation, a *one-sided test* is applied, and it is either concluded that there is no difference or that the mean for B is higher than the mean for A.

The decision whether to use a one-sided or two-sided test must be made before the experiment is carried out. In most cases, when there is no prior interest in a particular direction of the difference, a two-sided test is appropriate.

The method of analysis for comparing two sets of sensory data also depends on the type of sensory scale used (e.g. interval or ordinal), the underlying distribution (whether it can be considered to be approximately normal) and whether the two samples contain paired or independent observations.

4.1. Non-Parametric Comparison

Where the population of possible ratings for a product cannot be assumed to be normal, without an alternative specific distribution being postulated, a class of tests known commonly as non-parametric, or more correctly distribution-free, are widely used. Sprent (1981) gives an excellent introduction, more comprehensive texts are those by Siegel (1956), Conover (1980) and Daniel (1978), and a bibliography of such methods is given by Singer (1979).

4.2. Two Independent Samples, the Kolmogorov–Smirnov Two-Sample Test

This is a two-sided test of whether two *independent* samples come from the same population, or rather from populations having the same distribution.

TABLE 3
NON-PARAMETRIC TEST FOR TWO INDEPENDENT SAMPLES — KOLMOGOROV–SMIRNOV TWO SAMPLE TEST

Score	No. of assessors		Cumulative frequencies		Cumulative proportions		Difference[a]
	Panel 1	Panel 2	Panel 1	Panel 2	Panel 1	Panel 2	
1	0	0	0	0	0	0	0
2	1	2	1	2	0·02	0·05	0·03
3	5	6	6	8	0·12	0·20	0·08
4	10	20	16	28	0·32	0·70	0·38
5	20	10	36	38	0·72	0·95	0·23
6	12	2	48	40	0·96	1·00	0·04
7	2	0	50	40	1·00	1·00	0
Total	50	40					

[a] Regardless of direction.
D = maximum difference = 0·38
Critical levels of D:

at 5% significance level $1·36 \sqrt{\left(\frac{50+40}{50 \times 40}\right)} = 0·29$;

at 1% significance level $1·63 \sqrt{\left(\frac{50+40}{50 \times 40}\right)} = 0·35$;

at 0·1% significance level $1·95 \sqrt{\left(\frac{50+40}{50 \times 40}\right)} = 0·41$.

Therefore the distributions are significantly different at the 1% level.

A cumulative frequency distribution is obtained for each sample and the test statistic, D, is the maximum difference between corresponding cumulative frequencies. The method is shown in Table 3. If the sample sizes (panel sizes) in both cases, n_1 and n_2, are at least 40, if D is greater than $1·36\sqrt{[(n_1+n_2)/n_1 n_2]}$ the difference between the distributions is significant at the 5% level. (For 1% and 0·1% levels, replace 1·36 by 1·63 and 1·95 respectively.) For smaller samples, refer to tabulated values as in one of the texts mentioned above.

4.3. Two Paired Samples, the Wilcoxon Matched-Pairs Signed Ranks Test
The Wilcoxon test is a non-parametric test that two sets of paired observations come from different distributions (i.e. with different medians). The method is illustrated by the data in Table 4.

TABLE 4
HEDONIC RATINGS BY THE SAME PANEL ON TWO FRUIT JUICES

Assessor	Juice A	Juice B	Difference $A - B$	Rank	Positive ranks	Negative ranks
1	7	2	+5	11	11	
2	6	3	+3	8	8	
3	4	5	−1	2·5		2·5
4	6	4	+2	5·5	5·5	
5	5	6	−1	2·5		2·5
6	6	4	+2	5·5	5·5	
7	5	2	+3	8	8	
8	5	5	0			
9	7	3	+4	10	10	
10	5	6	−1	2·5		2·5
11	6	5	+1	2·5	2·5	
12	8	5	+3	8	8	
Median ratings	5·8	4·5				
Sums					58·5	7·5
T						7·5

For 11 non-zero differences, T should be 8 or less for a significance level of 5%.

Ranks are assigned to the differences (regardless of their sign) ignoring any cases where the difference was zero. Where there are a number of differences of the same value they each take the average of their ranks.

The sign of each difference is then attached to each rank, and the sums of positive and negative ranks taken separately. The test statistic, T, is the lower of these sums.

Where there is no difference between ratings for the two products, the positive and negative sums will be approximately equal, and T will be distributed with mean and standard deviation as follows:

$$\text{mean} = \frac{n(n+1)}{4}$$

$$\text{standard deviation} = \sqrt{\left[\frac{n(n+1)(2n+1)}{24}\right]}$$

where n is the number of non-zero differences (11 in Table 5).

For large panel size (greater than 25), it is appropriate to use a normal approximation by referring

$$\frac{T - \text{mean}}{\text{standard deviation}}$$

to tables of the standard normal deviate (see Appendix 1).

TABLE 5
STUDENT'S t-TEST
(Two independent samples, assuming normal distributions. Data are from two sets of cod, 2 and 14 days in ice, scored by an assessor using the odour score sheet of Shewan et al. (1953)).

	2-day	14-day
	8·5	4·5
	8·0	5·5
	9·0	5·0
	6·5	4·0
	7·5	6·0
	8·0	5·0
	8·0	5·0
	8·5	4·0
Mean	8·000	4·875
Sum of squares	4·000	3·375
Variance	0·571 4	0·482 1
Standard deviation	0·755 9	0·694 4

Variance ratio = 0·571 4/0·482 1 = 1·19.
Degrees of freedom = 7 and 7.
Variation ratio is therefore not significant at 5% level.
Therefore pooled variance may be used.
Pooled variance = 0·526 8

$$t = \frac{8 \cdot 000 - 4 \cdot 875}{\sqrt{[0 \cdot 526\ 8(\frac{1}{8} + \frac{1}{8})]}} = 8 \cdot 61 \text{ (14 degrees of freedom)}$$

For smaller panels, T should be compared with the critical value in tables for this test (e.g. Siegel (1956), Table G). For 11 non-zero differences, a value of T of 8 or less indicates a difference which is significant at the 5% level by a two-sided test. For the data in Table 4, T is 7·5, and it would be concluded that the acceptability of the juices was different.

4.4. Two Independent Normal Samples, Student's t-Test

Student's t-test is a test of equality of means of two normal distributions, and is illustrated by the data of Table 5. Generally if the number of observations in each set is quite small (less than 20) it is difficult to prove non-normality, but if there is evidence that the underlying distributions are not normal, it may be appropriate to use a non-parametric test of the type shown above.

The test also assumes that the two populations have the same variance.

To test this, the ratio of the larger of the sample variances divided by the smaller is compared with critical values of the F-distribution, tables of which are in most general statistical textbooks (e.g. Snedecor and Cochran, 1980). If the variances are concluded to be different, then this is evidence on its own that the populations are different without examination of the means.

To perform the test, two statistics are needed, the difference between the means and the standard error of the difference.

Where there are two samples of size n_1 and n_2 with variances s_1^2 and s_2^2 the pooled variance s^2 is given by

$$s^2 = \frac{(n_1 - 1)s_1^2 + (n_2 - 1)s_2^2}{n_1 + n_2 - 2}$$

Note that $(n_1 - 1)s_1^2$ and $(n_2 - 1)s_2^2$ are the sum of squares about each sample mean. When $n_1 = n_2$, $s^2 = (s_1^2 + s_2^2)/2$.

The standard error of the difference between the means is equal to

$$\sqrt{\left[s^2\left(\frac{1}{n_1} + \frac{1}{n_2}\right)\right]} \quad \text{or} \quad \sqrt{\left(\frac{2s^2}{n}\right)}$$

when $n_1 = n_2 = n$. The test statistic, t, is obtained by dividing the difference between the means by the standard error. For a two-sided test, in the case of the data of Table 5, the value of 8·61 with $n_1 + n_2 - 2 = 14$ degrees of freedom is significant at the 0·1% level.

4.5. Two Paired Samples, the Paired t-Test

The paired t-test, illustrated by the data of Table 6, tests the hypothesis that individual differences (e.g. for each sample the difference between scores given by assessor 1 and assessor 2) comes from a distribution (assumed normal) with mean equal to zero.

The mean and standard error of the mean are calculated for the differences, and the test statistic t is calculated:

$$t = \frac{\text{mean difference}}{\text{standard error of the mean difference}}$$

The number of degrees of freedom is one less than the number of pairs. With nine degrees of freedom for the data in Table 6, it may be verified by a two-sided test that the mean difference is not significantly different from zero, and there is therefore no evidence to suggest that the assessors have different thresholds.

TABLE 6
SCORES GIVEN ON AN INTENSITY SCALE BY TWO ASSESSORS TO TEN SAMPLES

Sample	Assessor 1	Assessor 2	Difference (1 minus 2)
1	7·5	6·0	+1·5
2	5·0	4·5	+0·5
3	5·5	6·0	−0·5
4	6·0	5·5	+0·5
5	7·0	6·0	+1·0
6	7·0	7·0	0
7	5·0	6·0	−1·0
8	7·5	8·0	−0·5
9	4·5	4·0	+0·5
10	6·5	5·5	+1·0
Means	6·15	5·85	+0·30
Sum of squares			5·60
Variance			0·622 2
Variance of mean			0·062 2
Standard error of mean			0·249
t			1·20
Degrees of freedom			9

5. COMPARISON OF MORE THAN TWO SETS

There are many situations in which several sets of observations are compared at the same time. The general class of statistical method is referred to as 'analysis of variance', although, as will be seen in due course, the name is not strictly accurate. As with comparisons of two sets in the previous section, there are versions which are appropriate only when assumptions of normality hold, while non-parametric versions should be used otherwise. It is not possible in the limited scope of this chapter to give details for more than a few situations, but the references given at the end of the chapter will allow the reader to proceed further.

5.1. One-Way Analysis of Variance

The one-way analysis is so-called because in the experiment only one 'factor', e.g. panel, treatment or product, changes. Either different products are assessed by the same expert or different panels assess the same product. For a two-way analysis, two factors would change, e.g. there might be different products and several assessors.

TABLE 7
ASSESSMENT BY ONE EXPERT OF SEVERAL PACKETS OF FROZEN PEAS FROM DIFFERENT RETAIL OUTLETS

Packet no.	Outlet				
	A	B	C	D	
1	11	9·5	10·5	10	
2	12·5	10	11	12	
3	10	8	9	9	
4	13	9		11·5	
5	11·5				
6	11				
Sum	69	36·5	30·5	42·5	178·5
Mean	11·50	9·12	10·17	10·62	10·50

The method is illustrated by the data of Table 7. In general algebraic form, if each of k sets (or outlets) indexed i has n_i observations, the jth being x_{ij}, with mean $\bar{x}_i$, there being N observations altogether, and the overall mean is $\bar{x} = (\sum_i \sum_j x_{ij})/N$ then the analysis of variance table is constructed as shown in Table 8.

The analysis of variance is a method for separating the total variation throughout the data (expressed by the sum of squares about the overall mean) into two components, one being the aggregated sum of squares of each observation about the mean of its set (the within-sets sum of squares) and the other representing variation of the means of all the sets about the overall mean (the between-sets sum of squares). The expressions for the sums of squares are shown in the most suitable form for understanding what they represent, although they are not the most efficient for calculating or computing purposes. Algorithms are given in Appendix 2.

For each component, the sum of squares is converted to a mean square (or variance) by dividing by the number of degrees of freedom. The variance ratio, or F (after R. A. Fisher) is obtained by dividing the between-sets mean square by the within-sets mean square, and is then compared with tabulated values of the F-distribution. If it is greater than the tabulated value then it is concluded that the set means are different, at the appropriate significance level.

Degrees of freedom need to be explained briefly. In general, for each component of variation there is one degree of freedom for every *independent* term within its sum of squares. The total sum of squares represents the variation of all observations *about the overall mean* $\bar{x}$. If there

TABLE 8
ONE-WAY ANALYSIS OF VARIANCE

Source of variation	Degrees of freedom	Sum of squares	Mean square	
Between sets	$k-1$	$SSB = \sum_i n_i(\bar{x}_i - \bar{\bar{x}})^2$	$MSB = SSB/(k-1)$	$F = MSB/MSW$
Within sets	$N-k = \sum(n_i - 1)$	$SSW = \sum_i \sum_j (x_{ij} - \bar{x}_i)^2$	$MSW = SSW/(N-k)$	
Total	$N-1$	$SST = \sum_i \sum_j (x_{ij} - \bar{\bar{x}})^2$		

MSB is an estimate of σ_W^2.
MSB is an estimate of $\sigma_W^2 + n^* \sigma_B^2$ where $n^* = (N^2 - \sum n_i^2)/N(p-1)$ or n if all values of n_i are equal (to n).

are N observations, only $N-1$ of them are independently represented in the sum of squares: once $N-1$ terms of the form $(x_{ij} - \bar{x})^2$ are known, the remaining one is immediately known since $N\bar{x} = \sum\sum x_{ij}$ and hence $\sum\sum (x_{ij} - \bar{x}) = 0$. There are similarly $k-1$ independent terms in the sum of squares between k sets, and for the within-sets sum of squares, the ith set contributes $n_i - 1$ degrees of freedom.

When referring to the tables of the F-distribution, the degrees of freedom along the top are for the mean square on the numerator, i.e. between sets, and those on the left-hand column are those for the divisor, i.e. within sets. The analysis of variance for the data of Table 7 is in Table 9. The F-value of

TABLE 9
ANALYSIS OF VARIANCE TABLE

Source of variation	Degrees of freedom	Sum of squares	Mean square	F
Between outlets	3	13·958 3	4·652 8	3·77
Within outlets	13	16·041 7	1·234 0	
Total	16	30·000 0		

3·77 is compared with tabulated values, with 3 and 13 degrees of freedom, of 3·41 (for 5% significance level) and 5·74 (for 1% level), and it may be concluded that variation between the means for the four sets is significant at the 5% level but not at the 1% level.

This conclusion means that the hypothesis that all the means are equal is rejected, and thus among *some* of the means there are differences. To find out which differences are significant, work out the least-significant difference (LSD) between two means:

$$\text{LSD} = t \sqrt{\left[s^2 \left(\frac{1}{n_1} + \frac{1}{n_2} \right) \right]}$$

where t is the tabulated value of the t-distribution with $N-k$ degrees of freedom, s^2 is the within-sets mean square, and n_1 and n_2 are the numbers of observations in the two sets whose means are to be compared. When the number of observations in every set is the same, LSD needs to be calculated only once.

It can be seen (Table 7) that, for example, the difference of 2·38 between the mean scores for outlets A and B is significant, being greater than the LSD for $n_1 = 6$ and $n_2 = 4$, which is 1·55.

When there are many comparisons the LSD should be used with some caution as comparisons within all pairs of means usually result in many more differences appearing significant than are implied by the significance level of the F-test. In those circumstances it is wiser to use a multiple comparison test such as the Newman–Keuls Q method (Newman, 1939; Keuls, 1952) or Duncan's multiple range test (Duncan, 1955).

The analysis of variance allows separate estimates to be obtained of the within-sets and between-sets variances. The within-sets mean square (MSW in Table 8) is an estimate of the within-sets variance $\sigma_W^2 + n^*\sigma_B^2$ where σ_B^2 is the between-sets variance (i.e. variance between the true means of all the sets) and n^* is equal to n if the same number of observations, n, has been taken from each set or a weighted mean of the values of n_i as formulated at the foot of Table 8 if they are different.

For the data of Table 7, $n^* = 4 \cdot 157$, so the estimated variances are $\sigma_W^2 = 1 \cdot 2340$ and $\sigma_B^2 = 0 \cdot 8224$.

Application of the analysis of variance assumes that, within each set, the observations are normally distributed with the same variance. If this is not the case, and particularly if the variance is not constant but differs in accordance with the mean, then the data should be transformed. For example if the standard deviations are roughly proportional to the mean, then a logarithmic transformation of the data will stabilise the variances and analysis of variance can then be performed on the transformed data.

5.2. Two-Way Analysis of Variance

For the two-way analysis of variance, two factors vary. In Table 10 the example used is one in which the rows represent different treatments (or storage conditions) and the columns represent assessors. Although it is primarily differences between the treatments that are of interest, it is anticipated that there will also be differences between the sensory thresholds of the assessors and hence between their levels of scoring. The two-way analysis of variance allows a component for each effect, treatment or assessors, to be extracted from the sum of squares. The method of analysis is as for the 'randomised block' design in experimental design texts, the assessors being the 'blocks'.

The requirements of the model are that the score x_{ij} given for treatment i by assessor j is a sum of four parts, thus:

$$x_{ij} = \mu + \alpha_i + \beta_j + \varepsilon_{ij}$$

where μ is an overall mean, α_i is the 'effect' of the ith treatment, β_j is the 'effect' of the jth assessor (his 'bias') and ε_{ij} is a random error term. It is

TABLE 10
DATA FOR TWO-WAY ANALYSIS

Treatment	Assessor					Sum	Mean	Effect
	A	B	C	D	E			
1	4·5	5	7	4·5	4·5	25·5	5·1	−2·2
2	7·5	8	6	9	8·5	39	7·8	+0·5
3	7·5	8	9	9	8	41·5	8·3	+1·0
4	6·5	6·5	6·5	6·5	7·5	33·5	6·7	−0·6
5	6	8	7	8	7	36	7·2	−0·1
6	6	8	7	9	7·5	37·5	7·5	+0·2
7	7	8	7	9·5	8	39·5	7·9	+0·6
8	7·5	8	6·5	8	8·5	38·5	7·7	+0·4
Sum	52·5	59·5	56	63·5	59·5	291	7·3	
Mean	6·6	7·4	7·0	7·9	7·4			
Bias	−0·7	+0·1	−0·3	+0·6	+0·1			

further assumed that the sum of the treatment effects α_i is zero, the assessors' biases β_j also add to zero, and the error terms ε_{ij} are normally distributed with zero mean and common variance.

Where there are r rows and c columns the analysis of variance table is constructed as in Table 11.

Within the expressions for sums of squares, the convention of dots and bars to represent means has been used, thus

$$\bar{x}_{i\cdot} = \left(\sum_j x_{ij}\right)\bigg/c \text{ is a row mean}$$

$$\bar{x}_{\cdot j} = \left(\sum_i x_{ij}\right)\bigg/r \text{ is a column mean}$$

and

$$\bar{x}_{\cdot\cdot} = \left(\sum_i \sum_j x_{ij}\right)\bigg/(rc) \text{ is the overall mean}$$

Also

$$x_{ij}^* = \bar{x}_{\cdot\cdot} + (\bar{x}_{i\cdot} - \bar{x}_{\cdot\cdot}) + (\bar{x}_{\cdot j} - \bar{x}_{\cdot\cdot})$$

represents the predicted value for the (i, j)th observation by the additive

TABLE 11
TWO-WAY ANALYSIS OF VARIANCE

Source of variation	Degrees of freedom	Sum of squares	Mean square	F
Between rows (treatments)	$r-1$	$\text{SSR} = c\sum (\bar{x}_{i.} - \bar{\bar{x}}_{..})^2$	$\text{MSR} = \text{SSR}/(r-1)$	$F_R = \text{MSR}/\text{MSE}$
Between columns (assessors)	$c-1$	$\text{SSC} = r\sum (\bar{x}_{.j} - \bar{\bar{x}}_{..})^2$	$\text{MSC} = \text{SSC}/(c-1)$	$F_C = \text{MSC}/\text{MSE}$
Error	$(r-1)(c-1)$	$\text{SSE} = \sum_i\sum_j (x_{ij} - x_{ij}^*)^2$	$\text{MSE} = \text{SSE}/\{(r-1)(c-1)\}$	
Total	$rc-1$	$\text{SST} = \sum_i\sum_j (x_{ij} - \bar{\bar{x}}_{..})^2$		

model. The resulting error terms $x_{ij} - x_{ij}^*$ sum to zero in each row and each column, and hence, although there are rc such terms, there are only $(r-1)(c-1)$ degrees of freedom for the error sum of squares as only $(r-1)(c-1)$ error terms are independent.

Each sum of squares is converted to a mean square by dividing by its corresponding number of degrees of freedom. The F-ratios for both rows and columns are obtained by dividing the appropriate mean square by the error mean square. Computation of the sums of squares by efficient algorithms is given by Appendix 2.

The method is illustrated with the data of Table 10 where samples of fish from eight different treatments were assessed by five trained assessors. From the analysis of variance (Table 12) it can be seen that variation between treatments is very significant (i.e. significant at the 1% level), the F-value having 7 and 28 degrees of freedom.

The level of significance of each source of variation is indicated by a commonly-used convention, by writing one, two or three asterisks after the F-ratio to indicate significance at the 5%, 1% or 0·1% level respectively (and where appropriate the letters NS if the variation is not significant at the 5% level).

The least significant difference between two treatment means is calculated by

$$\text{LSD} = t \sqrt{\left(\frac{2\text{MSE}}{c}\right)}$$

where t has $(r-1)(c-1)$ degrees of freedom and MSE is the error mean square.

A particular two-way analysis of variance model, in which it is assumed that the error variance is not constant but is different for each assessor, was

TABLE 12
ANALYSIS OF VARIANCE

Source of variation	Degrees of freedom	Sum of squares	Mean square	F
Between assessors (blocks)	4	8·600	2·150	
Between treatments	7	35·075	5·011	7·67**
Error	28	18·300	0·654	
Total	39	61·975		

Significance levels for F are indicated as follows:
** significant at 1% level.

developed by Ehrenberg (1950) and demonstrated in Shewan *et al.*, (1953). Its main purpose was in the training of assessors in an objective sensory descriptive scoring system, and should not be considered for anything simpler.

5.3. More Complex Designs

The types of analysis in Sections 5.1 and 5.2 are the two simplest among analyses for a wide variety of experimental designs in which the effects of different factors and their interactions can be studied at the same time. Texts such as Cochran and Cox (1957) or John and Quenouille (1977) are recommended. The warning should be repeated that the analysis of variance models and their consequent F-tests have underlying assumptions of additive effects, independence of errors and a common normal distribution of errors. It is only properly valid to contemplate such an analysis where a sensory panel produces precise objective data. In situations where even the calculation of a mean may not have validity, such a complex design and analysis should not be considered.

5.4. Non-Parametric Equivalents of Analysis of Variance

When assumptions for analysis of variance are clearly violated, it may be appropriate to consider non-parametric methods, for which further details are given in the References. The two commonest methods which are analogues of the one-way and two-way analysis of variance both depend on ranking the data.

In the Kruskal–Wallis one-way analysis of variance by ranks (Kruskal and Wallis, 1952) all the data (the N observations from k sets) are ranked as one series and a test statistic calculated from the k rank sums.

Friedman two-way analysis of variance by ranks (Friedman, 1937) is the equivalent of the randomised blocks method. For each assessor (in that example) the scores are ranked and the test statistic is based on the rank sums obtained by the separate treatments

There is a shortage of suitable non-parametric methods for more complex designs. Singer (1979) refers to a few cases of their use. Where there are no suitable methods, however, the only answer may be to perform an analysis of variance even though the required normality and linearity assumptions do not hold. Provided caution is exercised over interpretation of the analysis and the expression of significance levels, it may be the only way of gaining anything from the data until more appropriate methods are developed.

6. DATA IN TWO DIMENSIONS

It is often desired to investigate the relationship between two variables of which one or both may be a sensory assessment, e.g. quality score and time of storage of apples, flavour score and appearance, or sensory score and some chemical determination.

It is important, wherever possible, to plot the data in a scatter diagram, otherwise blind application of a computer program or calculator routine can lead to the wrong conclusions about the type of relationship. The most commonly used routines in the exploration of a relationship between two variables involve the estimation of the parameters of the 'best' straight line through the data (the regression line) and of the closeness of the points to that line (the correlation coefficient).

Some scatter diagrams which might occur are shown in Fig. 3. There is an excellent discussion of these and other types of 'regression' in Chatterjee and Price (1977), which is helpful in understanding what can and cannot be done. In the case of Fig. 3(d) there is clearly no dependence of one variable on the other, but in the others there is a relationship. Blind application of the usual programs would produce some claim of evidence for a relationship but as it assumes linearity the wrong conclusions would be reached in Figs 3(b) and (c).

6.1. Regression as Prediction

The most widely available method for estimating the linear relationship between two variables x and y is regression by 'least squares'. The model assumes that the n points (x_i, y_i) satisfy $y_i = \alpha + \beta x_i + \varepsilon_i$, where ε_i are normally distributed errors. The line $y = a + bx$, where a and b are the values of α and β respectively which minimise the sum of square of errors $\sum \varepsilon_i^2$, estimates the regression.

$$b = \frac{\sum (x_i - \bar{x})(y_i - \bar{y})}{\sum (x_i - \bar{x})^2}$$

$$a = \bar{y} - b\bar{x}$$

where $\bar{x}$ and $\bar{y}$ are the mean values of x and y.

This form of regression is used when at a number of fixed values of x (e.g. storage time) values of y are recorded, as in Table 13 and Fig. 4. It enables y to be predicted for a given value of x and is also the correct one to use for inverse prediction of x for a given value of y.

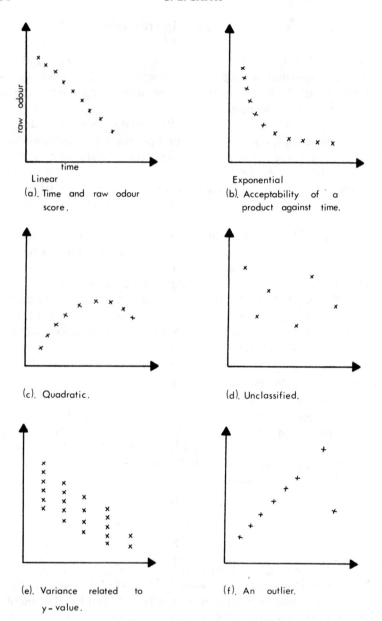

Fig. 3. Some examples of scatter diagrams illustrating relationships between two variables.

TABLE 13
MEAN RAW ODOUR SCORE FOR DIFFERENT STORAGE TIMES IN ICE

Days in ice (x)	Mean RO (y)
2	7·8
3	8·0
5	7·2
6	6·2
7	7·0
9	5·9
11	5·2
12	5·4
13	4·8
14	4·2
16	4·8
18	3·0

Slope (b) = $-0{\cdot}283$
Intercept (a) = $8{\cdot}53$

Formulae for standard errors of the estimates and predicted values can be found in suitable texts, e.g. Draper and Smith (1981) or Snedecor and Cochran (1980).

The mean square error, s^2, which loosely represents the variance about the line, is calculated by

$$s^2 = \frac{\sum(y_i - \bar{y})^2 - b^2 \sum(x_i - \bar{x})^2}{n-2}$$

The test for the slope being significantly different from zero is

$$t = \frac{b}{\sqrt{\left(\frac{s^2}{\sum(x_i - \bar{x})^2}\right)}}$$

where t, the Student's t-statistic, has $n-2$ degrees of freedom.

For the data of Table 13, $s^2 = 0{\cdot}183$ and $t = -11{\cdot}32$ with 10 degrees of freedom, which can be seen from tables of the t-distribution to be significant at the 1% level.

6.2. Regression in a Bivariate Population

When two variables x and y are jointly normally distributed in a population, it follows that the mean value of y for a given value of x is a

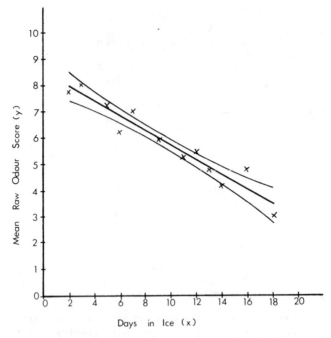

FIG. 4. Raw odour score versus days in ice.

linear function of x. This function is called the regression of y on x, and there is similarly a regression of x on y.

The method of calculating the coefficients for the regression of y on x is identical to that in the preceding section, although the basic philosophy is different. The calculations for the regression of x on y are performed in the same way with x replaced by y and vice versa in the formulae; thus

$$\text{slope } b_x = \frac{\sum (x_i - \bar{x})(y_i - \bar{y})}{\sum (y_i - \bar{y})^2}$$

$$\text{intercept} = \bar{x} - b_x \bar{y}$$

Unless all points lie on the same straight line, the two regressions will not represent the same line. The closeness of the relationship between the two variables is measured by the coefficient of correlation r, where

$$r = \frac{\sum (x_i - \bar{x})(y_i - \bar{y})}{\sqrt{[\sum (x_i - \bar{x})^2 \sum (y_i - \bar{y})^2]}}$$

Values for the correlation coefficient range from -1 (perfect negative correlation) when all the points lie on a line of negative slope, to +1 (perfect positive correlation) when all the points lie exactly on a line of positive slope. A correlation coefficient of 0 represents a situation in which there is no relation at all between the variables. Tables for testing the significance of a correlation coefficient are in many statistical texts.

6.3. Regression when Data Have Associated Error

The regression methods so far described assume that data are observations made without error. There may be situations where, for example, each observation represents a mean value for a batch based on data for a sample as in Connell *et al.* (1976). The observations used in the regression cannot be the actual batch means but are estimates with a degree of error, and allowance may need to be made in the estimation of the regression.

If the errors are only in the y-value, then the variance is a part of the error variance which otherwise measures only variation of the points about the fitted straight line.

If there are errors in the x-values, then the slope as calculated in the previous sections will be an underestimate of the slope of the regression of y on the true, rather than observed, x.

The linear functional relationship (Lindley, 1947) measures this true relationship. Estimation for a variety of error patterns is discussed in Sprent (1969).

6.4. Robust Regression

If there is evidence that deviations about the line are not normally distributed or that underlying distribution of the dependent variable is not normal, it may be worthwhile considering a 'robust' alternative. The method above, which minimises the sum of squares of deviations (the 'least-squares' method) is severely influenced by the occasional extreme observation (an 'outlier') and by the more serious departures from normality.

Robust regression methods are less affected by these two problems by minimising different functions of deviations such as the sum of the absolute values of the deviations or those in which the influence of an observation decreases in some way with its distance from the rest of the data.

The 'least-absolute deviations' line is the regression analogue of the median and is recommended in situations involving hedonic data, for example relating median acceptability scores for fish fingers to time of frozen storage. The coefficients of the line cannot be obtained by an immediate algebraic solution as with the least-squares method, but require an iterative solution. The algorithm is given in Armstrong and Kung (1978). For a comparison of the properties of some other robust methods see Ramsay (1977).

6.5. Confirming a Relationship

When a regression or correlation analysis has been performed, a scatter diagram with the regression line superimposed should be plotted. This will assist visual confirmation that a straight line was the most appropriate fit.

The pattern of the residuals can also be revealing. For each observation, the residual is the difference between the actual value y_i and the 'fitted' value $a + bx_i$, and these may be plotted against the corresponding values of x.

If the residuals have a symmetrical distribution with equal spread throughout the x-axis, then the correct regression has been fitted. If they have a banana-like shape, being mainly on one side of zero at each end of the x-axis but mainly on the other in the centre, this is evidence that a non-linear regression (perhaps quadratic) should have been fitted. The plot of residuals may also reveal the presence of an outlier, which should be tested and verified before perhaps repeating the analysis without that observation. Another possible pattern is one in which the spread of the errors changes with position on the axis. This is often associated with a non-linear relationship between the variables. Interpretation of residual plots is discussed in Chatterjee and Price (1977).

6.6. Non-linear Relationships

Fitting of a linear regression is clearly appropriate when there is evidence of a linear dependence of one variable on another, and can be used in less clear cases but only as an indication of the average change in one variable as the other one changes.

When there is evidence of a particular type of non-linear relationship (e.g. logarithmic, exponential, power law) some transformation may be appropriate which will make the relationship linear. (See Chatterjee and Price (1977) for a discussion on suitable forms.)

Unless data are very precise or there is an underlying reason for fitting a particular form of function, it is wise to choose a simple form. Often the data do not suggest a particular form and a scatter diagram, with perhaps points at successive values of x joined by line segments, is as much as can be done.

If there are a large number of points (of the order of 100 or more) the exploratory technique of robust smoothing (Cleveland, 1979) may help in examining the nature of a trend in a two-dimensional plot. This method, which was developed out of time-series analysis, enables a curve of local medians or other summary statistics of y at successive values of x to be drawn.

7. SAMPLING AND EXPERIMENTAL DESIGN

It is important to give some thought to the choice of experimental design before pursuing an investigation. It can be disappointing to complete an experiment and discover that no conclusions can be drawn because too small a panel was used or not enough samples were taken.

Where there is prior knowledge on variation between assessors, between samples etc., or reasonable assumptions can be made, a few simple calculations can give guidance on the amount of data which should be collected.

This section assumes that the normal distribution applies. If it is not appropriate, then the derived sample sizes would have to be used as very rough guides only.

7.1. Sampling

The following cases should illustrate how determination of the degree of sampling can be helped by some forethought.

A single set of observations — it is desired that the standard error of the mean score should be no more than a certain amount, d. Recall that if the standard deviation is s, then the standard error of the mean of n observations is $s/\sqrt{n}$. Thus $s/\sqrt{n}$ is to be no more than d then, exercising some algebra, we find that n should be not less than $(s/d)^2$. If the value of s is known, or can be assumed, from past experience, it can be substituted along with d in the expression. For example, if s is 2·0 and d is required to be 0·5 or less, then n should be at least 16. Note, however, that if d is to be halved, then n has to be multiplied by 4. Often little gain in precision can be achieved without a large increase in work or resources beyond a certain level.

Paired sets of observations—it is required that any mean difference greater than d will be significant at the 5% level. If the standard deviation of the n differences is assumed to be s, then since $t = $ mean difference$/(s/\sqrt{n})$, it follows that n should be at least $(ts/d)^2$. The appropriate value of t from tables of the t-distribution depends on n, having $n-1$ degrees of freedom, but an initial value of $t = 2$ would suffice before improving it as an idea of n emerges.

Two independent sets—it is required that the difference between their means will be significant if it is greater than some value of d. If it is assumed that the two sets have the same or pooled standard deviation s, then

$$t = \frac{\bar{x}_1 - \bar{x}_2}{\sqrt{\left(\frac{2s^2}{n}\right)}}$$

if each set comprises n observations. Then it follows that n should be at least $2(ts/d)^2$. The correct value of t should be that one with $2(n-1)$ degrees of freedom but $t = 2$ would suffice for initial estimates of the required sample size. Similar calculations would apply in comparisons of two out of several independent sets, where s^2 was the residual mean square from the analysis of variance.

7.2. Experimental Design

In general it is the case that the standard error of a mean value decreases with an increase in the amount of data from which the mean is calculated.

Suppose for example a mean colour score of a batch of frozen peas was to be determined. If resources permitted 12 observations to be made, would it be better to take one observation from each of 12 packets or more observations from fewer packets? If several assessors were available, should each assessor be given different samples from the rest of the panel, or could they all assess the same samples, producing the same amount of data with less material?

7.2.1. More Packs or Sub-Samples?

Suppose k packs are to be sampled, and from each of them n samples are taken. If the variance between packs is σ_B^2 and the variance between samples within a pack is σ_W^2, then the variance of the mean score is

$$\frac{\sigma_B^2}{k} + \frac{\sigma_W^2}{kn}$$

It can be seen that the variance of the mean is inversely proportional to k, the number of packs, but whether any appreciable improvement is gained by increasing the number of samples per pack depends on the relative values of the two components of variance.

When variance within packs is the larger component, an increase in n has almost the same effect as a similar increase in k, and a given total number of observations could be allocated in several ways while giving similar variances. When the variation between packs is the major component, however, there is virtually no change in the variance of the mean from an increase in the number of samples per pack. In such a situation, one sample should be taken from each of the full number of packs.

7.3. Sampling for a Two-Way Design

In quality control it is often the case that a mean score of some measurement of quality is required for a batch of some commodity. A panel of expert assessors may be available, and we wish to know how many samples to take, how many assessors to use and how best to allocate the work so as to have as low a variance as possible within the limits of available resources. If testing is non-destructive, as in the assessment of appearance, the same samples may be examined by more than one assessor. This is not possible with destructive sensory testing of, for example, toughness of meat, but it may be possible to allocate sub-samples so that assessors would in effect be assessing the same samples.

The variance components are: σ_S^2, the variance between samples; σ_A^2, the variance between assessors; and σ_0^2; the error variance. If the same n samples are assessed by m assessors, the variance of the mean is

$$\frac{\sigma_S^2}{n} + \frac{\sigma_A^2}{m} + \frac{\sigma_0^2}{mn}$$

The assessor variance enters in this expression only if assessors are chosen at random for each series of assessments. If the same assessors are used, then the assessor variance is not a random component and is removed for the expression. Thus with a fixed panel, the variance of the mean is

$$\frac{\sigma_S^2}{n} + \frac{\sigma_0^2}{mn}$$

If it is not possible for different assessors to examine the same samples, then the formula for the variance of the mean is different. If m assessors separately examine n samples *each*, the variance of the mean becomes $(\sigma_S^2 + \sigma_0^2)/mn$ with an additional term σ_A^2/m if the assessors are chosen at random. The last formula shows that with fixed assessors a given number of samples may be allocated in any way among them and the same variance results.

The best sampling procedure depends on the relative values of the three variance components. If the assessor's variance is larger than sample-to-sample variance, there is an advantage to be gained from sharing the work using fewer samples and more assessors, but when the sample-to-sample variance is the largest component, it is again important to take more samples.

If separate samples have to be taken for each assessor, the variance will be reduced in all cases by sharing them among the assessors. The size of advantage to be gained depends on the actual variances, and the calculations should be performed for each situation.

8. OUTLIERS

Estimation of parameters can be seriously affected by the presence among the observations of an outlier, a value that seems very unlikely when compared with the rest of the data. Barnett and Lewis (1978) and Hawkins (1980) give extensive reviews of methods for identifying outliers and the necessary treatment of data.

When an outlier is detected, it should not be rejected without investigation. It may have arisen through incorrect recording or transcription of data or a mistake in calculation. The appropriate course of action is then to correct the value and repeat the analysis. If the investigator is satisfied that the outlier arose because of some abnormality in the condition of a sample caused, for example by a taint or by faulty packaging, then it would be reasonable to discard that observation and repeat the analysis. If there is no explanation for the outlier, it may be evidence that the underlying distribution is not normal, in which case the data should have been transformed (perhaps by logarithms) or it may indicate that the wrong analysis was performed. It if finally appears that the rest of the data satisfy the criteria for the analysis undertaken and the suspicious observation is a genuine outlier it may be discarded, but it must be mentioned in the report on the analysis.

Tests which are often used are those by Dixon (1953), Grubbs (1969) or Grubbs and Beck (1972) for outliers in a single set of values, Lund (1975) for use in linear regression and Stefansky (1972) for residuals in an analysis of variance.

9. MULTIDIMENSIONAL DATA

Multidimensional data are collected in sensory analysis when, for example, different aspects of food quality such as appearance, colour, odour, texture and flavour are examined together (perhaps with several non-sensory attributes) or when a flavour profile, which may contain 30 or more components of sensory quality, is prepared.

To give details of the operation and interpretation of all the methods which could be used in the analysis of multivariate sensory data is beyond the scope of a part of one chapter so in the remainder of this chapter it is possible only to describe briefly the main types of method. The reader is directed to Chatfield and Collins (1980) for a general introduction or Piggott (1986) for a detailed account of their uses in food science, including sensory analysis. The exception is multidimensional scaling which is the subject of Chapter 10.

9.1. Presentation of Multidimensional Data

Multidimensional data can be presented in tabular form, with each column representing a different variable, but although this properly displays all the data, it gives no assistance in visualising any patterns.

For data of up to perhaps four dimensions, extensions of the two-dimensional scatter diagram can usefully convey all the information present in the raw data. Three dimensions can be plotted as in Fig. 5, with a vertical line, whose length corresponds to the value of the third variable,

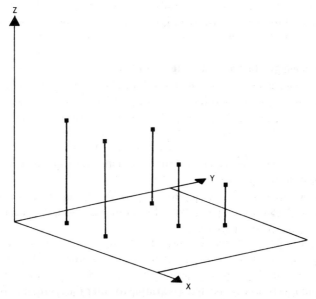

FIG. 5. A three-dimensional plot.

drawn from the appropriate position on the plane of the first two variables. The impression gained from such a plot depends, however, on which variable is chosen for the vertical axis.

A method of plotting four-dimensional data is to plot a point as if on a scatter diagram of the first and second variables and attach horizontal and vertical lines whose lengths correspond to the values of the third and fourth variables respectively. Again this conveys all the information but the visual impression depends on the ordering of the variables.

Andrews (1972) discusses some methods of representing data in many dimensions. Among these is the conversion of a multidimensional observation $(x_1, \ldots x_n)$ to a function

$$f(t) = x_1\sqrt{2} + x_2 \sin t + x_3 \cos t + x_4 \sin 2t + x_5 \cos 2t + \cdots$$

Another is the representation as a face (Chernoff, 1973) with the size and shape of each feature determined by the value of a different variable. Barnett (1981) gave an account of several attempts to tackle the problem of examining multidimensional data.

As no method of representing multivariate data has been universally accepted, there is more interest in methods of analysis which allow as much information as possible from the data to be plotted in fewer, if possible, two, dimensions. Among these are multidimensional scaling (Chapter 10) and principal components analysis, although, as will be seen later, that is not the main aim of the latter.

9.2. Dependence of One Variable on Others

Multiple regression is used as a method for investigating the dependence of one variable y on several others $x_1, \ldots, x_p$. The usual procedure involves the determination of coefficients $b_0, b_1, \ldots, b_p$ in the regression equation

$$y = b_0 + b_1 x_1 + \cdots + b_p x_p$$

using the same 'least-squares' criterion that was used in simple linear regression in Section 6.1.

It is possible to obtain the coefficients by the application of matrix algebra to the solution of a system of p simultaneous equations, to give the best fit using all the variables.

It usually happens that a fit which is almost as good can be obtained using fewer variables, and it is of interest to know which are the most important variables in the regression. *Stepwise* regression is usually performed, as follows: The first x-variable to enter the regression is that one which gives the best fit for the simple regression of y on a single variable, the next to enter is that one which gives the best improvement to the fit, and so on. A 'stopping criterion' is specified, so that if the degree of improvement or reduction in the error variance is less than a certain amount no more variables will be entered.

From most regression computer packages, regression statistics can be obtained for every stage at which a new variable is entered. In addition to the coefficients b_i and their standard errors, the partial correlation coefficients (correlation coefficients between y and each of the x-variables after removal of the effect on y of the other variables in the regression) are also usually displayed. It should be appreciated that the coefficient associated with a particular variable which has been entered into a regression will change when other variables are added.

The results of a multiple regression analysis should be interpreted with caution. The function specified is linear in the x-variables so faulty conclusions may be reached if y is non-linearly related to any of the x-variables. If any non-linearity is expected, then the appropriate variables should be transformed before the analysis. A plot of residual against fitted values of y will be helpful in detecting non-uniformity of the error variance or the presence of outliers; see Chatterjee and Price (1977).

9.3. Interrelationships Between the Variables

Two techniques are often used to examine interrelationships among the variables or to explore the underlying structure of the variables. Although they have differences in their motivation, they are often considered together.

The first is *principal component analysis* (PCA), whose purpose is to transform the original p variables $x_1, \ldots x_p$ into a new set of p *principal components* $y_1, \ldots y_p$ which are uncorrelated with each other and whose variances (in order of $y_1, \ldots, y_p$) decrease. Although all p principal components are required to reproduce exactly all the structure of the data, it is usually the case that a few will account for almost all the variation in the data, so in effect PCA is a dimension-reducing technique. Scatter diagrams of pairs of the first three or four principal components are often a sufficient display of the data.

Each principal component y_i is a linear combination of the original variables; thus

$$y_i = a_{i1}x_1 + \cdots + a_{ip}x_p$$

when the x-variables are scaled with zero mean and usually with variance equal to 1. The coefficients $a_{i1}, \ldots, a_{ip}$ are constrained so that

$a_{i1}^2 + \cdots + a_{ip}^2 = 1$. The first principal component y_1 is that linear combination of the x-variables, with that constraint on the coefficients, which has the highest variance. If the data points can be imagined to fill the volume of a hyperellipse then the first principal component would be in the direction of the major axis. The second principal component is the one which has the highest variance of all those uncorrelated with the first, the third has to be uncorrelated with the first two, and so on.

The vector of coefficients for the transformation from the original variables to each principal component is an eigenvector of the covariance matrix of the scaled original variables and the corresponding eigenvalue is the variance of that principal component. The importance of each principal component is indicated by its variance expressed as a percentage of the sum of the variances for all the principal components. Each coefficient a_{ij} measures the correlation between the ith principal component and the jth original variable.

The coefficients obtained will depend on the scale of the x-variables, and it is common practice to standardise all the variables to the same scale, i.e. a variance of 1. This is an important procedure if variables in the same analysis have different units, e.g. numbers on a sensory scale, milligrams of a chemical, items per kilogram, number of blemishes, but it should be noted that there may be occasions, perhaps in flavour profiling, where standardising would not be appropriate because differences in the ranges of variables were meaningful.

Factor analysis has been much in use by psychologists, and applied to sensory analysis by Harper (1956). For an examination of the statistical aspects of the method, see Lawley and Maxwell (1971).

The underlying idea is that each of the p variables $x_1, \ldots, x_p$ is dependent on a number (smaller than p) of factors which account for the covariance between the observed variables so that, apart from an error term, it can be expressed as a linear combination of them; thus

$$x_i = \sum_{r=1}^{k} a_{ir} f_r + e_i$$

The underlying variables f_r are the factors, the coefficients a_{ir} the factor loadings (a_{ir} being the loading of factor f_r on variable x_i) and e_i a residual representing sources of variation which only affect the variable x_i.

The method of estimation is initially similar to that of PCA in that loadings and variances of the factors are obtained from eigenvectors and eigenvalues of a dispersion matrix. Assuming the original variables have been standardised, the terms in the matrix apart from the diagonal terms are the correlations. However, instead of each entry on the main diagonal being 1, representing the correlation of each variable with itself, there is the 'communality' which reflects a reduction from the correlation with itself to take into account the specific error e_i. There are several interpretations in the literature of the communality and its estimation, but unless a value can be inserted the problem of estimating the factors is indeterminate.

There are also difference models for factor structure. Unlike principal components, they do not require to be uncorrelated, and can be 'obliquely' transformed to improve their loadings with the observed variables.

9.4. Methods for Investigating Configurations of Data

Among methods for displaying the data in fewer than the original numbers of dimensions, PCA (Section 9.3) and multidimensional scaling (Chapter 10) have already been mentioned.

A method which enables grouping among data points to be investigated is *cluster analysis* (see Everitt, 1981). There are many algorithms for clustering (Anderberg, 1973; Späth, 1980) for which much use has been found under the name of numerical taxonomy in bacteriology, zoology and botany, but applications have recently been developed in sensory analysis. Two such examples might be classification of assessors by their preferences for certain products and classification of cheeses by descriptions of sensory characteristics or evaluation of their similarities to each other.

The simplest method requires that a measure of similarity or dissimilarity (distance) between every object (assessor or cheese in the above examples) and every other object can be obtained. From the triangular matrix of similarities or dissimilarities a dendrogram can be constructed as in Fig. 6 in which those objects which have the highest similarities with each other are joined together and the similarity between two groups so formed is related to all the similarities between an object in one group and an object in the other.

Another method for exploring configurations, *procrustes rotation*, first developed by Green (1952) and Gower (1971), was applied by Banfield and Harries (1975) to compare visual assessments given by twelve judges. The method assumes that judges are more consistent among themselves than data would suggest once their interpretations of the scales are taken into account. Fitting by procrustes rotation involves dilation and rotation of the configuration of points from each assessor about a common mean until the configurations are as close together as possible. This method is applied to data obtained from Free Choice Profiling (Williams and Langron, 1984).

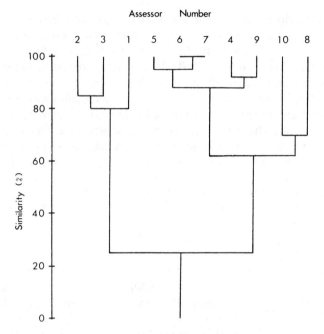

FIG. 6. An example of a dendrogram.

9.5. Comparison of Groups

When multivariate data have been obtained from two or more populations (e.g. products or species), it is often desired to determine whether or not their means are different.

Multivariate analysis of variance (MANOVA) permits this to be done. The method extends the one-dimensional ideas of analysis of variance by the use of dispersion matrices instead of simply the sums of squares. (In a dispersion matrix each diagonal element D_{ij} is a sum of products for the ith and the jth variables.) In place of the one-dimensional ratio between mean squares is a ratio of determinants of dispersion matrices, which is approximately distributed as a statistic known as Wilks's lambda (Λ). A significant value of the statistic indicates that the mean vectors for the populations are not all the same.

9.6. Discrimination

To examine how far it is possible to distinguish between members of different groups on the basis of their sensory data and to attempt to classify an unknown sample, *discriminant analysis* is performed.

The intention is to partition the multidimensional space into regions, one corresponding to each group, so that if an observation made on an individual falls into a particular region it is allocated to that group. The partitioning, by means of *discriminant functions*, is to be made in such a way as to minimise the degree of misallocation.

In the simplest case of two equally likely populations with different mean vectors but equal dispersion matrices, the discriminant function represents a hyperplane orthogonally bisecting the line between the means, and each unknown individual is allocated to the 'closer' population, i.e. the population from whose mean it has the lower *generalised distance*. This measure takes into account correlations between the variables, and is defined by

$$D^2(x,y) = \sum_{i=1}^{p}\sum_{j=1}^{p}(x_i - y_i)s_{ij}(x_j - y_j)$$

where s_{ij} is the (i,j)th term in the inverse of the common dispersion matrix, and x and y are the points $(x_1, \ldots, x_p)$ and $(y_1, \ldots, y_p)$ respectively.

With more than two populations, many discriminant functions are required, but essentially in the simple situation of equal probabilities and a common dispersion matrix the problem becomes that of choosing the population with the 'closest' mean.

The degree of discrimination between the populations is related to the probabilities of misclassifying. A crude method is to treat each of the original observations as new unknowns and note how many are classified where they belong, but as the observations are not independent of the discriminant functions this gives an incorrect estimate of the degree of discrimination. It can be calculated more correctly, given some knowledge of the distributions of the populations.

As with multiple regression, stepwise methods are often used to determine which of the variables are important in the discrimination.

10. CONCLUSIONS

In just one chapter it cannot be possible to give an exhaustive presentation of statistical methods appropriate to sensory analysis. Statistics applicable to some sensory methods such as difference testing and ranking have been given in the appropriate chapters. Users of statistics will doubtless find some of their favourite techniques absent in favour of some others which warrant consideration. Sensory analysis with its use of scales of varying degrees of sophistication and sometimes unknown underlying distributions has particular statistical problems most of which have only recently been critically tackled.

It is hoped that some encouragement has been felt to examine data before analysis and that this chapter has given the reader an indication of appropriate techniques to use. The list of references gives sources of greater detail for the methods than has been possible here. Statistical tables have not been included, as they can be found in many of the appropriate references.

REFERENCES

Anderberg, M. R. (1973). *Cluster Analysis for Applications*, Academic Press, London.
Andrews, D. F. (1972). Plots of high-dimensional data. *Biometrics*, **28**, 125–36.
Andrews, D. F., Bickel, P. J., Hampel, F. R., Huber, P. J., Roger, W. H. and Tukey, J. W. (1972). *Robust Estimates of Location: Survey and Advances*, Princeton University Press, Princeton, New Jersey.
Armstrong, R. D. and Kung, M. T. (1978). Algorithm AS 132: Least absolute value estimates for a simple linear regression problem. *Appl. Statist.*, **27**, 363–6.
Banfield, C. F. and Harries, J. M. (1975). A technique for comparing judges' performance in sensory tests. *J. Food Technol.*, **10**, 1–10.
Barnett, V. (Ed.) (1981). *Interpreting Multivariate Data*, J. Wiley & Sons, Chichester.
Barnett, V. and Lewis, T. (1978). *Outliers in Statistical Data*, J. Wiley & Sons, Chichester.
Chatfield, C. and Collins, A. J. (1980) *Introduction to Multivariate Analysis*, Chapman and Hall, London.
Chatterjee, S. and Price, B. (1977). *Regression Analysis by Example*, J. Wiley & Sons, Chichester.
Chernoff, H. (1973). The use of faces to represent points in k-dimensional space graphically. *J. Am. Statist. Ass.*, **68**, 361–8.
Cleveland, W. S. (1979). Robust locally weighed regression and smoothing scatterplots. *J. Am. Statist. Soc.*, **74**, 829–36.
Cochran, W. G. and Cox, G. W. (1957). *Experimental Designs* (2nd edn), J. Wiley & Sons, Chichester.
Connell, J. J., Howgate, P. F., Mackie, I. M., Sanders, H. R. and Smith, G. L. (1976). Comparison of methods of freshness assessment of wet fish. IV. Assessment of commercial fish at port markets, *J. Food Technol.*, **11**, 297–308.
Conover, W. J. (1980). *Practical Nonparametric Statistics* (2nd edn), J. Wiley & Sons, Chichester.
Daniel, W. W. (1978). *Applied Nonparametric Statistics*, Houghton Mifflin Co., London.
Dixon, W. J. (1953). Processing data for outliers. *Biometrics*, **9**, 74–89.

Draper, N. and Smith, H. (1981). *Applied Regression Analysis* (2nd edn), J. Wiley & Sons, Chichester.
Duncan, (1955). Multiple range and multiple F tests. *Biometrics*, **11**, 1–42.
Ehrenberg, A. S. C. (1950). The unbiased estimation of heterogeneous error variances, *Biometrika*, **37**, 247–57.
Ehrenberg, A. S. C. (1975). *Data Reduction*, J. Wiley & Sons, Chichester.
Everitt, B. (1981). *Cluster Analysis* (2nd edn), Heinemann, London.
Friedman, M. (1937). The use of ranks to avoid the assumption of normality implicit in the analysis of variance. *J. Am. Statist. Ass.*, **32**, 675–701.
Gower, J. C. (1971). Statistical methods of comparing different multivariate analyses of the same data. In: *Mathematics in the Archaeological and Historical Sciences*. F. R. Hodson, D. G. Kendall and P. Tautu (Eds). Edinburgh University Press, Edinburgh, pp. 138–49.
Green, B. F. (1952). The orthogonal approximation of an oblique structure in factor analysis. *Psychometrika*, **17**, 429–40.
Grubbs, F. E. (1969). Procedures for detecting outlying observations in samples. *Technometrics*, **11**, 1–21.
Grubbs, F. E. and Beck, G. (1972). Extension of sample sizes and percentage points for significance tests of outlying observations. *Technometrics*, **14**, 847–54.
Harper, R. (1956). Factor analysis as a technique for examining complex data on foodstuffs. *Appl. Statist.*, **5**, 32–48.
Hawkins, D. M. (1980). *Identification of Outliers*, Chapman and Hall, London.
John, J. A. and Quenouille, M. H. (1977). *Experiments: Design and Analysis* (2nd edn). Griffin, London.
Keuls, M. (1952). The use of 'Studentized range' in connection with an analysis of variance. *Euphytica*, **1**, 112–22.
Kruskal, W. H. and Wallis, W. A. (1952). Use of ranks in one criterion variance analysis. *J. Am. Statist. Ass.*, **47**, 583–621.
Lawley, D. N. and Maxwell, A. E. (1971). *Factor Analysis as a Statistical Method*, Butterworth, London.
Lindley, D. V. (1947). Regression lines and linear functional relationships. *J. Roy. Statist. Soc.*, B, **30**, 31–66.
Lindley, D. V. and Miller, J. C. P. (1953). *Cambridge Elementary Statistical Tables*, Cambridge University Press, Cambridge.
Lund, R. E. (1975). Tables for an approximate test for outliers in linear models. *Technometrics*, **17**, 473–6.
Neave, H. R. (1978). *Statistics Tables for Mathematicians, Engineers, Economists and the Behavioural and Management Sciences*, Allen and Unwin, London.
Newman, D. (1939). The distribution of range in samples from a normal population, expressed in terms of an independent estimate of standard deviation. *Biometrika*, **31**, 20–30.
Piggott, J. R. (1986). *Statistical Procedures in Food Research*. Elsevier Applied Science Publishers, London.
Ramsay, J. O. (1977). A comparative study of several robust estimators of slope, intercept and scale in linear regression. *J. Am. Statist. Ass.*, **72**, 608–15.
Shewan, J. M., Macintosh, R. G., Tucker, C. G. and Ehrenberg, A. S. C. (1953). The development of a numerical scoring system for the sensory assessment of the spoilage of wet white fish stored in ice. *J. Sci. Food Agric.*, **4**, 283–98.

Siegel, S. (1956). *Non-parametric Statistics for the Behavioral Sciences*, McGraw-Hill, London.
Singer, B. (1979). *Distribution-Free Methods for Non-Parametric Problems: A Classified and Selected Bibliography*, The British Psychological Society, Leicester.
Snedecor, G. W. and Cochran, W. G. (1980). *Statistical Methods* (7th edn), Iowa State University Press, Ames, Iowa.
Späth, H. (1980). *Cluster Analysis Algorithms*, Ellis Horwood, Chichester.
Sprent, P. (1969). *Models in Regression and Related Topics*, Methuen, London.
Sprent, P. (1981). *Quick Statistics*, Penguin Books, Harmondsworth.
Stefansky, W. (1972). Rejecting outliers in factorial designs. *Technometrics*, **14**, 469–79.
Tukey, J. W. (1977). *Exploratory Data Analysis*, Addision-Wesley, London.
Williams, A. A. and Langron, S. P. (1984). The use of free-choice profiling for the evaluation of commercial ports. *J. Sci. Food Agric.*, **35**, 558–68.

APPENDIX 1: SOME DISTRIBUTIONS AND TEST STATISTICS

The Normal Distribution

The distribution most often encountered in many fields of science is the *normal* or *gaussian* distribution. It is specified by its probability density function (pdf) which has the familiar bell-shaped curve,

$$f(x) = \frac{1}{\sigma\sqrt{(2\pi)}} \exp\left\{ -\frac{1}{2}\left(\frac{x-\mu}{\sigma}\right)^2 \right\}$$

This function is defined for a continuous variate, i.e. one which can take any value within an interval and is not restricted, for example, only to whole numbers. The parameters specifying the distributions are μ, the mean value of x, and σ, the standard deviation (π is the circular constant 3·14...). The function has its maximum, indicating the densest part of the distribution of x, at μ, and has its steepest slope at $\mu - \sigma$ and $\mu + \sigma$. The probability that the variate x lies within a particular interval (between a and b, say) is obtained by integrating the density function between the limits a and b.

Among many useful properties of the normal distribution is that, if the limits of the desired interval are expressed in units of the standard deviation from the mean, then the probability is independent of the values of μ and σ. Probabilities are given in general statistics texts (e.g. Lindley and Miller, 1953; Neave, 1978) for the so-called standard normal variate, the special case where $\mu = 0$ and $\sigma = 1$. A value of a variate is often referred to as a 'deviate'. Any normally distributed variable x can be converted to a standard normal variate z by subtracting the mean and dividing by the standard deviation; thus $z = (x - \mu)/\sigma$. Those tables generally display the integral from $-\infty$ to z or from 0 to z for a large range of values of z. Denoting $P(z_1, z_2)$ the probability for the interval (z_1, z_2) of the standard normal variate, required probabilities can be obtained from the tables as follows:

$$P(z_1, z_2) = P(-\infty, z_2) - P(-\infty, z_1)$$
$$P(z, \infty) = 1 - P(-\infty, z)$$
$$P(-z, 0) = P(0, z)$$
$$P(-z, z) = 2P(0, z)$$
$$P(z, \infty) = 0 \cdot 5 - P(0, z)$$
$$P(-\infty, -z) = P(z, \infty) = 0 \cdot 5 - P(0, z)$$

We often required the two-tailed probability,

$$P(-\infty, -z) + P(z, \infty) = 2P(-\infty, -z)$$
$$\text{or} \quad 2[1 - P(-\infty, z)]$$
$$\text{or} \quad 1 - 2P(0, z)$$

If n independent variables $x_1, \ldots, x_n$ are each normally distributed with mean μ and standard deviation σ, it follows that their mean $\bar{x} = (x_1 + \cdots + x_n)/n$ is normally distributed with mean μ and standard deviation $\sigma/\sqrt{n}$, and thus $(x - \mu)/(\sigma/\sqrt{n})$ has the distribution of z, the standard normal variate.

The t-Distribution

To convert from x to z requires the population standard deviation σ to be known. In practice, σ is not known and is estimated by s, the sample standard deviation.

The ratio of a normally distributed variate, whose mean is assumed to be zero, and its standard error based on f degrees of freedom has the t-distribution with f degrees of freedom. As f tends to infinity, t approaches z, the standard normal variate.

The F-Distribution

The ratio of two mean squares, which are estimates of variances assumed equal, has the F-distribution. If the mean squares have respectively f_1 and f_2 degrees of freedom, then F is said to have f_1 and f_2 degrees of freedom.

APPENDIX 2: ALGORITHMS FOR SUMS OF SQUARES ABOUT THE MEAN

Wherever a sum of squares about a mean is referred to below, the recommended method is the updating algorithm mentioned in Section 2.1 and Fig. 1.

One-Way Analysis of Variance

If the numbers of observations in the sets are different, calculate total sum of squares about the overall mean as above, the within-sets sum of squares by obtaining the sum of squares for each set about its mean, and the between-sets sum of squares is obtained by difference.

If the numbers of observations in the k sets are the same, n, calculate total sum of squares about the overall mean as above; the between-sets sum of squares is

$$\frac{1}{n} \sum_{i=1}^{k} (S_i - \bar{S})^2$$

where S_i is the sum of observations in the ith set and $\bar{S}$ is the mean of those sums—thus the between-sets sum of squares is obtained by calculating a sum of squares as above using the array of set sums and then dividing by the number of observations per set; the within-sets sum of squares is obtained by difference.

Two-Way Analysis of Variance

Calculate the total sum of squares about the overall mean as above using all the rc single observations; for the between-rows sum of squares divide the sum of squares calculated from the row sums by c, the numbers of columns; for the between-columns sums of squares divide the sum of squares calculated from the column sums by r, the number of rows; the error sum of squares is obtained by difference.

Linear Regression

The sum of squares about the mean for x and the sum of squares about the mean for y are both calculated as above.

The sum of products of x and y about their mean is calculated similarly.

If SP_i is the sum of products for the first i observations about the mean x- and y-values of the first i observations, x_i and y_i respectively, i.e.

$$SP_i = \sum_{r=1}^{i} (x_r - \bar{x}_i)(y_r - \bar{y}_i)$$

then

$$SP_i = SP_{i-1} + (x_i - x_{i-1})(y_i - y_{i-1})(1 - 1/i); \qquad SP_1 = 0$$

Chapter 10

PREFERENCE MAPPING AND MULTIDIMENSIONAL SCALING

H. J. H. MacFie

Agricultural and Food Research Council,
Institute of Food Research, Bristol, UK

and

D. M. H. Thomson

Department of Food Science and Technology,
University of Reading, UK

1. GENERAL INTRODUCTION

The last few years have seen growing interest in preference mapping methods. These procedures offer a way of superimposing either a preference vector or an ideal point, for each consumer involved in the study, on a multidimensional representation of properties of products or samples. By inspecting the various directions of preference amongst the samples, or points of optimum preference, sub-groups of consumers can be readily identified. These represent discrete behavioural groups which may be important to target when marketing products.

The main advantage of preference mapping over traditional affective methodology is that individual preferences are taken fully into account and not merely averaged.

Although the work of Schiffman *et al.* (1981) brought multidimensional scaling (MDS) procedures to a wider audience, they have not been widely applied in the sensory evaluation of foods. These methods can, and should, play a much more important role, as evidenced in recent studies on beef (Francombe and MacFie, 1985) and sweeteners (Thomson *et al.*, 1987).

The aim of this chapter is to show the use, analysis and interpretation of preference mapping and MDS procedures, in a sensory rather than a market research or psychological context. For more detailed information on how to design and run a study, and how to interpret the results, reference should be made to texts by Schiffman *et al.* (1981) and Coxon (1982).

2. PREFERENCE MAPPING

2.1. General Remarks
In the context of sensory evaluation of foods, affective tests typically take the form of the paired preference test, preference ranking or hedonic scaling (Stone and Sidel, 1985). In each case, data are usually aggregated for the purposes of statistical analysis. For example, with hedonic scaling, the mean hedonic score across subjects, along with a measure of spread, are calculated for each sample. Student's t-test or analysis of variance are then applied to determine significant differences among the samples.

The problem with these simple univariate analyses is the implicit assumption that all subjects exhibit essentially the same behaviour, and hence a single mean value or some other summary statistic is representative of all the subjects.

One solution is to inspect the distribution of data for multimodality, and to conduct separate analyses on the various sub-groups. However, a very large amount of data is usually required for all the separate modes to become apparent. A superior alternative is to use multidimensional preference mapping methods, since each subject is individually built into the statistical model.

In discussing preference mapping, it is necessary to distinguish between internal and external analysis (Carroll, 1972). With the former, the objective is to obtain a multidimensional representation of the stimuli, based solely on preference data. Using external analysis, the aim is to relate preferences to a multidimensional represenatation of the stimuli derived by other means.

2.2. Internal Preference Mapping—MDPREF
The simplest approach to modelling individual differences in preference is the vector model proposed by Tucker (1960). The set of stimulus points are embedded in a multidimensional space and each subject is represented by a vector in the space. The ordering of the projections of the stimulus points on to the vector gives the preference ranking of that subject. The cosine of the angle that a vector makes with the dimensions of the space is considered to be proportional to the relative importance of that dimension in the preference judgement.

An example from our own experience demonstrates the use of the vector model very effectively. The data (unpublished) were generated at the Torry Research Station, Aberdeen, and we are grateful to P. Howgate for permission to use them. Forty-eight subjects were asked to rate six types of

TABLE 1
MEAN PREFERENCE SCORE FOR SIX SUBJECTS ON SIX FISH PRODUCTS

Subject	White fish in parsley sauce	White fish fingers	Scad (good)	Scad (poor)	Cod mince fingers	Blue whiting fingers
A	7·7	7·3	7·4	6·0	7·3	8·0
B	5·0	5·2	6·2	3·7	5·3	5·0
C	7·5	6·6	5·0	4·0	6·3	4·5
D	7·8	6·8	5·6	4·7	6·0	5·7
E	6·5	6·3	6·1	6·5	5·3	3·9
F	5·7	6·8	6·0	6·6	7·2	4·7

fish or fish product on a hedonic scale: 1 = dislike extremely, 9 = like extremely. For brevity Table 1 shows the session means for only six subjects, A-F. The complete data were input to the MDPREF program (Chang and Carroll, 1968), and the two-dimensional solution, which accounts for 85·3% of the variation, appears in Fig. 1. The subjects appear as points on the unit circle and a preference ranking is obtained by drawing a line passing through the origin to a point and dropping perpendiculars from each stimulus point on to the line.

The horizontal dimension, around which most subjects cluster is clearly the conventional preference dimension. The familiar products project on to this dimension in the expected order and the unfamiliar products are well to the left with the reformed product of poor quality giving the lowest acceptability score. However, there are subjects who find the blue whiting and good quality scad nearly as acceptable as the white fish. Subjects A and B are the most extreme examples. The positions of Subjects E and F reflect the very low scores for blue whiting and the apparent preference for the poor quality scad shown by these subjects.

2.3. External Preference Mapping—PREFMAP

With this type of analysis, the preference data of a group of subjects is related to a multidimensional representation of the stimuli, usually derived from other 'non-preference' data. In sensory evaluation of foods, this would typically be the stimulus space defined by descriptive sensory assessment (see Chapter 7 by J. J. Powers), or by similarity or dissimilarity estimation in conjunction with multidimensional scaling (Section 3) or even from attitudinal data (McEwan and Thomson, in press). Alternatively, a

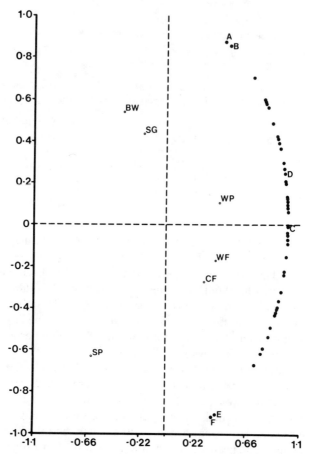

FIG. 1. MDPREF solution displaying configuration of six fish types. WP: white fish in parsley sauce. WF: white fish fingers. SG: good scad. SP: poor scad. CF: cod mince fingers. BW: blue whiting. Subjects are displayed as points or letter A–F on the unit circle.

stimulus space derived from physical/chemical measurements could be used.

If the preference data are measured at the interval level, simple multiple regression of the preference scores of each subject on to the external measurement is used. This is equivalent to the vector model discussed in the previous section. Separate regressions are obtained for each subject if necessary. If the data are ordinal, nonmetric regression procedures can be used (Carroll and Chang, 1967).

When a subject's preference scores cannot be fitted satisfactorily by the vector model, the ideal-point model may be appropriate. This model postulates that there is some optimum combination of the external measurements (i.e. sensory characteristics or physical parameters) which represent the ideal point, or most preferred stimulus, for a particular subject. His or her preference for any other stimulus is therefore directly related to its similarity to the ideal point. This is in direct contrast to the vector model, which assumes a single direction of increasing preference throughout the space. Carroll (1972) showed that the vector model can be formulated as a special case of the ideal point model in which the ideal point is a long way from the space of the stimuli and so the distances can be well represented by projections of the stimuli on to a line. Bennett and Hays (1960) developed the original multidimensional ideal-point model. This type of analysis is also called multidimensional unfolding.

Carroll (1972) proposed two extensions of the unfolding model. In the first generalisation ideal points are found for each subject as before but the subjects are also permitted to weight the dimensions of the external space differently. In the most general model of all, the assumption that the same basic set of dimensions are involved in the judgements of all individuals is relaxed. Each subject is allowed to rotate the reference axes of the external space and then to weight the new dimensions differently.

The PREFMAP algorithm that is mounted under various names in the MDS packages (see Section 7) is based on this hierarchy of models. Each model is fitted to each subject and an analysis of variance constructed that enables the best fitting model to be selected and tested for significance.

These concepts are very well illustrated by Schiffman *et al.* (1981) in their example on raspberry-flavoured liqueurs. Nine subjects were asked to give judgements of dissimilarity between the flavours of seven liqueurs that varied in sugar and alcohol content and the amount of flavouring. Judgements of overall preference were obtained. The PREFMAP program was used to test the goodness of fit of the vector and ideal-point models in relating the preference ratings of each subject to the two-dimensional INDSCAL solution (see Section 5.2.1.) of the flavour judgements.

The results are reproduced in Fig. 2. The circled numbers represent the seven liqueurs and the two INDSCAL axes were shown to reflect the variation in sugar and alcohol content respectively. The preference vectors of each subject are shown, but only the ideal points of Subjects 1,7 and the average subject (A) appear in the squares. Subjects 2, 5, 6, 8, 9 were best fitted by a vector model and Subjects 3 and 4 were not well fitted by either model. This suggests that these subjects were either inconsistent in their

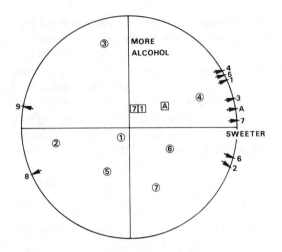

FIG. 2. Preference vectors (arrows), ideal points (boxed numbers indicates subjects) and stimuli (circled numbers) obtained by PREFMAP on raspberry liqueur data (Schiffman et al., 1981).

judgements, or that the principal sensory characteristics which discriminate amongst the liqueurs in the INDSCAL space were not the most important determinants of preference for these subjects.

Most of the preference vectors are in the direction of more sweetness and more alcohol. Subjects 8 and 9 prefer less sweetness, whilst Subjects 2 and 6 prefer less alcohol. Schiffman et al. (1981) conclude that liqueur 4 is probably close to the optimum composition.

2.4. Guidelines on the Conduct and Analysis of Preference Mapping Studies

2.4.1. Conduct
One practical point to bear in mind is that for internal preference mapping (MDPREF) each subject has to judge each of the samples. Missing value routines for eigenvector analyses are available (e.g. Wold et al., 1982) but are not widely implemented. For external preference mapping (PREFMAP) this is not so critical since the scores of each subject are fitted separately. The key point with PREFMAP is that the samples used to achieve the external configuration *must be identical* to those used in the preference trial.

Another important, but often neglected, aspect is to test for and balance out the effect of order of assessment. In our experience simple randomisation of order is inadequate. We have adapted an algorithm given by Williams (1949) which produces designs balanced for the effect of a single preceding treatment every n replications if n is even or $2n$ replications if n is odd. Copies are available on request from the first author.

Regarding the number of questions, experimenters should define carefully what sort of preference score they are seeking. If it is a consumer preference which is often instinctive and immediate, do not surround the question with a large battery of analytical questions! We favour as few as possible (ideally one) but no more than four questions on each product.

Another point to consider is that there are two ways to ask a preference/hedonic question. One can score on a scale of dislike-to-like (hedonic scale) or either side of a point that represents the ideal product (e.g. too salty or not salty enough). Our practice has been to use the dislike-to-like scale. This method is particularly compatible with the vector model in MDPREF and PREFMAP and thus we tend to recover most but not all individuals as conforming with this model.

To monitor the level of discrimination shown by subjects we recommend that about 20% or 40 individuals (whichever is the greater) are given half the samples twice. If no significant one way variance ratios of between to within sample differences are obtained, the samples may not be perceptually different to the particular section of the population you have selected.

2.4.2. Analyses
(a) Internal preference mapping

The first step is usually to centre the scores of each individual to zero and scale them to unit variance.

The second is to carry out a one-way analysis of variance to check for significant order effects. If the result is positive the residuals from the analysis can be used for subsequent analyses.

Internal preference mapping is carried out next. Important output that should be collected includes: a detailed summary of the cumulative variance accounted for each subject over the first seven principal axes; graphs of stimuli and subjects relative to all pairwise combinations of the first three principal axes. In our GENSTAT implementation of MDPREF the assessors are not placed on the unit circle as in Fig. 1, but placed a distance corresponding to the variance accounted for by the two dimensions of the graph. The graphs are usually reproduced with subject number replaced by other classifications such as age or brand usage etc.

(b) External preference mapping
The external configuration is usually taken from the first two or three principal component dimensions of a sensory profile or collection of instrumental data. Although it is not essential, zero correlation between the various dimensions of the external configuration simplifies the arithmetic if PREFMAP is by regression rather than a specialised package (see next section). We usually fit the ordinary vector model (Phase 4), the circular ideal point model (Phase 3) and the elliptical ideal point model (Phase 2); we then examine the variance explained by each model and record the number of subjects showing a satisfactory fit.

Graphs of vector positions and ideal points are output for those individuals that are well fitted.

2.4.3. Computing Aspects

Programs and packages are listed at the end of the chapter. However it is not difficult to carry out *metric* preference mapping using most statistical packages.

Internal preference mapping is carried out by finding the left and right eigenvectors of the matrix with n rows, one for each subject, and p columns, one for each sample. Wold *et al.* (1982) give a simple algorithm.

External preference mapping is really a series of regression models fitting a preference vector y of length p to some external variables x_1, x_2, x_r.

In Phase 1 we fit

$$y = b_0 + \sum_{i=1}^{r} b_i(x_i - \bar{x}_i)$$

In Phase 2, the circular ideal point model, one new term is introduced. This is the sum of the squared coordinates. Now we fit

$$y = b_0 + \sum_{i=1}^{r} b_i(x_i - \bar{x}_i) + b_{r+1} \sum_{i=1}^{r} (x_i - \bar{x}_i)^2$$

The ideal point is then found by differentiation to be at coordinates

$$q_a = -b_a/2b_{r+1}$$

for each of the external axes $1 \ldots a$.

In Phase 3, the elliptical ideal point model, a separate regression coefficient is found for each squared term. Thus we fit

$$y = b_0 + \sum_{i=1}^{r} b_i(x_i - \bar{x}_i) + \sum_{i=1}^{r} b_{r+i}(x_i - \bar{x}_i)^2$$

and the ideal point now has coordinates

$$q_a = -b_a/2b_{r+a}$$

Readers who wish to pursue this further should consult Schiffman *et al.* (1981) or Coxon (1982).

3. MULTIDIMENSIONAL METHODS IN THE SENSORY EVALUATION OF FOODS

The three main techniques which can be used to derive a representation of stimuli (i.e. food products or experimental samples) in multidimensional space are:

1. Descriptive profiling using a consensus vocabulary (sometimes known as consensus profiling) (see Chapter 7 by J. J. Powers);
2. Free choice profiling, where each subject uses his or her own private vocabulary (Williams and Langron, 1984);
3. Similarity estimation in conjunction with multidimensional scaling (MDS).

A multidimensional representation, however obtained, is merely a 'picture' of the stimuli in one or more dimensions, where those stimuli that are located proximally are likely to be perceptually similar, whilst those which are spaced apart are different. From this it is usually possible to deduce the nature of the main discriminating features amongst the samples. Such information may be extremely useful in fundamental research, product development and marketing.

With the two descriptive techniques mentioned above, subjects either collectively or individually derive a list of words to qualitatively describe the sensory characteristics perceived in the stimuli. The magnitude of each characteristic, in each sample, is then quantified using either a category or line scale. As a consequence, an attribute x-stimulus data matrix is obtained for each subject. Where a consensus vocabulary is used, principal component analysis (PCA) is used to derive a representation of the stimuli in multidimensional space (Jolliffe, 1986). With free choice profiling, generalised Procrustes analysis (GPA) is used to obtain the multidimensional representation (Arnold and Williams, 1986).

In the application of the (dis)similarity/MDS approach, subjects merely estimate overall (dis)similarity within all possible pairs of the stimuli under investigation. This gives rise to a simple (dis)similarity matrix, usually one for each subject. The matrices are submitted to MDS, to yield a spatial representation of the stimuli in n dimensions.

There are several reasons for choosing (dis)similarity estimation in conjunction with MDS rather than other forms of descriptive sensory analysis. Using descriptive sensory methods, it is necessary to assume that the relevant sensory characteristics of the stimuli can be adequately expressed in words. Previous experience with other food systems (e.g. Thomson, 1981) has shown that subjects often have great difficulty in finding meaningful qualitative expressions to describe the sensory characteristics of unusual or complex perceptual phenomena. Multidimensional scaling avoids this problem because subjects merely quantify overall perceptual differences between samples. With descriptive sensory methods the perceptual attributes of foods are compared in terms of their component sensory characteristics. By so doing, the overall impression created by the sensory characteristics of the food may not be taken fully into account. The well known Gestalt maxim, 'the whole is more than the sum of its parts', would seem to be particularly appropriate in this context. This problem is avoided when using MDS. Another advantage is that, unlike PCA or GPA, non-metric MDS methods (see Section 4) make minimal distributional assumptions about the data.

Williams and Arnold (1985) and Tunaley (1988) compared the multidimensional representations obtained when consensus profiling, free choice profiling and (dis)similarity estimation/MDS procedures were applied to the same set of stimuli. In each study, the stimulus configurations obtained by the three methods were very similar. However, when using the MDS approach, where no qualitative descriptors are obtained, interpretation of the stimulus space is sometimes more difficult.

4. WHAT IS MULTIDIMENSIONAL SCALING?

The basic data for most multidimensional scaling (MDS) analyses is an interdistance matrix, which is a triangular (or lower half) matrix with one

TABLE 2
MATRIX OF INTERDISTANCES BY ROAD BETWEEN 12 UK TOWNS

	Aberdeen	Aberystwyth	Cardiff	Dover	Edinburgh	Exeter	Glasgow	Hull	Liverpool	London	Norwich
Aberystwyth	427										
Cardiff	484	100									
Dover	552	284	226								
Edinburgh	115	312	369	436							
Exeter	555	196	119	243	440						
Glasgow	142	313	370	459	44	441					
Hull	337	214	224	232	222	278	245				
Liverpool	327	100	164	268	211	235	213	128			
London	488	212	154	72	373	170	394	168	197		
Norwich	471	270	236	153	356	277	378	143	215	111	
Penzance	663	304	227	355	548	112	549	386	343	281	389

row and one column pertaining to an object. Each entry in the matrix gives the distance between the objects associated with the corresponding row and column. The interdistance matrix (travelling by road) between 12 towns in the UK is shown in Table 2.

In its simplest form MDS attempts to obtain a graphical representation of the objects such that the interdistance matrix obtained from their relative positions on the graph closely approximates the original interdistances. A graphical representation consists of axes (usually called dimensions) that are perpendicular (orthogonal) to each other and each point has a coordinate for each dimension. The task of MDS is therefore to obtain a set of coordinates for each object.

The coordinates recovered by the metric MDS technique, principal coordinates analysis, henceforth referred to as PCO (Torgerson, 1952; Gower, 1966) were used to draw the two-dimensional plot in Fig. 3. This figure displays 96% of the variation and is clearly a very good approximation to the true configuration.

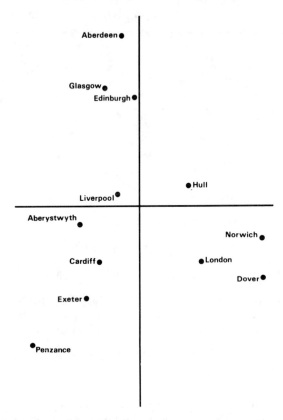

FIG. 3. Metric multidimensional scaling representation of interdistances between 12 UK towns.

Note that PCO actually produced coordinates relative to 11 dimensions. This is because, mathematically, it is possible for n points to be in a space that requires up to $n-1$ dimensions to define it. In this example the towns can be considered to be in two-dimensional space, and we are therefore quite content with Fig. 3.

Principal coordinate analysis assumes that the space in which the objects lie is metric. A simple definition of a metric space is one in which for any three points A, B and C in the space:

$$\text{distance AB} \leq \text{distance AC} + \text{distance CB}$$

This may seem a perfectly reasonable assumption to make, but consider the distance matrix in Table 3 which represents the pairwise difference

TABLE 3
A NON-METRIC DISSIMILARITY MATRIX

A				
B	5			
C	1	7		
D	4	3	7	
	A	B	C	D

scores given by a hypothetical subject between four stimuli—A, B, C, D. The metricity condition is broken twice:

$$BA + AC = 6 \quad \text{but} \quad BC = 7$$
$$CA + AD = 5 \quad \text{but} \quad CD = 7$$

The interdistance matrix is therefore non-metric.

An understandable reaction by an experimenter faced with such data would be to argue that these scores are inadmissible and should therefore be discarded. However, if any number $C \geq 2$ is added to each of the scores then the metricity condition is satisfied for all triplets. So the scores for another subject who perceived exactly the same difference pattern but happened to use large numbers would be admitted to the analysis!

The realisation that many non-metric distance matrices had metric counterparts with distances in exactly the same rank order led psychometricians in the 1950s and 1960s to search for methods of representing non-metric matrices in metric spaces. A number of methods were developed, notably Shepard (1962a,b), but by far the most elegant and widely adopted algorithm was proposed by Kruskal (1964a,b).

The task of the MDS algorithm may be stated formally as follows:

Define δ_{ij} as the given dissimilarity between objects i and j ($i, j = 1$ to N). The number of dimensions of the solution is set by the experimenter to be K. Then each object is to be represented in a new metric space by coordinates x_{ik} ($k = 1$ to K). The resulting metric interdistance matrix (d_{ij}) is calculated as

$$d_{ij}^2 = \sum_{k=1}^{K}(x_{ik} - x_{jk})^2$$

The advance made by Kruskal was to define as explicit badness of fit measure between the δ_{ij} and the d_{ij} which he termed STRESS. This was defined as

$$\text{STRESS} = \left[\frac{\sum_i \sum_j (d_{ij} - \hat{d}_{ij})^2}{\sum_i \sum_j d_{ij}^2} \right]^{1/2}$$

where the $\hat{d}_{ij}$ are distances that are closest to the d_{ij} but still maintain the original ranking of the δ_{ij} values. Another STRESS formula (STRESSFORM 2) is also used as a standard form. This differs only in the form of the denominator which is $\sum_i \sum_j (d_{ij} - \hat{d})^2$ as opposed to $\sum_i \sum_j d_{ij}^2$ (Here $\hat{d}$ is the average of d_{ij} values).

To show how effectively the non-metric MDS algorithm recovers structure, the 66 interdistances between the 12 UK towns were ranked

TABLE 4
MATRIX OF INTERDISTANCE RANKS OF 12 UK TOWNS INPUT TO NON-METRIC MDS

	Aberdeen	Aberystwyth	Cardiff	Dover	Edinburgh	Exeter	Glasgow	Hull	Liverpool	London	Norwich
Aberystwyth	53										
Cardiff	60	3									
Dover	64	38	26								
Edinburgh	7	40	47	55							
Exeter	65	17	8	31	56						
Glasgow	10	41	48	58	1	57					
Hull	43	22	25	28	24	36	32				
Liverpool	42	4	14	33	19	29	21	9			
London	61	20	13	2	49	16	53	15	18		
Norwich	59	34	30	12	46	35	50	11	23	5	
Penzance	66	49	27	45	62	6	63	51	44	37	52

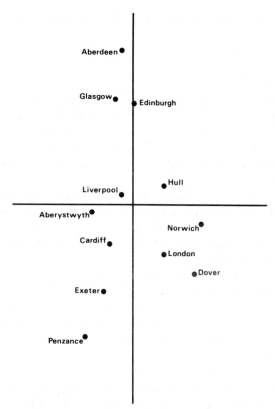

FIG. 4. Non-metric multidimensional scaling two-dimensional representation of interdistance rankings between 12 UK towns.

(Table 4). The ranks were input to the non-metric program KYST-2A (see Section 7).

The two-dimensional configuration thus obtained appears in Fig. 4 and is virtually indistinguishable from Fig. 3 obtained using the original interdistances. The accuracy of recovery of structure from purely ordinal data led workers to conclude that a complete set of difference rankings can only emanate from a fairly specific configuration. This was not realised until the advent of non-metric MDS algorithms and gave workers additional confidence in the use of ordinal data for analysis by MDS. The other implication of these results is that a satisfactory recovery may be possible from an incomplete interdistance matrix. This is discussed further in Section 6.3.

5. MDS METHODS AND CASE STUDIES

5.1. Two-way Scaling

5.1.1. The Basic Model
The basic non-metric model and algorithm have been introduced in Section 4. However, there are three further aspects that require elaboration before the reader can sensibly interpret a non-metric MDS computer output. These are the principle of least-squares monotone regression, the definition of adequate STRESS values, and the determination of suitable dimensionality.

In two-way scaling, a single interdistance matrix is obtained and a graphical representation is required. It is called two-way because the matrix is effectively a two-way table. The matrix is assumed to be symmetric, although there are MDS programs that will handle asymmetric matrices. These are not discussed here.

Least-squares monotone regression (Kruskal, 1964b) is used at the end of each iteration to calculate the STRESS (goodness) of fit of the reconstructed distances d_{ij} to the observed dissimilarities δ_{ij}. The regression estimates distances $\hat{d}_{ij}$ that are closest to the d_{ij} *but are in the same rank order as the* δ_{ij}. The plot of d_{ij} (vertical axis) versus δ_{ij} is termed the Shepard scatter diagram.

The Shepard diagram for the final iteration of the two-dimensional non-metric MDS of the 12 UK city interdistance rankings appears in Fig. 5. The least-squares monotone regression line has been superimposed. Only the points denoted by a D give $\hat{d}_{ij}$ that are not equal to d_{ij}. These plots can also be used to identify erroneous or atypical interdistances and should always be inspected.

There are two approaches to defining a satisfactory STRESS value. The first is to use the formula derived by Spence (1979) using Monte Carlo simulations of randomly generated configuration. This is calculated as

$$\text{STRESS} = a_0 + a_1 K + a_2 N + a_3 \ln K + a_4 (\ln N)$$

where K is dimensionality, N is number of stimuli, and the a_1 are fixed constants with values $a_0 = 524 \cdot 25$, $a_1 = 33 \cdot 80$, $a_2 = -2 \cdot 54$, $a_3 = -307 \cdot 26$ and $a_4 = 588 \cdot 35$. In this formula *STRESS is measured in units of 0·001*. The calculated value of STRESS for $N = 12$ and $K = 2$ is 0·227. The value of 0·024 obtained for the non-metric fit is thus well within this value. This is to be expected since this was an error-free matrix.

The second, and more traditional, way of defining a suitable STRESS

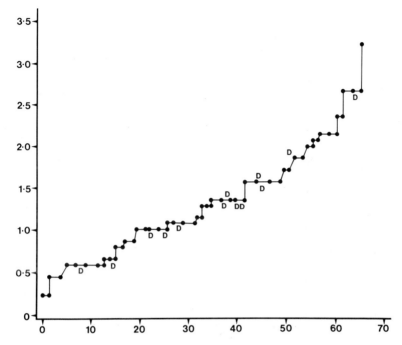

FIG. 5. Shepard scatter plot of recovered interdistances from non-metric multidimensional scaling (vertical axis) versus interdistance rankings. Least squares monotone regression line has been superimposed.

value is the rule of thumb given in the MDS(X) documentation (see Section 7) on the non-metric program MINISSA. This considers:

$$S < 0 \cdot 01 \text{ excellent}$$
$$0 \cdot 01 < S < 0 \cdot 05 \text{ good}$$
$$0 \cdot 05 < S < 0 \cdot 10 \text{ fair}$$
$$0 \cdot 10 < S < 0 \cdot 15 \text{ moderate}$$
$$0 \cdot 15 < S \quad \text{poor}$$

where S is the value of STRESS.

Suitable dimensionality is usually defined as the lowest number of dimensions that gives a satisfactory STRESS value. In practice many workers plot the STRESS values for the MDS solutions against the number of dimensions (K) and select the number at which STRESS starts to flatten off. The choice of dimensionality (K) is quite clear cut for our example, as shown by the stress versus dimensionality plot in Fig. 6.

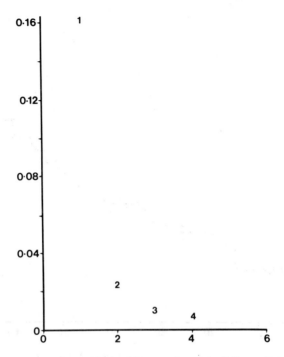

FIG. 6. Plot of STRESS (vertical axis) versus number of dimensions used in non-metric MDS solution.

5.1.2. Case Study—Ekman Colour Data

This analysis is selected not only because it is one of the first published examples of MDS (Shepard, 1962b) but it is also a good example of the effectiveness of a low dimensional solution.

Ekman (1954) obtained measures of similarity between 14 colours that varied only in hue. All possible pairwise comparisons were made, the results averaged over subjects and scaled such that 0 implied no similarity at all and 1 indicated an identical pair. Ekman treated this similarity data as representing correlation coefficients between the 14 stimuli and used factor analysis. This is a metric technique which obtains an orthogonal set of axes that most effectively explains the data. These axes are then rotated to directions that are easier to interpret. Ekman concluded that five factors (not all uncorrelated) were necessary to account for the similarity measures.

The two-dimensional non-metric MDS solution achieved by Shepard (1962b) is reproduced in Fig. 7. This configuration bears a close

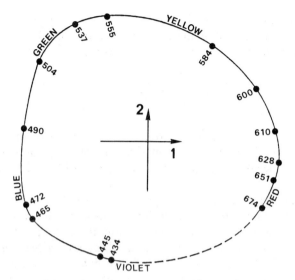

FIG. 7. Two-dimensional representation of 14 colours varying in hue as obtained by non-metric multidimensional scaling (Shepard, 1962b).

resemblance to the conventional colour circle and leads to a more parsimonious summary of the data. Shepard concluded that it was more appropriate to treat the similarity measures as proximity measures at the ordinal level rather than as correlation coefficients.

5.2. Multi-Way Scaling

5.2.1. The Weighted MDS Model

The previous section described two-way analysis, where a multidimensional representation is derived from a single data matrix. Under these circumstances, each datum in the matrix is either a single measurement, or a mean value across subjects, replicates or both. When the data are averaged in this way, the influence of individual differences or inter-replicate differences is lost. Three-way scaling, where a separate matrix is input for each subject, was developed by Carroll and Chang (1970), specifically to account for individual differences in pairwise judgements. In four-way scaling, separate matrices for each replicate and each subject can be input.

Carroll and Chang (1970) postulated that individuals might perceive the same basic configuration of stimuli (the consensus configuration) but give different weight to the underlying dimensions in evaluating pairwise differences. Thus, extending the notation of Section 4, a dissimilarity score d_{ijm} between stimuli i and j for subject m would be modelled as:

$$d_{ij}^2 = \left[\sum_{k=1}^{K} W_{km}(x_{ik} - x_{jk})^2\right]^{1/2}$$

where x_{ik}, $k = 1 - K$ are the coordinates of stimuli i relative to the k-dimensions of the underlying configuration. The W_{km} are the *weights*, sometimes called *saliences*, that subject m applies to differences along the K-dimensions.

The algorithm proposed by Carroll and Chang (1970), and implemented in the INDSCAL program, is quite different from the two-way non-metric algorithm of Kruskal (1964b). The dissimilarity matrices are first made metric by adding suitable constants, as discussed in Section 4 of this chapter. In the first iteration a set of coordinates x_{ij}, where $i = 1 - N$, $k = 1 - K$ (K fixed by experimenter), are obtained for an initial set of weights $W_{im} = 1$, The x_{ik} are then fixed and a new set of weights calculated to give optimum improvement in fit. These weights are then fixed and the process repeated until convergence is obtained.

The extension to four-way scaling simply introduces another set of weights v_{km} and the basic model is now written as:

$$d_{ijm} = [\sum W_{km}V_{km}(x_{ik} - x_{jk})^2]^{1/2}$$

There are two very important differences between two-way and multi-way MDS solutions. First, the underlying dimensions of the consensus configuration of the stimuli are not necessarily independent, although the axes are usually presented as perpendicular in any plots. Dimensional independence can be examined by recording the correlations between dimensions from the computer output. The second difference is that because the associated weights relate specifically to the selected dimensions of the stimulus space, it is not possible to rotate this space to directions of more convenient interpretation, without recalculating the weights. This would inevitably lead to a less than optimum solution.

The weighted model of Carroll and Chang (1970) has been successfully applied to a wide range of stimuli (for further examples see Wish and Carroll, 1974). The individual differences scaling program INDSCAL, written by these authors, is still widely used but almost every MDS package contains programs capable of performing metric or non-metric multi-way scaling. Schiffman *et al.* (1981) (a highly recommended text for MDS users) report that metric and non-metric solutions do not differ very much in practice.

5.2.2. Case Study—Perception of Electrical Stimuli

The objective of the experiment was to examine the application of the INDSCAL algorithm to pairwise comparisons of electrical stimuli that varied in both frequency and voltage (MacFie and Thomson, 1981). Nine stimuli were created according to the design shown in Fig. 8 and nine subjects were required to give a number reflecting the overall perceived difference within each pair of stimuli. No constraints on magnitude or direction of the scores were imposed. No formal training was given, and the subjects were deliberately kept unaware of the precise nature of the stimuli and object of the experiment. Five replicates were obtained, and there were no missing values. The first two replicates were regarded as preliminary to permit subjects to become consistent in their use of number, and were not used in the INDSCAL analysis.

The two-dimensional configuration of stimuli obtained is shown in Fig. 9. This was obtained using four-way INDSCAL analysis, which is available as part of the MDS(X) series of programs (see Section 7). The correlations of the computed matrices using the subject and replicate weights varied from 0·79 to 0·91, which was considered satisfactory.

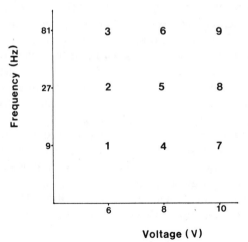

FIG. 8. Plot of the nine electrocutaneous stimuli against frequency and voltage. Coordinates of frequency are plotted after transformation of logarithm to the base 3.

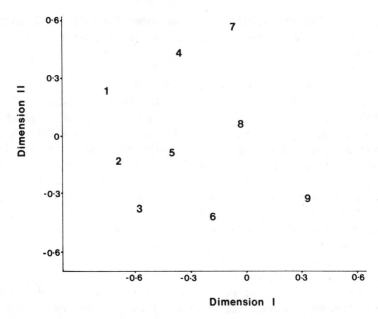

FIG. 9. Two-dimensional representation of electrocutaneous stimuli in the group space given by a four-way INDSCAL. Numbering of stimuli as in Fig. 4.

The weights for the nine subjects (1–9) and the three replicates (A, B, C) are plotted in Fig. 10. The positions of A, B, C, which represent the three replicates are positioned close together in the top right-hand corner of the plot. The square of the distance of each point from the origin is roughly proportional to the percentage of variation accounted for in the INDSCAL solution for the individual or replicate to which that point refers. Thus replicates A, B, C are well recovered and also weight equally and similarly on the two dimensions of the stimulus space.

Comparing Figs 8 and 9 indicates that the dimensions of the solution can be interpreted as voltage (horizontal) and frequency (vertical) although there is evidence of interaction between the two components on the size of the distances. This is discussed further by MacFie and Thomson (1981). The correlation between the dimensions is small (-0.06). The direction of the stimuli along the dimensions is arbitrary; it is the internal structure that is relevant.

The pattern of subject weights is of interest in that three categories emerge. Subjects 1, 2, 3, 6 and 9 weight heavily on the voltage dimension, Subjects 4 and 5 weight voltage slightly more than frequency, and Subjects

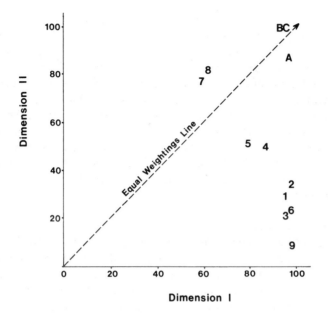

FIG. 10. Two-dimensional configuration of Replicates 3, 4 and 5 (denoted by A, B and C, respectively) and subjects (denoted by digits 1–9), defined by their relative weightings on the group space (Fig. 5).

7 and 8 weight frequency more than voltage. It is pertinent to note that the last two subjects were both physicists and probably more familiar with the concept of frequency in an electrical impulse.

It was concluded that the INDSCAL model provided a satisfactory fit and gave an easily interpretable solution. In retrospect it is possible that operating on the logarithm of the given dissimilarities would have produced a 'squarer' configuration. This is recommended by Ramsay (1982), but his work was subsequent to our analysis. However, potential users of INDSCAL are urged to bear this in mind.

6. GUIDELINES FOR THE DESIGN OF MDS EXPERIMENTS

6.1. Choice of Stimuli

When the experimenter has no control over the stimuli, for example having to select from a commercial product range, stimuli should be selected to span the perceptual space under investigation. Equally, stimuli should be perceptually different or comparisons will be wasted. A pilot study is usually essential. It is also advisable to make as many physical or chemical measurements as possible. By relating the perceived configuration to these measurements it is often possible to interpret the perceptual dimensions in meaningful terms such as toughness, saltiness, bitterness, etc.

When the experimenter is able to manipulate and even create the stimuli the design should reflect the known or suspected perceptual dimensions. For example with two dimensions the regular 3×3 square shown in Fig. 8 is useful, although a 4×4 design is preferable. With three dimensions a $3 \times 3 \times 3$ cube may be selected and the practice is easily extended to higher-order designs.

6.2. Number of Stimuli
The design of the experiment is usually severely constrained by this factor.

This is because there are $n(n-1)/2$ possible pairwise comparisons between n stimuli. In Section 6.3. we discuss various ways in which the number of pairwise comparisons can be reduced, but the number of stimuli selected is ultimately governed by the suspected dimensionality of the data. For adequate recovery of a single-subject single-replicate experiment Kruskal and Wish (1978) recommend a minimum of 9 stimuli for two dimensions, 13 for three, and 17 for four. With multiple-subject multiple-replicate designs it seems logical to suppose that these numbers can be trimmed slightly. This is considered in the next section.

Regarding the likely dimensionality of a perceptual space, this does seem to vary according to the sense involved. Three dimensions is the usual expectation for taste and odour, although Schiffman and Erickson (1971) have proposed a fourth. As many as seven dimensions have been suggested for colour perception by Chang and Carroll (1980). Getty et al. (1979) derived a five-dimensional model from similarity judgements of complex sounds and validated the model by predicting the performance of subjects in identifying the sounds.

6.3. The Number of Paired Comparisons
The fact that $n(n-1)/2$ pairwise judgements are required for a complete design frequently deters experiments from the MDS approach. For example a relatively small experiment involving 12 stimuli and 10 subjects would require 660 judgements to be made! However, it is possible that all $n(n-1)/2$ comparisons may not be necessary for adequate recovery (Young and Cliff, 1972), and since most non-metric MDS algorithms can operate

with incomplete dissimilarity matrices, the solution may be to select a subset.

Spence and Domoney (1974) compared a number of possible designs for incomplete sampling of a single dissimilarity matrix and concluded the following:

1. The most effective sampling strategy was a cyclic design based on a partially balanced incomplete block structure. These designs have the property that each stimulus appears equally often (balance) and it is not possible to divide the selected pairwise comparisons so that they refer to mutually exclusive subsets of stimuli (connectedness). Clatworthy (1973) gives tables to allow such designs to be constructed, and the method is discussed at length by David (1963).
2. Random selection of the pairwise comparisons (with checks for connectedness) proved almost as effective as cyclic designs.

Regarding the number of comparisons that could safely be deleted from the design, Spence and Domoney examined the correlation between recovered and true configurations by Monte Carlo simulation methods. Figure 11 shows their results for 40-point (stimuli) configurations containing zero and medium error levels, the curve of correlation against percentage remaining dissimilarities flattens off at about 40%. It was concluded that in practice the ratio of the number of comparisons to be

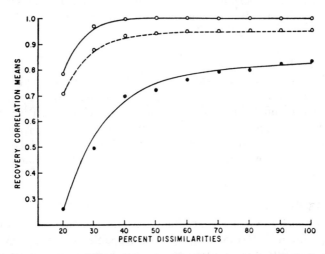

FIG. 11. Recovery correlation means vs percentage remaining dissimilarities for randomly generated configurations of 40 points (Spence and Domoney, 1974).

retained to the 'degrees of freedom' $m(n-1) - m(m-1)/2$ should exceed 3·5 (m = number of dimensions, n = number of stimuli).

More recently, Whelehan *et al.* (1987) found that up to 40% of the distances in a complete design could be excluded, provided that replicate measurements indicate that error levels are not severe. Van Trijp and Thomson (1986) studied the influence of missing values on the data from the electric shock experiment reported in Section 5.2.2. In spite of the apparent reproducibility of the three replicates (A, B and C, Fig. 10) the stimulus configuration was subject to change when even a small proportion of the data was omitted.

7. MDS PROGRAMS PACKAGES AND DOCUMENTATION

7.1. Bell Laboratories Packages
A suite of 18 programs including KYST-1, INDSCAL, MDPREF and PREFMAP is available from Bell Laboratories, Computing Information Library, 600 Mountain Avenue, Murray Illill, New Jersey 07974, USA. In the UK it is available from SIA Computer Services, Ebury Gate, 23 Lower Belgrave Street, London SW1W 0NW.

We have no experience of using these programs but have used KYST-2A, a more recent version. This program is well documented in 'How to use KYST-2A, a very flexible program to do multidimensional scaling' by J. B. Kruskal, K. W. Young and J. B. Seery, available from Bell Laboratories at the address above.

7.2. ALSCAL
ALSCAL (alternating least-squares algorithm) is an extremely comprehensive and flexible program developed by Forrest W. Young and is available from him (with documentation) at L. L. Thurstone Psychometric Laboratory, University of North Carolina, Chapel Hill, North Carolina 27514, USA. Alternatively, a similar program PROCALSCAL is available as part of the SAS package. The UK distributor is SIA Computer Services (address in Section 7.1).

7.3. MDS(X)
MDS(X) is a library of ten programs including: a non-metric two-way MDS program MINISSA, INDSCAL (capable of up to seven-way analysis), MDPREF and PREFMAP. It was developed in the UK by Professor Tony Coxon, Peter Davies and Charles Jones at University

College, Cardiff. We have used these programs successfully and the documentation is good, particularly for PREFMAP. Care must be taken in setting up the data input. One criticism is that the two-way scaling program, in our version, will not handle missing values. We use KYST-2A as a backup. MDS(X) is distributed by Program Library Unit, University of Edinburgh, 18 Buccleuch Place, Edinburgh EH8 9LN, UK.

7.4. MULTISCALE

This is the 'odd man out' of MDS packages as it uses maximum likelihood estimation to obtain parameter estimates. Ramsay (1982) describes the principles. A good discussion of the problems in this approach is appended to that paper. A unique aspect of MULTISCALE is the provision of confidence regions around the stimuli points. The program and excellent documentation is distributed by Scientific Software Inc., PO Box 536, Mooresville, Indiana 46158, USA.

7.5. GENSTAT

A GENSTAT program that performs both internal and external metric preference mapping is available from the first author.

REFERENCES

Arnold, G. M. and Williams, A. A. (1986). The use of generalized Procrustes techniques in sensory analysis. In: *Statistical Procedures in Food Research*, J. R. Piggott (Ed.), Elsevier Applied Science, London, pp. 233–53.
Bennett, J. F. and Hays, W. I. (1960). Multidimensional unfolding: Determining the dimensionality of ranked preference data. *Psychometrika*, **25**, 27, 43.
Carroll, J. D. (1972). Individual differences in multidimensional scaling. In: *Multidimensional Scaling: Theory and Applications in the Behavioural Sciences, Vol. 1*, R. N. Shepard, A. K. Romney and J. B. Nerlave (Eds), Seminar Press, New York, pp. 105–55.
Carroll, J. D. and Chang, J. J. (1967). Relating preference data to multidimensional scaling solutions via a generalization of Coombs' unfolding model. Paper presented at meeting of Psychometric Society, Madison, Wisconsin.
Carroll, J. D. and Chang, J. J. (1970). Analysis of individual differences in multidimensional scaling via an N-way generalization of 'Eckart-Young' decomposition, *Psychometrika*, **35**, 283–319.
Chang, J. J. and Carroll, J. D. (1968). How to use MDPREF, a computer program for multidimensional analysis of preference data. Unpublished report, Bell Telephone Laboratories.

Chang, J. J. and Carroll, J. D. (1980). Three are not enough. An INDSCAL analysis suggesting that colour space has seven (± 1) dimensions. *Color Res. Appl.*, **5**, 193–206.

Clatworthy, W. H. (1973). *Tables of Two Associate Class Partially Balanced Designs*, National Bureau of Standards, Applied Mathematics Series.

Coxon, A. P. M. (1974). The mapping of family composition preferences: a scaling analysis. *Soc. Sci. Res.*, **3**, 191–210.

Coxon, A. P. M. (1982). *The User's Guide to Multidimensional Scaling*. Heinemann Educational Books, London.

David, H. A. (1963). The structure of cyclic paired comparison designs. *J. Austral. Math. Soc.*, **3**, 117–27.

Dravnieks, A. (1982). Odour quality: Semantically generated multidimensional profiles are stable. *Science*, **218**, 799–801.

Ekman, G. (1954). Dimensions of colour vision. *J. Psychol.* **38**, 467–74.

Francombe, M. A. and MacFie, H. J. H. (1985). Dissimilarity scaling and INDSCAL analysis in the study of flavour differences between normal pH and DFD beef. *J. Sci. Food Agric.* **36**(8), 699–708.

Getty, D. J., Swets, J. A., Swets, J. B. and Green, D. M. (1979). On the prediction of confusion matrices from similarity judgements. *Percept. Psychophys.*, **6**, 1–19.

Gower, J. C. (1966). Some distance properties of latent root and vector methods used in multivariate analysis. *Biometrika*, **53**, 325–38.

Harries, J. M. and MacFie, H. J. H. (1976). The use of a rotational fitting technique in the interpretation of sensory scores for different characteristics. *J. Food Technol.* **11**, 449–56.

Jolliffe, I. T. (1986). *Principal Component Analysis*. Springer-Verlag, New York.

Kruskal, J. B. (1964a). Multidimensional scaling by optimising goodness of fit to a nonmetric hypothesis. *Psychometrika*, **29**, 1–27.

Kruskal, J. B. (1964b). Nonmetric multidimensional scaling: A numerical method. *Psychometrika*, **29**, 28–42.

Kruskal, J. B. and Wish, M. (1978). *Multidimensional Scaling*, Sage Publications, Beverley Hills.

McEwan, J. A. and Thomson, D. M. H. (1988). An investigation of the factors influencing consumer acceptance of chocolate confectionery using the repertory grid method. In: *Food Acceptability*, D. M. H. Thomson (Ed.), Elsevier Applied Science Publishers, London.

MacFie, H. J. H. and Thomson, D. M. H. (1981). Perception of two-component electrocutaneous stimuli. *Percept. Psychophys.*, **30**, 473–82.

Ramsay, J. O. (1982). Some statistical approaches to multidimensional scaling data. *J. R. Statist. Soc. A.*, **145**, 285–312.

Schiffman, S. S. and Erickson, R. P. (1971). A psychological model for gustatory quality. *Physiol. Behav.*, **7**, 617–33.

Schiffman, S. S., Reynolds, M. L. and Young, F. W. (1981). *Introduction to Multidimensional Scaling: Theory, Methods and Applications*, Academic Press, New York.

Shepard, R. N. (1962a). The analysis of proximities: multidimensional scaling with an unknown distance function I. *Psychometrika*, **27**, 125–40.

Shepard, R. N. (1962b). The analysis of proximities: multidimensional scaling with an unknown distance function II. *Psychometrika*, **27**, 219–46.

Spence, I. (1979). A simple approximation for random ranking STRESS values. *Multivar. Behav. Res.* **14**, 355–65.
Spence, I. and Domoney, D. W. (1974). Single subject in complete designs for nonmetric multidimensional scaling. *Psychometrika*, **39**, 469–90.
Stone, H. and Sidel, J. L. (1985). *Sensory Evaluation Practices*. Academic Press, London.
Thomson, D. M. H. (1981). An investigation of non-verbal sensory techniques for evaluating meat flavour. Doctoral thesis, University of Bristol.
Thomson, D. M. H., Tunaley, A. and van Trijp, H. C. M. (1987). A reappraisal of the use of multidimensional scaling to investigate the sensory characteristics of sweeteners. *J. Sens. Studies*, **2**, 215–30.
Torgerson, W. S. (1952). Multidimensional scaling: I, theory and method. *Psychometrika*, **17**, 401–19.
Tucker, L. R. (1960). Intra-individual and inter-individual multidimensionality. In: *Psychological Scaling: Theory and Applications*, H. Gulliksen and S. Messick (Eds), J. Wiley & Sons, New York, pp. 155–67.
Tunaley, A. (1988). Development and application of techniques to investigate the perceptual characteristics of sweeteners. Doctoral thesis, University of Reading.
van Trijp, H. C. M. and Thomson, D. M. H. (1986). Unpublished work. Department of Food Science and Technology, University of Reading.
Williams, A. A. and Arnold, G. M. (1985). A comparison of the aromas of six coffees characterized by conventional profiling, free-choice profiling and similarity scaling methods. *J. Sci. Food Agric.*, **36**, 204–14.
Williams, A. A. and Langron, S. P. (1984). The use of free choice profiling for the evaluation of commercial parts. *J. Sci. Food Agric.*, **35**, 558–68.
Williams, A. A., Baines, C. R., Langron, S. P. and Collins, A. J. (1981). Evaluating taster's performance in the profiling of foods and beverages. In: *Flavour '81*. P. Schreier (Ed.), Walter Gruyter, New York, pp. 83–92.
Williams, E. J. (1949). Experimental designs balanced for the estimation of residual effects of treatments. *Austral. J. Sci. Res. Ser. A.*, **2**, 149–68.
Wish, M. and Carroll, J. D. (1974). Applications of individual differences scaling to studies of human perception and judgement. In: *Handbook of Perception, Vol. 2*, E. C. Carterette and M. P. Friedman (Eds), Academic Press, New York.
Whelehan, O. P., MacFie, H. J. H. and Baust, N. G. (1987). Use of individual differences scaling for sensory studies: simulated recovery of structure under various missing value rates and error levels. *J. Sens. Studies*, **2**, 1–8.
Wold, S., Albano, C., Dunn, W. J., Esbensen, K., Hellberg, S., Johannson, E. and Sjostrom, M. (1982). Pattern recognition: finding and using regularities in multivariate data. In: *Food Research and Data Analysis*, H. Martens and H. Russwurm (Eds), Applied Science Publishers, London.
Young, F. W. and Cliff, N. (1972). Interactive scaling with individual subjects. *Psychometrika*, **37**, 385–415.

INDEX

Acceptance testing
  affective methods for, 301, 302
  analytical methods for, 301–2
  consumer panel testing, 302–15
  field testing, 308–15
  home testing, 307–8
  institutional testing, 308–15
  laboratory testing, 303–7
  method of measurement, 305
  selection of samples, 304–5
Adaptation
  chromatic, 108–9
  olfactory, 44–5
  taste, 15
Adhesiveness
  definition of, 83
  instrumental measurement of, 93
Affective tests, 301, 302, 382
Ageusic patients, 18
Aging, olfaction affected by, 49–50
Alkaline taste, 8–9, 16
  test solutions used, 193
ALSCAL computer program, 406
Alternative forced choice (AFC)
    procedures, 136
  see also Duo tests; Triangular tests
American Society for Testing and
    Materials (ASTM)
  assessor training guidelines, 191,
    192, 193
  flavour amplitude definition, 198
  vocabulary proposed, 208–9

Amplitude, flavour assessment, 198–9
Analysis of variance (ANOVA), 176,
    348–56
  advantages of, 228
  flavour profiling interaction, 218–19
  multivariate analysis, 225–9, 372
  one-way analysis, 348–52
  product difference determined by,
    223
  two-way analysis, 352–6
Anatomical structures
  olfactory system, 26–30, 37–40
  taste system, 2–8
  vision system, 105–6
Anosmic people, 1, 8, 42
Appearance
  definition of, 104
  factors affecting, 110
  measurement of, 109–26
  texture assessment affected by, 88
Appetisers
  preference testing studies, 284, 286,
    297
  sensory perception of, 224
Arthur D. Little Company, flavour
    panelists in, 197
Artificial Daylight lighting, 116, 118,
    122
Assessors
  analytical ability of, 190, 191, 302
  classification by cluster analysis,
    233, 234, 235

Assessors—contd.
  elimination of results, 236, 239
  evaluation of performance, 212–15, 225–6
  meaning of term, 158
  selection of, 192–3, 203–4, 306
  sensory leaders for, 197, 199, 200, 206
  trained assessors, evaluation of performance, 215–16, 227–9
  training of, 191–6, 204, 306
Association diagrams, 243, 250
Astringency, 191
  test solutions used, 193, 194
Attribute rating, 85, 175, 301, 306
Australia, consumer panels in, 310
Austria, consumer panels in, 310
Autonomic nervous system, olfaction affected by, 47–8

Bacon, opacity of, 119–22
Baked beans, preference testing of, 285, 291, 300
Bartlett's test, 244
Beans, green
  preference testing of, 285, 291, 298
  sensory assessment of, 231–2
Beefburgers/hamburgers
  preference testing of, 284, 290, 298
  texture profile analysis of, 84–6
Beer
  flavour profiling for, 240
  preference testing of, 287
  product differences in, 224–5
  sensory evaluation of, 245
Belgium, consumer panels in, 310
Berry note (in wine), 204
Beverages
  mouthfeel characteristics of, 71–2
  preference testing of, 285, 286–7, 296, 297, 300
  see also Beer; Coffee; Fruit-based beverages; Tea; Vegetable juices; Whisky
Binomial probability tables, assessor rating using, 214–15, 216

Binomial test, 137
Biting processes, 74
Bitter taste, 10
  electric response characteristics for, 14, 17
  test solutions used, 193
  tongue areas responsive to, 12
Bloodborne odours, 46
Blueberry–whey beverage, green, 232, 234
Body-care products, sensory assessment of, 195, 222
Bowman's glands, 29, 30
Box-and-whisker plots, 338
Brazil, consumer panels in, 310
Bread products, preference testing of, 285, 287–8, 297
Breakfast, preference testing studies, 284, 288, 297
Brewing industry, flavour vocabulary used by, 207
Brightness, meaning of term, 113
Brittleness
  definition of, 83
  instrumental measurement of, 93
Brussels sprouts, preference testing of, 285, 291, 298, 300

Cabbage, preference testing of, 285, 291, 298, 300
Cakes, preference testing of, 294, 299
Canada, consumer panels in, 310
Canonical analysis, 253
Caramel odour, 35
Carbonated beverages, preference testing of, 287, 297
Carrots
  flavour difference scoring data, 168, 169
  preference testing of, 285, 291, 292, 300
Category rating, food acceptance testing, 301, 306
Category scaling, 81, 165–8
  comparison with other methods, 174, 175

Cauliflower, preference testing of, 285, 292, 300
Central tendency, 179
Checklists, food attitude testing, 273
Cheese
 flavour profiles of, 224, 227
 product differences in, 224
 sensory assessment of, 228–9, 230, 233
 texture assessment of, 81
Chemical analysis, texture assessment supplemented by, 96
Chemical senses, definition of, 1
Chemicals, odour of, 188, 209
Chewiness
 definition of, 83
 instrumental measurement of, 93
Chewing processes, 74–5
Chi-square test, 137
Chicken meat
 preference testing of, 284, 289, 297
 texture assessment of, 81
Chicken patties, flavour profiling for, 229
Cholecystokinin (CCK), olfaction affected by, 47
Chorda tympani, 7, 9
Chroma, meaning of term, 113
Chromatic adaptation, 108–9
CIELAB colour space system, 112, 114
 meat colours depicted by, 118
CIELUV colour space system, 112, 114
 meat colours depicted by, 118
Cluster analysis (CA), 229–36, 371
Coefficient of
 kurtosis, 340, 341
 skewness, 340, 341
 variation (CV), 337
Coffee
 flavour profiling of, 196
 preference testing of, 285, 287, 297
Cohesiveness
 definition of, 83
 instrumental measurement of, 93

College students, consumer studies among, 284–5, 317, 318–19
Colour
 absorption characteristics, 114–15
 matching, 107
 opponent mechanisms, 107–8
 scatter characteristics, 114–15
 specification, 111–12
 temperature, 111
 terminology used, 113–14
 tridimensional representation of, 104
 vision, 104–9
Colourfulness, meaning of term, 113
Commission Internationale de L'Eclairage (CIE) system (of colour measurement), 104
 application of, 124–5
 uniform colour spaces proposed, 112–13
Computer programs
 cluster analysis, 236, 238
 multidimensional scaling, 406, 407
 preference mapping, 406, 407
 regression analysis, 369
 see also ALSCAL; INDSCAL; KYST; MDPREF; MDS(X); MINISSA; MULTISCALE; PREFMAP; PROCALSCAL; VARCLUS
Cones and rods (in eye), 106
Confidence intervals, 341, 342
Confidence limits, 342
Configurational analysis methods, 371
Consensus profiling, 190–203, 389, 390
 see also Flavour profile method; Quantitative descriptive analysis
Consumer panels
 comparative cost of, 307
 food acceptance testing, 302–15
 home testing, 307–8, 310–11
 in-house/laboratory, 304
 major panels listed worldwide, 310–11
 texture assessment, 80
 see also Panel testing

Cookies, preference testing of, 294, 299
Correlation analysis, comparison with factor analysis, 241
Correspondence analysis, 253
Crisp foods, texture assessment of, 78–9, 81
Cross-modality matching, 162, 163
Cultural differences, sensory testing, 194, 248–9, 250

Data analysis
 descriptive methods, 212–25
 flavour profiling, 212–53
 scaling methods, 176, 181
Data comparison
 more than two sets, 348–56
 non-parametric methods, 343, 356
 one-way analysis of variance used, 348–52
 two independent normal samples, 346–7
 two independent samples, 343–4
 two paired samples, 344–6, 347–8
 two sets of results, 342–8
 two-way analysis of variance used, 352–6
Data presentation, 335, 336–41
 food preference testing, 284–5, 296, 297–9
 multidimensional data, 367–8
 preference lists, 284, 296
 quartile distribution graphs, 296, 299
 winners/losers lists, 296, 297–9
Degrees of freedom
 meaning of term, 349, 351
 variance analysis affected by, 249, 251, 351
Delphi technique, 211
Dendrograms, 371, 372
Denmark, consumer panels in, 310
Descriptive methods
 compared with other methods, 301
 data sets evaluated, 212–25
 practicalities, 203–12

Descriptive methods—*contd.*
 procedures described, 189–203
 statistical analysis for, 225–53
 texture assessment, 81–8
 variations between, 210–13
Descriptive terms
 flavour profile, 205–10
 effectiveness of, 216–18
 see also Vocabulary
Desserts, preference testing of, 285, 293, 294–5, 296, 299
Diet measurement
 reliability of methods, 321, 322
 validity of methods, 321, 322, 325
 variables in, 315–20
Dietary record, 317, 320
Difference testing
 advantages/disadvantages of, 131
 compared with rating scale methods, 164
 popularity of, 131
 texture assessment, 81
Differential sensitivity, meaning of term, 133
Dilution methods, 301, 306
Dimensionality, MDS analysis, 397, 398
Dimethyl sulphide, odour of, 209
Discriminability
 scaling by, 162–3
 stimulus difference related to, 140–1
Discriminal dispersions, 143, 159–60
Discriminal processes, 142, 160
Discriminant analysis (DA), 176, 236, 247–9, 252, 372–3
Discriminant functions, 373
Discriminatory testing, texture assessment, 80–1
Diseases, odours caused by, 188
Duncan's range test, 223, 352
Duo tests, 134, 146, 147, 149
Duo–trio tests, 134, 146, 147, 301, 306

Ekman colour data, MDS analysis of, 398–9

INDEX

Elasticity
  definition of, 83
  instrumental measurement of, 93
Elderly people, olfactory perception of, 49–50
Electric taste, 16–17
Electrical stimuli
  MDS analysis of, 401–3
  taste qualities elicited by, 17
Electrogustometry, 16, 17
Electromyography, mouth movements studied using, 74, 75
Entrées, preference testing of, 284, 288–90, 296, 297–8
Europe, consumer panels in, 310, 311
Evaporated milk, appearance of, 124–6
Experimental design, 178–80, 181, 336, 364–5
  MDS experiments, 403–6
    choice of stimuli for, 403–4
    number of paired comparisons used, 404–6
    number of stimuli used, 404
Expert, meaning of term, 158
Expert trained panel
  purpose of, 302
  see also Assessors
External preference mapping, 383–6
  computer aspects, 388–9
  conduct of, 388
  see also PREFMAP computer program
Eye, anatomy of, 105

F-distribution, 349, 350, 351, 377
  flavour profile assessors, 213, 214, 226, 227
  wine flavour descriptors, 217–18
Factor analysis (FA), 176, 240–2, 370–1
  benefits of, 245–6
  caution required, 246
  comparison with correlation analysis, 241, 371
  orthogonal versus oblique rotation effects, 242–7

Fechner's Law, 141, 142, 159
Feedback forms, food acceptance testing, 309, 312–13
  validity of, 314
Female cycle, olfaction/taste sensitivity affected by, 10, 48
Field food acceptance testing, 308–15
  forms used, 309, 312–15
  interview technique, 308–9, 315
Filiform papillae, 2, 3, 12
Fish
  preference testing of, 284, 288, 297
  texture assessment of, 84
Fisher variance ratios, 349
  see also F-values
Flavour, definition of, 25
Flavour profile method (FPM), 190–201
  amplitude assessment in, 198–9
  assessment described, 199
  compared with other methods, 202, 203, 301, 306
  cost of training, 199–200
  first described, 187
  interaction in, 195–6
  leader for assessors, 197, 199, 200
  logic behind, 190–1
  product reference compounds used, 196–7
  selection of assessors for, 192–3
  statistical analysis applied to, 211
  subjective decisions in, 197–8
  training of assessors for, 191–6
  typical profile illustrated, 201
Flavourists, 188, 208
Fluorescent lights, meat appearance under, 115–18
Foliate papillae, 3, 4, 12
Food acceptance, 300–15
  see also Acceptance testing
Food action rating (FACT) scale, 171, 282, 301, 306
Food attitudes
  interviews, 271–3
  meaning of term, 269
  measurement of, 270–300
  preference testing used, 278–300

Food attitudes—*contd.*
  questionnaires, 271–3
    design of, 273–8
    scales for, 274–6
Food frequency checklist, 317, 318–19, 321
Food habits
  chronological outline of, 268
  consumer studies of, 267–331
  meaning of term, 267
  organisation of, 268
Food intake
  meaning of term, 269
  measurement of, 315–25
  see also Intake studies
Food preferences, *see under* Preference studies
Food selection, meaning of term, 269
Food waste, 270, 325, 327–31
  see also Waste studies
Forced-choice procedures, 136
Fractionation (scaling) method, 162
Fracturability
  definition of, 83
  instrumental measurement of, 93
France, consumer panels in, 310
Free-association vocabulary, 209, 211
Free-choice profiling, 209–10, 229, 371, 389, 390
Freedom, degrees of. *See* Degrees of freedom
Frequency scales, food preference testing, 275, 279–80
Fresh meat
  appearance of, 110
  oxidation of, 115–19
Friedman two-way analysis, 177, 356
Fruit juices
  light scattering from, 122–3
  preference testing of, 284, 286–7, 297
  sweetness–viscosity interaction in, 245
Fruits, preference testing of, 293, 296, 299, 300
Fungiform papillae, 2–3, 12

Gas chromatography
  odours analysed by, 251, 252
  smelling of effluents from, 188, 208
General Foods texture profile, terms used, 83, 84
General Foods Texturometer
  interpretation of curves from, 93
  principle of, 91
  typical curves from, 92
Generalised Procrustes analysis (GPA), 390
Genetics, taste perception, 18
GENSTAT computer program, 387–8, 407
Germany, consumer panels in, 310
Glutamate, taste sensation of, 15, 194
Grape jelly, flavour profile testing of, 200, 201
Great Britain, consumer panels in, 311
Groups, data, comparison of, 372
Guessing, probabilities of, 136–7, 138, 153
Gumminess
  definition of, 83
  instrumental measurement of, 93
Gymnemic acid, 15

Hard palate, texture assessed using, 72
Hardness
  definition of, 83
  instrumental measurement of, 93
Hedonic scaling, 169–71
  category labels used, 170
  contrasted with other scales, 171
  food acceptance/preference testing, 274, 278, 301, 306
  statistical analysis of, 338–9
Histogram, hedonic scores, 339
Home methods
  acceptance testing, 307–8
  intake studies, 316, 317, 320
  waste estimation, 328
Hormones, olfaction affected by, 47–9
Hotelling's T^2 test, 244

Hue
  meaning of term, 113
  two-dimensional representation of, 399
Hunger, olfaction affected by, 46–7
Hunt model (for colour appearance), 108
Hurvich model (of photopigment absorption), 108
Hypogeusic patients, 18

Ice cream, preference testing of, 287, 295, 299
In-house food acceptance testing, 303–7
Individual differences scaling program, 400, 406
INDSCAL computer program, 385, 400–3, 406
Inherent patterning phenomenon (in olfaction), 36
Institute of Food Technologists (IFT), testing methods listed by, 301, 305, 306
Institutional food acceptance testing, 308–15
Institutional methods
  acceptance testing, 308–9, 312–15
  preference studies, 282–300
  waste estimation, 327–31
Instron Universal Testing Machine, texture assessment using, 92, 93, 94
Insulin, olfaction affected by, 46–7
Intake studies
  diet history method used, 320, 321
  diet measurement variables, 315–20
  examples of research, 324–6
  frequency checklist used, 317, 318–19, 321
  household aggregate method, 316
  methods used, 320–4
  pantry inventory used, 321
  recall methods, 316, 317, 321, 324–6
  record methods, 317, 321, 324–6
  weighed intake method used, 321

Interaction
  assessor–assessor, 195–6
  flavour profiling, 195–6
  product–assessor, 195, 218–20, 222
  segregation of, 223
  segregation of, 223
  statistical analysis of, 218–20
  texture–taste, 245
Internal preference mapping, 382–3
  analysis for, 387
  computer aspects, 388
  conduct of, 386
  see also **MDPREF** computer program
International Standards Organization (ISO), sensory assessor training guidelines, 194
International System of Flavour Terminology, 240
Interval estimation, 341–2
Interval scale, meaning of term, 157
Interval scaling, 81, 82
Interviews
  food acceptance testing, 308–9, 315
  food attitude testing
    characteristics of, 271
    design of questions, 277–8
    scales used, 274–6
    structured/unstructured questions used, 272–3
Intravascular olfaction, 46
Ireland, consumer panels in, 311
Italy, consumer panels in, 311

Japan, consumer panels in, 311
Just-noticeable-difference (JND) concept, 141, 159

Kolmogorov–Smirnov two-sample test, 343–4
Kramer analysis, 177
Kruskal MDS algorithm, 393–4
Kruskal–Wallis one-way analysis, 356
Kubelka–Munk (reflectance) theory, 114, 115
  applications of, 119, 121, 122, 126

INDEX

Kurtosis, coefficient of, 340, 341
KYST computer programs, 395, 406

Laboratory food acceptance testing, 303–7
Lamb, roast, preference testing of, 284, 289, 300
Least-significant difference (LSD), calculation of, 351, 355
Least-squares criterion regression analysis, 357, 368
Light scatter, 114–15
  orange juice, 122–3
  protein denaturation as cause, 126
Lightness, meaning of term, 113
Line scaling, 202–3
Linear regression analysis, 357, 360
  sum-of-squares calculated, 378–9
Liquers, preference mapping for, 385–6
Liquids
  instrumental measurement of, 93–4
  sensory assessment of, 71–2, 76–7
Lower senses, definition of, 1

McCormick & Company, flavour profile assessors in, 199–200
Magnitude estimation (ME), 173–4
  compared with other methods, 174–5
  criticisms of, 174
  first developed, 162, 173
  food acceptance testing, 301, 306
  multidimensional approach, 163
  texture assessment, 82
Mail surveys
  consumer studies using, 307, 310–11, 320, 324
  food acceptance testing by, 308, 310–11
Masking techniques, 179
  texture assessment, 79, 88–9
Mastication processes, 26, 73–4
  texture perception during, 72–7

MDPREF computer program, 383, 384, 386, 387, 406
  GENSTAT implementation of, 387–8, 406
MDS(X) computer programs suite, 401, 406–7
Mean square error (MSE), calculation of, 359
Meat
  appearance of, 110, 115–22
  lighting for, 115–16, 118
  preference testing of, 284, 289, 297, 300
  see also Bacon; Beefburgers...; Chicken...; Fresh meat; lamb
Mechanical testing instruments, 93–4, 95
  fingers/jaw/teeth/tongue used as, 73, 77
Median, definition of, 337
Memory, fading of, 138, 152, 278
Metallic taste, 8, 194
Metmyoglobin
  oxidation of, 115, 117
  reflectance spectrum for, 116
Military personnel, consumer studies among, 286–300, 309, 320, 322–3
Milk and milk products
  appearance of, 111, 124–6
  preference testing of, 285, 297, 300
MINISSA computer program, 397, 406
Miraculin, 15–16
Mixture effects, olfaction affected by, 42–4
Monte Carlo simulation methods, 396, 405
Mouth
  physiological function of, 25–6
  texture perception in, 72–7
Mouthfeel characteristics, 72–7
  beverages, 71–2
Mucosa
  olfactory processes in, 28, 29, 30
  taste processes in, 5

Multidimensional data, 366–73
  presentation of, 367–8
Multidimensional scaling (MDS)
  procedures, 163–4
  design of experiments, 403–6
  example of use, 401–3
  explanation of, 390–5
  multi-way scaling, 399–403
  two-scaling, 396–9
  weighted model used, 399–401
Multidimensional unfolding, 385
Multiple alternative forced-choice
  procedures, 148–51
Multiple choice questions, food
  attitude testing, 273
Multiple comparison procedures,
  168–9, 175–6
Multiple discriminant analysis
  (MDA), 247
Multiple regression analysis, 368–9
Multiple scaling, texture assessment,
  82
MULTISCALE computer program,
  407
Multivariate analysis (MVA), 176
  sensory profiling applications, 176,
    211, 225–53
  texture assessment application, 97
Multivariate analysis of variance
  (MANOVA), 225–7, 372
Multivariate data, presentation of,
  367–8
Multiway scaling
  example of MDS analysis, 401–3
  weighted MDS model used,
    399–401
Munsell (colour hue) system, 108, 109,
  112
Myoglobin, oxidation of, 115

Nasal cavity, innervation of, 27
Nasal cycle, 48
Nasal patency, 47
Netherlands, consumer panels in, 311
Neural coding of taste, 12–13

Neural pathways
  olfactory system, 30–1, 37–8, 39
  taste system, 7–8, 9
  vision system, 105, 106
Noise, used in texture assessment, 79,
  88–9
Nominal scale, meaning of term, 156
$trans$-2-Nonanal, odour of, 209
Non-linear relationships, 362–3
Non-parametric methods, variance
  analysis, 343, 356
Non-sensory methods
  sensory data related to, 96–7, 306
  texture assessment using, 89–96
Normal distributions, 144–5, 376–7
Normal probability graph paper, 340
Normality tests, 340
Null hypothesis, 343

Oddity choice test, 136
Odours
  bloodborne, 46
  clearance of, 30
  detection of, 25
  disease-caused, 188
  inhalation rates, effect on
    perception, 41
Oestrogen, olfaction affected by, 48–9
Olfaction
  adaptation processes in, 44–5
  aging effects on, 49–50
  anatomical studies of, 36–7
  autonomic nervous system, effects
    of, 47–8
  chromatographic model of, 36
  functional-group theory of, 34–5
  hormonal influences on, 47–9
  hunger effects on, 46–7
  hydrogen bonding theory of, 33–4
  infrared theory of, 33
  inherent patterning phenomenon in,
    36
  mixture effects on, 42–4
  oxidative theory of, 34
  physiological studies of, 35–6

Olfaction—*contd.*
  puncturing theory of, 33
  sniffing effects on, 40–1
  stereochemical theory of, 34, 35
  structure–activity relationships in, 32–7
  taste supplemented by, 8
  vibration theory of, 33, 34
Olfactory bulbs, structure and function of, 37–40
Olfactory epithelium
  basal cells in, 28, 29
  function of, 30–2
  microvillar cells in, 28
  receptor cells in, 28
  structure of, 28–30
  supporting cells in, 28, 29
Olfactory modulators, 40–50
One-dimensional data, summarising of, 336–40
One-way analysis of variance, 348–52
  sum-of-squares calculated, 378
  see also Univariate analysis
Opacity, 110
  bacon products, 119–22
Opponent colour mechanisms, 107–8
Orange juice, light scattering from, 122–3
Oranges, preference testing of, 293, 299, 300
Ordinal scale, meaning of term, 156–7
Outliers, 366
  assessors as, 229, 236, 239
  effect of, 339
  scatter diagram for, 358, 362
Oxymyoglobin
  oxidation of, 115, 117
  reflectance spectrum for, 116

Paired comparison procedures
  food acceptance testing, 301, 306
  inadequacy of, 132–3
Paired comparisons, Thurlstone theory of, 148–9, 160, 161
Paired preference methods, 301, 306
Paired t-test, 347–8

Panel, meaning of term, 158
Panel testing
  flavour profile, 190–1, 197–8
  olfaction, 41
  texture assessment, 80, 88–9
  see also Assessors; Consumer research; Flavour profile method
Panelists. *See* Assessors
Pasta dishes, preference testing of, 284, 289–90, 297–8
Pathological states, taste perception affected by, 18
Peaches, preference testing of, 293, 300
Percentiles, meaning of term, 338
Perfumers, 188, 208
Physical response, sensory relationship with, 142–5
Pi(π)-mechanisms, retinal vision, 107
Pies, preference testing of, 285, 294–5, 299
Pigments
  fresh meat, 115–19
  mixing of, 104
  retinal, 106
Plate waste, 325
  see also Waste studies
Polyhedral testing, 135
Potassium carbonate, taste of, 9
Potassium chloride, taste of, 10, 11
Potato products, preference testing of, 285, 291, 298
Power law concept, 162, 163
Preference mapping, 381
  computing aspects, 388–9, 406, 407
  external preference mapping, 383–6
  guidelines to conduct and analysis of, 386–9
  internal preference mapping, 382–3
Preference studies
  examples of resultant data, 282, 284–300
  forms used, 281, 283
  home-use scale used, 282
  scales used, 278–80, 282

PREFMAP computer program, 385, 386, 387, 406
Principal component analysis (PCA), 176, 369–70
  descriptive methods, PCA used, 239–40, 252–3
  texture assessment using, 88
Principal coordinates analysis (PCO), 391–2
PROCALSCAL computer program, 406
Process development
  expert panels used, 302
  test methods used, 306
Procrustes analysis, 209, 210, 236, 371
Product development
  expert panels used, 302
  test methods used, 306
Product differences, statistical analysis of, 223–5, 251
Profile attribute analysis, 197
Profiling techniques
  flavour, 190–201
  standards used, 85, 175
  standards used with, 175
  texture assessment, 84–8, 175
Psychometric relationships, 113–14, 139
Psychophysical relationships, 113, 139–40
Puddings, preference testing of, 285, 294–5, 299

Quality control, test methods used, 251–2, 306
Quantitative descriptive analysis (QDA), 172, 201–3
  compared with other methods, 202, 203, 301, 306
  line scaling used, 202–3
  selection of assessors for, 203–4
  training of assessors for, 202, 203, 204
Quantitative sensory profiling (QSP)
  data sets evaluated, 212–25
  definition of, 189

Quantitative sensory profiling (QSP)—contd.
  practical methods, 203–12
  procedures available, 189–203
  selection of assessors for, 203–4
  statistical analysis for, 225–53
  training of assessors for, 202, 203, 204
  variations between methods, 210–13
  vocabulary used, 205–10
  see also Flavour profile method (FPM)
Quartermaster Food and Container Institute (QMFCI) food preference scale, 278–80, 282
Quartiles
  definition of, 337–8
  food preference, 296
Questionnaires
  acceptance testing, 309, 312–13
  attitude testing, 271–3
  characteristics of, 272
  design of questions, 277–8
  scales used, 274–6
  structured/unstructured questions used, 272–3
  intake studies, 317, 318–19, 321

Randomised block design, 352
Range, definition of, 337
Range–frequency effects, 179–80
Range tests, 223, 352
Rank sum tables, 177
Ranking, 176–8
  advantage of, 177, 181
  compared with rating scale methods, 164
  food acceptance testing, 301, 306
  meaning of term, 157, 176–7
  practical procedure for, 177–8
  texture assessment, 81
Rating scale methods
  advantages of, 164
  arbitrariness of, 165
  food attitude/acceptance testing, 273, 301, 306

Rating scale methods—*contd.*
  texture assessment, 84, 85
Ratio scales, 81–2
  food preference/acceptance testing, 282, 301
  meaning of term, 157
Recall method, food intake studies, 317, 321, 325
Receptor cells
  olfactory system, 28, 30–2
  taste system, 5
  vision system, 106
Reference compounds, flavour profiling, 196, 204, 227
Reflectance characteristics, food surfaces, 114
Regression
  bivariate population, 359–61
  errors treated, 361
  prediction by, 357–9
  robust regression, 361–2
Relative-to-ideal rating procedure, 172–3
Replication, 221–3
Retina, anatomy of, 105–6
Rhodopsin, 106
Rice, steamed, preference testing of, 285, 291, 300
Road distance table, multidimensional scaling applied to, 391–5
Round-table discussions, flavour vocabulary, 208, 211
Rye/wheat breads
  sensory assessment of, 235, 236
  sensory characteristics of, 223

Saccharin, flavour affected by, 243, 244, 246
Salad dressings, preference testing of, 293, 298
Salads, preference testing of, 284, 292–3, 296, 298
Salty taste, 10
  electric response characteristics for, 14, 17
  test solutions used, 193

Salty taste—*contd.*
  tongue areas responsive to, 12
Sampling, 363–4
  replication of, 221–3
  two-way design, 365–6
Sandwiches, preference testing of, 284, 290, 298
Sanitary pads, sensory attributes for, 195
Saturation (in colour), meaning of term, 113
Scales
  acceptability amount, 276
  category scales, 165–8
  comparison of different scales, 174–5
  definition of, 156
  food acceptance/preference testing, 274–6
  hedonic scaling, 169–71
  magnitude estimation used, 173–4
  multiple comparison procedure used, 168–9
  practically used scales, 164–74
  preference frequency, 275, 279–80
  relative-to-ideal rating used, 172–3
  standards used with, 175–6
  unstructured scales, 171–2
Scaling, 81–2
  data analysis used, 176
  direct methods of, 162
  historical background to, 141, 159–64
  indirect methods of, 161–2
  meaning of term, 156
Scatter diagrams, 357, 358, 362
Scoring, meaning of term, 157–8
Seafood, preference testing of, 284, 288, 297
Sensory difference testing, 131–53
  reasons for use, 131
Sensory difference tests, 134–8
  duo test, 134
  duo–trio test, 134
  oddity choice test, 136
  reasons for use, 131
  tetrad test, 135

Sensory difference tests—*contd.*
   triangular test, 135
   Type 1 and Type 2 errors considered, 137
Sensory receptor cells, 4, 5
Sensory response behaviour, basic principles of, 132–4
Sensory texture profile analysis, 84–8
Shepard scatter plot, 396, 397
Signal detection theory (SDT), 149, 161
Significance level, meaning of term, 137
Similarity estimation methods, 389, 390, 398
Skewness, coefficient of, 340, 341
Smell
   sense of, 25–50
   see also Olfactory...
Sniffing, olfaction affected by, 40–1
Soapy taste, 8–9, 16
Sodium chloride, taste of, 10–11
Somatosensory information, taste supplemented by, 8
Sound, texture perception affected by, 78–80, 89
Soups, preference testing of, 284, 286, 296, 297
Sour taste, 10
   electric response characteristics for, 14, 16, 17
   test solutions used, 193
   tongue areas responsive to, 12
Soy sauce, sensory evaluation of, 248–9
Spain, consumer panels in, 311
Spearman's rank correlation procedure, 229
Spectrum™ flavour profiling system, 206
Spider-web diagrams, product differences illustrated by, 223–5
Spirits, sensory evaluation of, 251–2
Spreadability tests, 78
Springiness
   definition of, 83
   instrumental measurement of, 93

Standard errors, 342
Standardisation (of data), 220–1
Standards
   scaling, 175–6
   texture assessment, 85
Starchy foods, preference testing of, 284, 285, 287–8, 289–90, 296, 297–8
Statistical analysis, 178, 335–79
   descriptive methods, 225–53
   sensory difference tests, 137–8, 151–2
Stepwise discriminant analysis (SDA), 230, 247
Stepwise regression analysis (SRA), 253, 368–9
Stereochemical theory of olfaction, 34, 35
Stevens Compression Response Analyser, 93, 95
Stickiness test methods, 77–8
Stimulus–Organism–Response (S–O–R) model, 147
Storage testing, methods used, 306
STRESS (badness-of-fit) measures, 394, 396–7
Structure–activity relationships, olfactory system, 32–7
Structured questions, food attitude testing, 272–3
Student's t-test, 346–7
Sucrose
   flavour affected by, 243, 244, 245, 246
   taste of, 10, 193
Sum of squares
   algorithms for, 337, 378–9
   definition of, 337
Summary statistics, 336
Sweden, consumer panels in, 311
Swedish Natural Colour system, 108
Sweet taste, 10
   electric response characteristics for, 14, 17
   test solutions used, 193
   tongue areas responsive to, 12

Sweeteners, tea flavour affected by, 243, 244
Sweetness rating, 165–6
Switzerland, consumer panels in, 311

Taste
  blindness, 18
  buds, 3, 4–6
    nerve supplies for, 7
    number of, 3, 5
    structure of, 4–6
  cells
    adaptation of, 15
    primary processes in, 13–16
  definition of, 25
  enhancers, 15–16
    see also Glutamate
  genetics of, 18
  neural coding of, 12–13
  neural pathways of, 7–8, 9
  olfaction contribution to, 8
  perception, pathological states of, 18
  pores, 4, 5
  qualities, 8–9
  receptor cells, 2
    electric responses of, 14
  sense of, 1–18
  somatosensory contribution to, 8
  texture assessment affected by, 88
  threshold sensations of, 9–12
  see also Flavour
Tea
  flavour profiles of, 195, 220, 226
  preference testing of, 285, 287, 297, 300
  sweetening of, 243, 244
Teeth, texture assessed using, 73
Telephone data collection, consumer studies using, 311, 320, 321, 324
Terminal nerve, chemosensitivity of, 26
Test statistic (t), 346, 377
Tetrad tests, 135

Textural characteristics, classification of, 70–2
Texture, definition of, 69, 95–6
Texture assessment
  appearance effects, 88
  chemical analysis used, 96
  descriptive tests used, 81–8
  discriminatory tests used, 80–1
  general comments on procedures, 88–9
  geometrical characteristics proposed, 70, 82–3
  instrumental methods used, 89–95
    compression tests, 90
    cutting tests, 90
    elasticity tests, 93, 95
    empirical methods, 89–90
    extrusion tests, 90
    flow/mixing tests, 90
    imitative instruments, 91–5
    puncture/penetration tests, 89–90
    shear tests, 90
    vibrational techniques, 95
    viscous-elasticity tests, 93–4
  mechanical characteristics proposed, 70, 82, 83
  mouthfeel characteristics, 72–7
  non-oral methods, 77–8
  non-sensory methods used, 89–96
  relationship of sensory and non-sensory data, 96–7
  sounds, effect of, 78–80
  structural studies used, 95–6
  taste effects, 88
  terminology used, 69–72
Texture profile analysis, 84–8, 175, 301, 306
  instrumental measurement, 91
Texturometers, 91–2
Thorn De Luxe Natural (DLN) lighting, 117, 118
Three-dimensional plot, 367–8
Threshold testing
  food acceptance, 301, 306
  texture assessment, 80–1

INDEX

Thresholds
  olfactory, effects of inhalation rates on, 41
  taste substances, 10
Thurlstone theory (of paired comparisons), 148–9, 160, 161
Tomato paste, appearance of, 110, 125
Tongue
  liquid perception using, 71, 77
  taste areas of, 2–4, 11, 12
Toothpaste, flavour profiling for, 211
Training (of) assessors
  flavour profiling, 191–6, 204
  texture assessment, 80, 84
Translucency, 110
  freshly cut meat, 119
Triangular tests, 135, 146, 301, 306
Trichromacy, 107
Trigeminal chemical stimulation, 26–7, 42–3
Tristimulus colorimetry, 104
Two-dimensional data, 357–63
Two-way analysis of variance, 352–6
  sum-of-squares calculated, 378
Two-way experimental design, sampling for, 365–6
Two-way scaling
  example of MDS analysis, 398–9
  MDS method used, 396–9

UK
  consumer panels in, 311
  food preference studies, 300
Univariate analysis, 176, 218–19, 223, 228
  see also Multivariate analysis (MVA); One-way analysis of variance
Unstructured questions, food attitude testing, 272–3
Unstructured scales, 171–2, 175
US Armed Forces, food preference studies, 285, 286–95, 297–9, 300

USA, consumer panels in, 311

Vallate papillae, 3–4, 12
Vanillin, sensory characteristics of, 206
VARCLUS program, 236, 238
Variance, definition of, 336
Variation, coefficient of, 337
Vegetable juices, preference testing of, 284, 286, 297
Vegetables
  flavour reference compound, 204
  preference testing of, 285, 291–2, 296, 298, 300
Vibrational methods, texture assessment, 95
Visco-elasticity
  instrumental measurement of, 93–4
  sensory assessment of, 76–7
Viscometers, liquid texture assessed using, 94
Viscosity, definition of, 83
Vision, definition of, 103–4
Vocabulary
  flavour profiling, 205–10
  food acceptance/preference testing, 274–6, 277, 279, 280, 282, 283
  texture assessment, 85
Volodkevich Bite Tenderometer, 91, 97
Von Ebner's rinsing glands, 3, 4

Ward's minimum variance cluster analysis, 235, 236, 237
Warner–Bratzler shear test, 75–6, 90
Waste studies, 326, 327–31
  aggregate plate waste measurement methods, 327–8
  individual plate waste measurement methods, 328–9
  plate waste methods listed, 329
  visual estimation methods, 329–31
Weber's Law, 141, 159
Whisky, sensory evaluation of, 245, 252–3

Wilcoxon matched-pairs signed ranks test, 344–6
Wilk's lambda, 372
Wine, flavour vocabulary used, 207, 209, 217–18

Winners/losers (food preference) lists, 296, 297–9

Yams, texture profile analysis of, 87